LANDSCAPE ECOLOGY in AGROECOSYSTEMS MANAGEMENT

Advances in Agroecology

Series Editor: Clive A. Edwards

Soil Ecology in Sustainable Agricultural Systems,
  Lijbert Brussaard and Ronald Ferrera-Cerrato
Biodiversity in Agroecosystems,
  Wanda Williams Collins and Calvin O. Qualset
Agroforestry in Sustainable Agricultural Systems,
  Louise E. Buck, James P. Lassoie, and Erick C.M. Fernandes
Agroecosystem Sustainability: Developing Practical Strategies,
  Stephen R. Gliessman
Interactions Between Agroecosystems and Rural Communities
  Cornelia Flora
Structure and Function in Agroecosystem Design and Management
  Masae Shiyomi and Hiroshi Koizumi

Advisory Board

LANDSCAPE ECOLOGY in AGROECOSYSTEMS MANAGEMENT

Edited by Lech Ryszkowski

CRC Press
Taylor & Francis Group
Boca Raton London New York

CRC Press is an imprint of the

Cover Art

Top: Example of damage resulting from erosion and pollution in a beet field in Erlon, northern Parisian basin. (From F. Tellier, June 1999.)

Bottom: Shaded composition of General Chłapowski's Agro-ecological Park (Greater Poland), Satellite Image SPOT HRV-17, June 1997.

Library of Congress Card Number 2001052416

Library of Congress Cataloging-in-Publication Data

Landscape ecology in agroecosystems management / edited by Lech Ryszkowski.
     p. cm. -- (Advances in agroecology)
     Includes bibliographical references (p.).
     ISBN 0-8493-0919-0 (alk. paper)
     1. Landscape ecology. 2. Agricultural ecology. 3. Ecosystem management. I. Ryszkowski, Lech. II. Series.

QH541.15.L35 L347 2001
577.5′5--dc21
                                              2001052416

CRC Press
Taylor & Francis Group
6000 Broken Sound Parkway NW, Suite 300
Boca Raton, FL 33487-2742

© 2002 by Taylor & Francis Group, LLC
CRC Press is an imprint of Taylor & Francis Group, an Informa business

First issued in paperback 2019

No claim to original U.S. Government works

ISBN-13: 978-0-367-45505-7 (pbk)
ISBN-13: 978-0-8493-0919-9 (hbk)

The Editor

Lech Ryszkowski, Ph.D., is Professor of Natural Sciences and Director of the Research Centre for Agricultural and Forest Environment of the Polish Academy of Sciences in Poznań, Poland. He graduated from the Faculty of Biology and Earth Sciences in 1955 and completed his Ph.D. on coypu population structure in 1963. In 1971, he received his Doctor of Sciences from Warsaw University for his studies on the role of small mammal communities in energy flows of temperate forest ecosystems. For 2 years, he was the student of Eugene Odum at the University of Georgia in Athens. Dr. Ryszkowski became an Associate Professor in 1971 and a Full Professor nominated by the State Council in 1985.

Dr. Ryszkowski's major interests are the functional role of animals in agroecosystem energy flow and matter cycling, and energy flow and matter cycling at the level of agricultural landscape with emphasis on the role of ecotones (e.g., mid-field shelterbelts or stretches of meadows) in control of elements cycling in rural areas. His main scientific achievements focus on guidelines for the protection of rural environment and for the biological conservation and management of agricultural landscapes. Particular emphasis is placed on the possibilities of control over nonpoint pollution sources by the formation of biogeochemical barriers such as woodlots shelterbelts, grasslands, and water bodies differentiating agricultural landscape.

Dr. Ryszkowski is the head of the International Group for Agroecosystems and a member of many ecological organizations. He was a member of the Board of INTECOL (1992 to 1998) and has organized many international symposia. As a member of the Countryside, Wildlife and Landscape Working Group of the Council of Europe in Strasbourg, Dr. Ryszkowski presented two reports on landscape and biological diversity in the Central and Eastern European countries and was the Polish organizer of the symposium on landscape diversity in rural areas for the Council of Europe.

Dr. Ryszkowski has published 354 papers in Polish and international journals. He is the editor or co-editor of 18 books.

Contributors

Ewa Arczynska-Chudy
Research Centre for Agricultural
  and Forest Environment
Poznań, Poland

Stanisław Bałazy
Research Centre for Agricultural
  and Forest Environment
Poznań, Poland

Gary W. Barrett
Institute of Ecology
University of Georgia
Athens, Georgia, U.S.A.

Alina Bartoszewicz
Department of Soil Science
Agricultural University
Poznań, Poland

Jacques Baudry
INRA
Rennes, France

Françoise Burel
Laboratoire d'Evolution des Systemes
  Naturels et Modifiés
Université de Rennes I
Campus de Beaulieu
Rennes, France

Susan R. Crow
Carl Vinson Institute of Government
and
College of Agricultural
  and Environmental Sciences
University of Georgia
Tifton, Georgia, U.S.A.

Frank Eulenstein

Marc Galochet
Center of Biogeography-Ecology
  (UMR 8505-CNRS/ENS)
Saint-Cloud, France
and
University of Paris IV — Sorbonne
Paris, France

Bärbel Gerowitt
Georg-August-Universität
Landwirtschaft und Umwelt
Göttingen, Germany

Vincent Godard
Center of Biogeography-Ecology
  (UMR 8505-CNRS/ENS)
Saint-Cloud, France
and
University of Paris VIII — Saint-Denis
Saint-Denis, France

Hanna Gołdyn
Research Centre for Agricultural
  and Forest Environment
Poznań, Poland

Micheline Hotyat
Center of Biogeography-Ecology
  (UMR 8505-CNRS/ENS)
Saint-Cloud, France
and
University of Paris IV — Sorbonne
Paris, France

Janusz Jankowiak
Research Centre for Agricultural
  and Forest Environment
Poznań, Poland

Jerzy Karg
Research Centre for Agricultural
and Forest Environment
Poznań, Poland

Andrzej Kędziora
Department of Agrometeorology
Agricultural University
im A. Cieszkowskiego
Poznań, Poland
and
Research Centre for Agricultural
and Forest Environment
Poznań, Poland

David Kleijn
Department of Environmental Sciences
Nature Conservation and Plant Ecology
Group
Wageningen University
Bornsesteeg, Wageningen
The Netherlands

Krzysztof Kujawa
Research Centre for Agricultural
and Forest Environment
Poznań, Poland

Didier Le Coeur
Laboratoire d'Ecologie et Sciences
Phytosanitaires
Ecole Nationale Superieure
Agronomique de Rennes
Rennes, France

Richard Lowrance
Southeast Watershed Research
Laboratory
Agricultural Research Service

Jon Marshall
Marshall Agroecology
Barton, Winscombe
Somerset, U.K.

Andrzej Mizgajski
Educational Center for Environment
Protection and Sustainable
Development
A. Mickiewicz University
Poznań, Poland

Camilla Moonen
Scuola Superiore Santa Anna
Pisa, Italy

Janusz Olejnik
Department of Agrometeorology
Agricultural University
im A. Cieszkowskiego
Poznań, Poland
and
Department of Land Use Systems
and Landscape Ecology
Center for Agricultural Landscape
and Land Use Research (ZALF)
Müncheberg, Germany

Maurizio Paoletti
Dipatimento di Biologia
Università Degli Studi di Padova
Padova, Italy

Lech Ryszkowski
Research Centre for Agricultural
and Forest Environment
Poznań, Poland

Laura E. Skelton
Institute of Ecology

Lech Szajdak
Research Centre for Agricultural
 and Forest Environment
Poznań, Poland

Barbara Szpakowska
Research Centre for Agricultural
 and Forest Environment
Poznań, Poland

George Thomas
Department of Agriculture
 and Food Studies
Seale-Haync Faculty
University of Plymouth
Newton Abbot, Devon, U.K.

Armin Werner
Department of Land Use Systems
 and Landscape Ecology
Center for Agricultural Landscape
 and Land Use Research (ZALF)
Müncheberg, Germany

Stanislas Wicherek
Center of Biogeography-Ecology
 (UMR 8505 CNRS/ENS)
Saint-Cloud, France
and
Member, Board of Directors
 and Vice-Chairman
European Institute of Sustainable
 Development
http://www.iedd.org-www.iedd.net

Peter Zander
Department of Socioeconomics and
Department of Land Use Systems
 and Landscape Ecology
Center for Agricultural Landscape
 and Land Use Research (ZALF)
Müncheberg, Germany

Irena Zyczynska-Baloniak
Research Centre for Agricultural
 and Forest Environment
Poznań, Poland

Contents

CHAPTER **1**

The Functional Approach to Agricultural Landscape Analysis

Lech Ryszkowski

CONTENTS

OVERVIEW

In response to a growing need to bridge aspects of geography and ecology, Troll (1950, 1968) coined the term "landscape ecology," which was adopted as a new scientific discipline. According to Troll (1968), the landscape can be studied in terms of its morphology, classification, and changes in time (history), as well as the functional relationships between its components, which he called landscape ecology. Troll also considered that problems of landscape protection as well as management should be included in geographical analyses of landscapes. Although these fields of landscape study are closely interrelated, the first three depend on a geographical approach, whereas the second two are ecological. In the 1950s, when Troll first introduced the term *landscape ecology*, the main concerns of ecology were focused

by the International Biological Programme in the late 1960s and the first half of the 1970s. These studies broadened the functional analysis of ecosystems or landscapes, thereby providing insights into driving forces for dynamic processes (energy fluxes causing the transformation of structures) and also introduced new options for the control of environmental threats brought about by the intensification of agriculture and other human impacts on ecosystems.

Other studies published at the same times as Troll's papers also stressed the importance of an ecological approach to landscape analysis, especially for problems of landscape management. For example, Wodziczko (1948) in Poland strongly emphasized that the maintenance of biological balance in landscapes was an essential goal of protection or restoration activities. Wodziczko also introduced the term *physiotactics* for the ecological management of landscapes, which in 1953 Odum called *ecological engineering*.

EMERGENCE OF THE DISCIPLINE

The progressive interest in landscape studies eventually resulted in the emergence of landscape ecology as a separate scientific discipline incorporating both geographical and ecological studies with landscape planning and management aspects. By the 1980s, landscape ecology was an established discipline, marked by the organization of the International Association for Landscape Ecology (IALE) in 1982 and the publication of books summarizing its scope and goals. Benchmark publications include those of Naveh and Lieberman (1984) and Forman and Godron (1986). Naveh and Lieberman stressed the role of human impact on landscape structures and functions and proposed ways for restoring degraded landscapes. Thus, their appreciation of landscapes reflected the approach of both the landscape designer and the environment protectionist as well as the social geographer. Forman and Godron emphasized the ecological aspects of landscape function. One very interesting result of their functional analysis was the proposition of the landscape stability principle, which indicates the importance of landscape structural heterogeneity in developing resistance to disturbances, recovery from disturbances, and total system stability. This principle was an outstanding contribution to general ecological theories identifying the importance of relationships among the various components of the landscape in maintaining resistance to external threats.

SCALE OF LANDSCAPE ECOLOGY

The geographical approach reflected in analyses of land-use changes (e.g., Turner et al. 1990, Brouwer et al. 1991), which has always played an important role in

forming the natural basis for landscape function are interwoven with the economic, social, cultural, and political spheres of human activities. As a result of such interactions, the landscape structure may be changed in ways that in turn influence landscape functions. The interactions between natural and human-induced processes are fundamental to understanding conflicts between ecological and human systems (Haines-Young and Potschin 2000).

The scale of landscapes, and to a lesser extent ecosystems, is such that changes in their component structure have implications for modifications or resistance to threats to their function by human activities. Therefore, the landscape constitutes a suitable ecological unit for management of environmental threats as well as a functional unit that can reflect interactions of ecological and human aspects of processes. A comprehensive analysis of a landscape enables not only the identification of ecological and human-induced processes in a holistic way but also may help detect the sources of threats to the sustainable development of agriculture or other human activities.

The optimization of human activities directed toward specific goals in coordination with properties of natural systems is a fundamental assumption of successful environmental protection. As Haines-Young and Potschin (2000) articulated this question, the ecological services or resources are exploited according to their values to society, and such values observed by society cannot be resolved unless their relations to ecological functions are understood. A complete understanding of these natural relationships should also include the historical aspect, which is how human values are redefined in terms of ecological functional knowledge. In this context, human needs should be accommodated to ecological processes rather than subjugated to them.

FEEDBACK MECHANISMS

The recognition of feedback mechanisms between the functions of landscapes and the exploitation of goods or services needed by human society should form a first step in developing strategies for sustainable development of the countryside. In many situations, deterioration of the environment caused by human activities was observed, but the causal relationships underpinning such phenomena were often poorly recognized. As a result, there was a widespread belief that human economic activities cannot be coordinated with protection of nature. However, with increasing recognition of a landscape's basic processes, such as energy fluxes, organic matter cycling, and mechanisms of their management, there is a growing conviction that it is the way in which natural resources are used, not the fact that they are exploited, that has led to environmental degradation (Ryszkowski 2000).

The argument that the less an ecosystem or landscape is changed by human

evaluation of alternative technologies of production, with objectives that seek to optimize production and environmental protection and to meet social needs at the same time. It cannot be stressed enough that this goal cannot be achieved on a scale smaller than the landscape, and that landscape ecology therefore becomes a very important pillar for the implementation of sustainable development of agriculture.

OPTIONS EXPLORED IN THIS BOOK

The results of various studies presented in the subsequent chapters of this book further document this idea. Some of the results, described more fully in the various chapters, are provided here to highlight the options for the management of threats to the environment and enable development of guidelines for the implementation of an environmentally friendly landscape management.

The successful management of agricultural landscapes depends on a full recognition of the relationships between those processes and structures that maintain the whole system. Concerns are often focused on simplified input and output analyses that consider the reacting system as a black box. This approach does not allow for any modification of environmental threats by rearrangement of system components, which could increase inner compensation processes that can amend unwanted effects. Chapter 2, relying on long-term interdisciplinary studies on agricultural landscapes carried out by the Research Centre for Agricultural and Forest Environment in Poznań, Poland as well as on a vast quantity of literature, evaluates the functional changes of agroecosystems compared to natural ecosystems. Because of structural simplifications in agroecosystems, brought about by the obvious need to increase yields, the cultivated fields are characterized by a lower tie-up of organic matter cycling, which can result in increased leaching or wind loss of chemicals from cultivated fields. Farmers can moderate the intensity of various material-dispersing processes through properly applied tillage technologies, but they are unable to eliminate them entirely regardless of whether they use integrated or organic farming systems. Combining on-farm environmentally friendly technologies with the structuring of associated landscapes with various stretches of permanent vegetation can provide a more successful elimination of environmental threats. Thus, protection activities carried out at the landscape level can enhance environmentally friendly technologies applied within the farm.

The functions that are of the utmost significance for landscape management, such as partition of solar energy for evapotranspiration and air and soil heating, are rarely considered in landscape studies. The newer methods of heat balance estimates under field conditions, relying on empirical relationships between solar energy fluxes and meteorological characteristics described by Kędziora and co-workers (see Chapters 3 and 4) have opened new frontiers for the management of water cycling

convection processes, can modify mesoscale atmospheric circulation, which in turn influences regional climatic phenomena.

The other important function of landscape studies is the control of diffuse aspects of water pollution caused by leaching of chemical compounds from the soil of cultivated fields. Information presented in Chapter 5 provides new insights on control mechanisms of diffuse pollution operating in permanent vegetation strips (biogeochemical barriers) located in agricultural landscapes. Thus, for example, biological recycling of nitrogen within the biogeochemical barrier plays a very important role in the efficiency of its retention. When rates of organic matter decomposition overcome the rates of nitrogen uptake by organisms, the biogeochemical barrier changes from a sink to a source of nitrogen pollution in the water reservoirs of a landscape. Thus, management of biogeochemical barriers is essential if their water-purifying functions are to be retained. The indication that shelterbelts located in upland parts of watersheds can slow rates of nitrous oxide release to the atmosphere, compared to that occurring in adjoining cultivated fields, is an another example that the disclosure of the inner mechanisms operating in the biogeochemical barriers can provide options for the management of organic matter cycles at the landscape level.

The implementation of agricultural landscape management for optimization of biomass production with environmental protection should be stimulated by a policy encouraging farmers to undertake certain actions. Chapter 6 discusses U.S. policies for the implementation of riparian buffer strips as a means of diffuse pollution control, emphasizing both the achievements and the obstacles.

Chapter 7 examines studies concerning waterborne transport of organic compounds in the ground water under the agricultural landscape and provides an almost pioneering analysis of factors influencing those processes. The formation of metal-organic complexes, the influence of the biogeochemical barrier on processes of chemical speciation, which causes change in the ionic forms of chemical compounds, shows that water transport of chemicals is a very complicated phenomenon. Solubility, hence the toxicity to organisms of mineral compounds, could be changed enormously when the relevant inorganic ions are bound to organic compounds, which usually occurs when ground water seeps through the root system of the biogeochemical barrier. Consequently, for effective control of waterborne migration of heavy metals through a landscape, it is necessary to recognize the chemical forms of their transport.

Chapters 8 and 9 report on studies showing that semi-natural habitats, maintained in the agricultural landscape, can constitute important refuge for many plants and animals, thereby supporting biological diversity in farmland. Thus, the introduction of a mosaic landscape, composed of cultivated fields and semi-natural habitats, to some extent compensates for the negative effects exerted by agriculture on the biota. The effects of stimulating biodiversity can appear very quickly when a new shelterbelt is planted, and in 2 to 3 years a very high level of biodiversity is achieved (Chapter 8)

   Agriculture-induced changes in environment are discussed in Chapters 10 and 11. Comparison of results presented in Chapters 10 and 11 shows some common threats for both regions discussed (the Wielkopolska and the Northern Parisian Basin), such as erosion processes or ground water pollution, which were created by the intensification of agriculture. The different climatic conditions of the regions produced different threats related to induced changes in water cycling. In the Wielkopolska, because of relatively low precipitation and high evapotranspiration, the drainage of cultivated fields resulted in an overall drying of the region (Chapter 10). In France, rapid removal of water from cultivated fields under moist Atlantic climatic conditions resulted in an increase in the severity of floods (Chapter 11). These areas illustrate clearly that the effects of introduced agricultural technology (drainage) can lead to different environmental effects if the regimes of matter cycling (water) in the whole system are different.

   Chapter 12 shows the methods of forecasting threats that can be induced by various trends in changes in agricultural land use. This approach is very important when a precautionary principle is applied to prevent actions that could deteriorate environmental conditions, and the causal link between such activities and the impact has not been fully confirmed. Thus, for example, some scenarios of the agricultural development proposed by the European Union should, according to Werner's and Zander's analysis, lead to environmental threats. By applying the proposed procedures we have the opportunity to correct potential trends of development before the real environmental impacts are observed.

   The protective role of shelterbelts against wind erosion and severe climatic phenomena has been known for a long time. Chapter 13 discusses the synthesis of the various environmental services of shelterbelts and reviews methods ensuring their efficiency for management of environmentally negative effects of agriculture. The methods of planting trees to increase the value of shelterbelts for environmental protection, thereby ensuring their multifunctionality, are presented. Methods of shelterbelt classification using satellite images are examined in Chapter 14. Such methods enable surveys of shelterbelts to be made in large areas of the countryside.

   The prospects of agricultural landscape ecology development are discussed in Chapter 15. Barrett and Skelton conclude that although science will yield new environmentally friendly agricultural technologies, sustainable development depends on the abilities of farmers to utilize the achievements of that science. Thus, efficient methods of modern knowledge dissemination are of the utmost significance for rural landscape management.

OUTLOOK

   The brief review of recent achievements in ecology presented in this book clearly

and decision makers' growing acknowledgment of the usefulness of landscape ecology for implementation of sustainable development in rural areas. Detailed documentation of that situation is presented in Chapter 16.

REFERENCES

Brouwer F. M., Thomas A. J., Chadwick M. J., Eds., 1991. *Land Use Changes in Europe,* Kluwer Academic Publishers, Dordrecht. 528 pp.
Forman R. T. T., Godron M., 1986. *Landscape Ecology,* John Wiley & Sons, New York. 619 pp.
Haines-Young R. H., Potschin M. B., 2000. Multifunctionality and value, in *Multifunctional Landscapes,* J. Brandt, B. Tress, G. Tress, Eds., Centre for Landscape Research, Roskilde, 111–118.
Naveh Z., Lieberman A. S., 1984. *Landscape Ecology — Theory and Application.* Springer-Verlag, New York. 356 pp.
Odum E. P., 1953. *Fundamentals of Ecology.* W.B. Saunders, Philadelphia. 384 pp.
Ryszkowski L., 2000. The coming change in the environmental protection paradigm, in *Implementing Ecological Integrity,* P. Crabbe, A. Holland, L. Ryszkowski, L. Westra, Eds., Kluwer Academic Publishers, Dordrecht, 37–56.
Troll C., 1950. Die geographische Landschaft und ihre Erforschung. *Studium Generale* (Heidelberg) 3: 163–181.
Troll C., 1968. Landschaftsökologie, in *Pflanzensoziologie und Landschaftsökologie,* R. Tuxen, Ed., Junk, The Hague, 1–21.
Turner II B. L., Clark W. C., Kates R. W., Richards J. F., Mathews J. T., Meyer W. B., 1990. *The Earth as Transformed by Human Action.* Cambridge University Press, Cambridge. 713 pp.
Wodziczko A., 1948. Na straży przyrody (On guard of nature) [in Polish]. *Państwowa Rada Ochrony Przyrody,* Kraków.

CHAPTER 2

Development of Agriculture and Its Impact on Landscape Functions

Lech Ryszkowski and Janusz Jankowiak

CONTENTS

INTRODUCTION

It seems obvious that human activities in pursuit of safe food supplies, more comfortable housing, resource exploitation, easier transport of people and goods, and so on have transformed nature. Long ago, when population density was low and people lived in small groups separated by vast ranges of wilderness, visible impact on environment appeared only locally, and degraded sites eventually regenerated after their use was abandoned. Ecologically, primeval people differed from other mammals by only one very important skill — use of fire — which sometimes had remarkable impact on the local landscape. The extinction of the large flightless birds about 50,000 years ago in Australia may be interpreted as the result of habitat destruction by fires ignited by primeval humans (Miller et al. 1999). Restricting living activities to a confined area for long periods of time resulted in landscape transformations evi-

caused a decline in their size, a possible sign of overexploitation. Ecologically, such impacts are similar to the broad category of animal influences on habitat, such as overgrazing of plant cover by ungulates confined to an island.

Improvement of paleolithic technology led to development of tools that increased work efficiency and increased human options to change landscapes. To ensure prosperity, people subjugated nature for profits that enabled higher standards of living. A crucial event in the subjugation of the countryside was the beginning of agriculture about 10,000 years ago in both the Fertile Crescent and China (rice cultivation) and later in other parts of the world (Pringle 1998). The accumulation of surplus food enabled large settlements and work division, which led to development of complex social organization. The transformation from the hunter-gatherer lifestyle was a slow process, lasting millennia. Inhabitants of southern Egypt used grinding stones as well as mortars and pestles and harvested wild barley and other grass seeds with stone or bone sickles nearly 18,000 years ago (Boyden 1992). In the Fertile Crescent 13,000 years ago, people used sickles for harvesting wild cereals and cultivated some cereals, but at the same time many were still hunters and gatherers (Boyden 1992, Pringle 1998). In large early neolithic settlements excavated in Anatolia, Turkey — such as Asikli, founded about 10,000 years ago — people lived mostly by hunting and gathering. In a large settlement of 2000 families founded about 1000 years later, in Catalhöyük, inhabitants ate both wild and cultivated plants (Balter 1998). About 9000 years ago the main crops, such as einkorn wheat, emmer wheat, barley, lentil, and pea, were domesticated (Lev-Yaduni et al. 2000), and the first successful human attempts to subjugate landscape were made.

The progress of agriculture was achieved by trial and error with the focus on yield increase. The scientific basis for plant cultivation and animal husbandry was developed only about 150 years ago, while the environmental consequences of agriculture were recognized only during the last 30 years. Attempts at agricultural landscape management for sustainable development of rural societies are just now being undertaken. The consequences of the agricultural revolution, which is the cornerstone for development of civilization, included formation of a complex social organization that relied on class differentiation. The important feature of this transformation was full-time craft specialization, which in the long run provided positive feedback mechanisms for subjugation of nature, and led to the creation of modern technologies that changed the face of the Earth. At the same time, peasants could focus activities only on farming and became more and more dependent on other class services (protection, trade, production of tools). That change enhanced their non-nomadic lifestyle, the activities of which in turn caused accumulated impact on the environment. For example, soil erosion started to appear in the Mediterranean hillsides as a result of forest clearing, continuous cultivation of large areas, or overgrazing. Such soil degradations were noted by Homer in the 9th century B.C. (Boyden 1992)

processes, and cultivation skills and irrigation techniques of peasants ensured constant yields. Under such conditions, the agriculture provided such surplus food supplies that the great urban states developed in Sumerian Mesopotamia, ancient Egypt, and other river civilizations. Their sustainability was endangered by salinization or by wars destroying the irrigation systems and exterminating the peasant population.

In upland areas, peasants had to face the problem of nutrient replenishment. Conversion of forests or grasslands into cultivated fields was accomplished in various ways in the different regions of the world, but one can distinguish a general trend of agriculture intensification. Slash-and-burn agriculture occurred at the stage of development when population density was very low and vast areas of pristine ecosystems were available for conversion. The next step, alternating two-field system (crop, fallow), was made possible by the use of farm animal manure, which intensified land use for crop production. The third step relied on the three-field rotation, in which on one field an overwintering crop was followed by a spring crop and then left fallow. When mineral fertilizers became readily available, the fallow field was eliminated from the crop rotation pattern. The cultivation of cereals, row crops, and pulse crops in rotation without intensive application of chemicals is usually referred to as traditional rotational agriculture. In this system, the recycling of nutrients sustaining soil fertility is maintained by rotation of pulse crops (especially legumes) and organic fertilizers produced within the farm, although some inorganic nutrient inputs can constitute a factor of production intensification. This method was the most intensive form of cultivation before industrialized agriculture gained momentum in the second half of the 20th century, and brought with it intensive use of mineral fertilizers and pesticides, more highly productive cultivars, and total mechanization of tillage.

The driving force for agricultural development has been continuously increasing input of energy, at the beginning by human labor, draft animals, and fire, then by harnessed wind and water energy in mills and application of more efficient tools produced by craftsmen. When those craftsmen were replaced by engineers who applied scientific achievements, chemical, electrical, and nuclear energy sources became available for increasing agricultural production and shielding crops from pests, pathogens, harmful climatic conditions, and other obstacles to effective cropping.

Boyden (1992) pointed out that the ratio of energy inputs for cropping to energy output (energy value of produced product) in primitive farming societies ranged from 1:15 to 1:20. The same ratio in modern agriculture in the U.K., Holland, or the U.S. ranges from 1:0.5 to 1:0.7. Thus, without very substantial input of energy for management modern agriculture cannot provide high yields.

Ryszkowski and Karg (1992) estimated that under conditions of moderately intensive agriculture the greatest contribution to energy subsidies provided by farmers for cereal cultivation was made by industrial fertilizers. With input of 60 kg

situation examined were spent on goods produced outside the farm. The maintenance of such a situation was facilitated by the human population harnessing more and more energy fluxes. Agriculture became closely tied to activities of other sectors of society. Reciprocal and close relationships between agriculture and industry, trade, and other human activities recently attained its full expression in industrialized agriculture. Human labor was replaced by work of machines, regeneration of nutrients by inorganic fertilizers, biological control of pests and pathogens by use of pesticides, water problems by drainage and irrigation, and so on (Edwards et al. 1993). Thus, many obstacles to crop production are under the farmer's control in modern agriculture, and agroecosystems become to a large extent controlled by humans. It is not suprising, therefore, that agriculture has substantial impact on structure and function of the landscape. The impacts of agriculture on landscape features are connected with shielding crops against detrimental factors on the one hand, and with the environmental consequences of applied technologies on the other hand.

IMPACTS OF AGRICULTURAL PRACTICES ON LANDSCAPES

During the long period of agriculture development, focus was first directed on practical guidelines for keeping stable yields; then, when scientific grounds for production were discerned, the spectacular growth of yield was achieved. With the exception of the last two or three decades, agricultural activity was valued as purely beneficial for humanity, and almost nobody thought that farmers' activities could threaten the environment. Conversion of forests or grasslands into cultivated fields changed, of course, the structure of landscape as well as endangered the existence of some plant and animal species. But until industrialized agriculture was developed and global environmental problems were caused by industry and urbanization, the nature conservancy programs assumed that it was possible to shield biological diversity from human influences in national parks, reserves, or other protected habitats. When threats to the environment attained global expression (Hannah et al. 1994, IUCN 1980, Lubchenco 1998, Pimentel and Edwards 2000, Ryszkowski 2000, Vitousek et al. 1997a,b, Watson 1999, and others) the need for reconciliation of human activities with protection of nature was recognized. The conviction is slowly growing that it is the way in which natural resources have been used, not the fact that they are exploited, that has resulted in environmental degradation. A new paradigm for nature protection is proposed that maintains that nature should not be separated from socioeconomic activities but that instead the two should be reconciled (Ryszkowski 2000). The management of landscape processes can be a useful approach to achieving this goal.

While increasing production, farmers subsidize energy to simplify plant cover directly by input of fertilizers, pesticides, etc., or indirectly by changing water cycling and decreasing holding capacities of soils for chemical compounds. In addition, agricultural activity often leads to decrease of humus contents. Increased use of farm machinery enables not only stronger impact on soil but also land surface leveling, modification of water drainage, etc., which causes changes in geomorphological characteristics of terrain. These effects of farm activity result in the development of a less complex network of interrelations among the components of agroecosystems. As a consequence of this simplification, relationships among agroecosystem components are altered, so that there is less tie-up in local cycles of matter. Thus increased leaching, blowing off, volatilization, and escape of various chemical components and materials from agroecosystems should be expected (Ryszkowski 1992, 1994, Ryszkowski et al. 1996).

Many environmentally significant effects of agriculture intensification are connected with the impoverishment or simplification of the agroecosystem structure. However, to obtain high yields, farmers must eliminate weeds, control pests and pathogens, be assured that nutrients are easily accessible only for cultivated plants during their growth, increase mechanization efficiency, etc. Therefore, agricultural activity aiming at higher and higher yields leads inevitably to the simplification of agroecosystem structure, which in turn causes further environmental hazards. Such ecological analysis leads to a conclusion of major significance for the sustainable development of rural areas: applying intensive means of production, farmers cannot prevent such threats to the countryside as leaching, blowing-off, volatilization of various chemical compounds which causes increased diffuse pollution of ground and surface waters, evolution of greenhouse gases (N_2O, CO_2, CH_4), and water or wind erosion. Biological diversity is also impoverished. Clearly, although farmers can moderate the intensity of these processes through careful selection of crops and tillage technologies, they are unable to eliminate them entirely, regardless of the farming system — industrialized, integrated, or organic.

These ecological implications can be evidenced by the analysis of the agriculture development in the European Union (EU) within the framework of the Common Agricultural Policy (CAP). The main objective of the CAP was to ensure agricultural self-sufficiency for countries suffering food shortages during World War II. To fulfill this goal, the CAP was proposed in the late 1950s and started to operate in the late 1960s. Guaranteed high prices were fixed each year for selected products, such as cereals, beef, and milk products. Quotas were later imposed and guaranteed prices were lowered. Other mechanisms were grants for land improvement, customs protection of agricultural commodities, low-interest loans for investment, and well-organized extension of research services, among others. This protectionist policy encouraged intensive agricultural technologies, which resulted in a substantial increase in production. For example, the average yield of wheat for all countries of the EU rose during the period 1975 to 1991 from 3.2 t·ha⁻¹ to 4.9 t·ha⁻¹, that is, by

the difference is much higher if one compares yield per hectare with the leading EU wheat producers such as the Netherlands (7.0 t ha^{-1}), France, Belgium, the U.K., Denmark, and Ireland, where the average wheat production was about 6.0 t ha^{-1} (Stanners and Bourdeau 1995).

The success of the CAP led to two constraints: one economical and the second one concerning environmental problems. Advances in agricultural technology led to overproduction, and surpluses were stored, which introduced a new burden on the EU economy. The annual rate of increase in agricultural production was 2 to 3% until the mid-1980s, while the rate of increase in consumption was about 0.5% (Laude 1996). Production surpluses accumulated, causing storage problems. To address this situation, proposals to increase production were introduced, such as production quotas, set-aside programs, and lower frontier protection.

These issues and other economic problems led to reforms of the CAP in 1992. Price-supporting mechanisms were lowered, while emphasis was placed on decoupling support for farmers from production by tying gains to levels more balanced with market demand. Environmental concerns were also taken into account in the new CAP policy. The negative environmental effects of the previous CAP policy are now quite well recognized (Stern 1996).

Intensification of agriculture caused environmental threats in the EU. The CAP-supported increase in farm sizes was related to more efficient use of labor hours and equipment as well as to lower costs for cultivation of large fields not segmented by shelterbelts, open drainage ditches, etc. Lack of permanent plant cover and large open fields exposed to wind and precipitation increased soil erosion processes. Presently in Europe, about 115 million ha suffer from water erosion and 43 million ha suffer from wind erosion. Water erosion is most severe in the Mediterranean region while wind erosion causes serious damage in the southern Ukraine and regions close to the northwest coast of the Caspian Sea (European Environment Agency 1998). Agricultural land comprises up to 70% of total territory of the Ukraine, and arable land cover is 55% of the country area (Voloshyn et al. 1999). Thus, the majority of the countryside in the Ukraine has been almost totally converted into arable land that is tilled with heavy machinery. The water balance is tight, and the central and southern parts of the country suffer water deficits. About 43% of the total arable land is threatened by soil erosion (wind and water), and in productive regions of black soils this figure rises to 70% (Voloshyn et al. 1999). According to estimates of Medvedev and Bulygin (1996), about 500 million metric tons of soil is annually lost from cultivated fields. Because of erosion processes, the potential fertility of soil is lowered by 1.2 times in weakly eroded soils, by 1.4 times in moderately eroded soils, and by 1.6 times in strongly eroded soils. In recent years, because of economic crisis, lower doses of fertilizers and pesticides are used and the intensity of mechanization has decreased. Nevertheless, problems of water, erosion, and soil pollution are still significant.

DEVELOPMENT OF AGRICULTURE AND ITS IMPACT ON LANDSCAPE FUNCTIONS   15

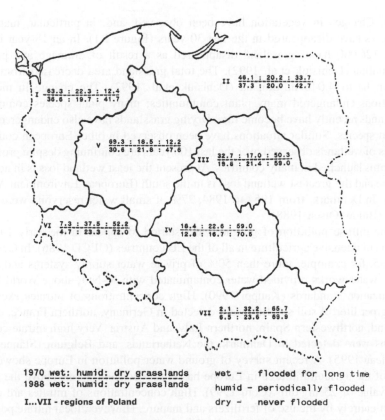

1970 wet: humid: dry grasslands	wet –	flooded for long time
1988 wet: humid: dry grasslands	humid –	periodically flooded
I...VII - regions of Poland	dry –	never flooded

Figure 2.1   Change in Polish grasslands area 1970–1989. Over a period of 18 years, approximately 126,000 ha of grasslands were lost (3%) because of drainage and peatbog reclamation (from 4,166 million ha to 4,040 million ha). (Modified after Denisiuk et al. 1992.)

mean annual precipitation is 717 mm if corrections for sampling error estimates are considered. The uncorrected value of annual precipitation was estimated at 599 mm (Gutry-Korycka 1978). In the central part of Poland, about 80% of precipitation is used for evapotranspiration (Kosturkiewicz and Kędziora 1995). The situation indicates a very tight water balance, and even a small variation in the ratio of precipitation to evapotranspiration caused by climatic conditions could have a great ecological and economic impact. Low runoff resulting from tight water balance delimits the area of surface water shortage, which amounts to 120,000 km², that is, 38% of the total area of Poland. In this area, located in the Central Plain, annual water runoff is less than 2 dm⁻³·s⁻¹·km⁻² (Kleczkowski and Mikulski 1995). Climatic conditions have not changed over the last centuries, and, therefore, the increasing water shortage

1991). Changes in vegetation have been observed, and, in particular, many wet meadows have disappeared in the last 30 years (Figure 2.1). In an 18-year period, about 126,000 ha of grassland disappeared as a result of drainage and peatbog reclamation (Denisiuk et al. 1992). The total grassland area decreased from 4.166 million ha to 4.040 million ha (Denisiuk et al. 1992). The change in moisture conditions endangered many plant communities; many plant species common to grasslands recently have become rare. Drying grasslands have also endangered many animal species. Similar situations have been observed in other European countries. Losses of wetlands observed over the last 100 years are continuing despite protection programs launched in many countries. At present the least wetland loss is in northern Europe and the greatest wetland loss is in the south (European Environment Agency 1998). In Denmark, from 1954 to 1984, 27% of small water reservoirs were eliminated (Bülow-Olsen 1988).

The diffuse pollution of ground water started to appear in the early 1980s in regions of intensive agriculture in all of the EU countries (OECD 1986). In Germany, in 1985, for example, more than 50% of private water supply systems and 8% of public water works provided water contaminated with $N-NO_3^-$ above World Health Organization standards (Kauppi 1990). High concentrations of nitrates exceeding 50 mg per liter of soil solution were detected in Germany, northern France, eastern England, northwestern Spain, northern Italy, and Austria. Very high nitrate concentrations were detected in Denmark, the Netherlands, and Belgium (Stanners and Bourdeau 1995). A recent survey of ground water pollution in Europe showed that 87% of the agricultural area in Europe has nitrate concentrations above the guide-level value of 25 $mg \cdot dm^3$ (Com 1999). High concentrations of nitrates are caused almost entirely by the use of fertilizers and manure. However, local nitrate pollution is caused by municipal or industrial effluents. Thus, in the late 1980s it appeared that modern intensive agricultural practices had brought with them threats to the environment and that the CAP of the EU should be changed by introduction of more environmentally friendly technologies.

This situation proved, almost experimentally, that agriculture cannot be considered only in terms of plant and animal production and economy, but that the environmental aspect of these activities should also be taken into account.

The other environmental threat brought by agriculture intensification is soil salinization, which appears in the Mediterranean and southeast Europe (Hungary, Romania, the Ukraine, Russia). Salinization appears presently in an area of about 29 million ha (Szabolcs 1991). Irrigation is the main anthropogenic factor influencing salt accumulation under semiarid or arid climatic conditions. Other environmental changes caused by agriculture intensification include loss of soil organic matter, acidification, and soil compaction (Van Lynden 1995).

Recognition that environmental threats are increasing forced EU administrators to change the Fifth Action Programme on the Environment (5EAP) to tackle growing

proposed to promote extensive production methods, set aside of parcels, and codes of good agricultural practice. But the integration of environmental concerns into agricultural practice still presents many difficulties.

Nevertheless, environmental issues have become a major concern of the CAP, and a strong focus is placed on the need to integrate farming practices with environment protection "to safeguard the environment and preserve the countryside" (Com 1999). According to the new CAP formulated by the Commission of the European Communities, sustainable agriculture development is founded on five main objectives: competitiveness of production, food safety and quality, fair standards of living for the agricultural community with stabilized incomes, protection of environment, and job opportunities for farmers and their families (Com 1999). The new approach to agriculture development presented by the CAP is shown by the emphasis on holistic treatment of all processes driving agriculture. The landscape approach is clearly recommended, therefore, because it enables identification and integrated analysis of all processes. From this standpoint, "policy choices can be more easily made to express the desired direction for development" (Com 1999). Thus, landscape analysis was recognized by policy makers as a useful tool.

AGRICULTURAL LANDSCAPE MANAGEMENT FOR CONTROL OF ENVIRONMENTAL THREATS

Efficient higher control of environmental threats from agriculture could be achieved by structuring agricultural landscape with various nonproductive components such as hedges, shelterbelts, stretches of meadows, riparian vegetation, and small ponds. Therefore, any activity to maintain or increase landscape diversity is important not only for aesthetics and recreation, but even more so for environment protection and, by the same token, for the protection of living resources in the countryside (Ryszkowski 1994, 1998).

The above considerations lead us to conclude that activities aimed at optimizing farm production and environment as well as protecting biodiversity should be carried out in two different but mutually supportive directions. The first component involves actions within the cultivated areas whose objective is to maintain the possibly high level storing capacities of soil and to preserve or improve its physical, chemical, and biological properties. Such actions include agrotechnologies, which increase humus resources or counteract soil compaction, and rely on differentiated crop rotations. Important effects of humus resources augmentation would be improved water storage capacity, more intensive processes of ion sorption, etc. Integrated methods of pest and pathogen control and proper application of mineral fertilizers adapted to crop requirements and to chemical properties of soil help facilitate the

The second component of the integration program of farm production and environment protection is the management of landscape diversity. It consists of such differentiation of the rural landscape so as to create various kinds of so-called biogeochemical barriers, which restrict dispersion of chemical compounds in the landscape, modify water cycling, improve microclimate conditions, and ensure refuge sites for living organisms. In landscapes having a mosaic structure, greater amounts of fertilizers can be applied without subsequent water pollution problems (Ryszkowski 1998a) than can be used in a homogenous structure composed of arable fields only. This conclusion is very important for the program of sustainable development of the countryside. Implementation of those ecological guidelines into the integrated agriculture policy will help develop new environment-friendly agrotechnologies that at the same time enable intensive production balanced with the ability of natural systems to absorb side effects of agriculture without being damaged. By saving natural capital of resilience capacities of the landscape, farmers will increase competitiveness of their farms in a manner similar to that described by Jacques Delors, president of the European Commission, in a white paper (1993) that maintained that improved environmental performance in an industry could increase its competitiveness in the world market.

Such a policy requires redefining ideas accepted up to now. Emphasis on increased production and its economical protection without much respect for interrelations of processes and interests should be changed to a more holistic approach that includes landscape issues. The heart of the dilemma at the national level is the failure of economies to elaborate efficient ways to incorporate environmental costs into proposals for rural area development.

Recent progress in agroecology and especially studies on agroecosystems and rural landscape functions such as energy flows, matter cycling, and biodiversity maintenance have shown that the following threats to environment and protection of living resources cannot be controlled only at the farm level but must also be curbed by management of the landscape structure diversity:

- Decreased water storage
- Increased pollution from nonpoint sources
- Soil erosion
- Impoverishment of plant and animal communities

The protective activities within the farm can only moderate generation of those threats (e.g., by reasonable use of fertilizers, regardless whether organic or industrial). Because of interconnectivity of water fluxes in phreatic aquifer of the whole watershed, one can observe widespread ground water migration of chemicals leached from soil of cultivated fields, or, due to large ranges of biota dispersion, the protective activities within the small farm are not sufficient to achieve successful biodiversity

- Increase the water storing capacities in the landscape by introducing mid-field afforestation networks, forming small water reservoirs with controlled outlets, and enhancing humus resources.
- Develop efficient technical plants for sewage treatment combined with control measures of diffuse pollution by biogeochemical barriers at the watershed level. Technical and ecological systems of pollution control should enhance their efficiencies by reciprocal support.
- Increase erosion control by propagating anti-erosive tillage practices and crop rotation patterns, as well as by structuring landscape with mid-field afforestations and strips of grassland.
- Augment application of organic fertilizers and crop rotation patterns, enriching soil in organic matter.
- Protect or restore refuge habitats for biota and ensure their connectivity by networks of ecological corridors.
- Adjust the application of artificial fertilizers to levels balanced with cultivar needs and introduce biochemical barriers to control spread of unused plant nutrients.
- Apply the integrated plant protection methods enhancing effectiveness of biological control by introducing refuge sites for predators.

Higher control efficiency of environmental threats from agriculture could be achieved, therefore, by structuring landscape with various nonproductive components such as shelterbelts, hedges, stretches of meadows, riparian vegetation strips, small mid-field ponds or wetlands, and so on.

Natural compatible structures that assist in controlling matter cycles in agricultural landscape are of great importance for enhancing the countryside's resistance to degradation. Various plant cover structures such as hedges, shelterbelts, stretches of meadows, and riparian vegetation strips are of special interest. They can be easily planted. The provision of such biogeochemical barriers in the landscape is not expensive, and they could provide economic benefits (e.g., timber, herbs, honey, etc.) as well as fulfill some societal needs (hunting, photography, mushroom and berry picking, etc.). Very important from the point of ecological engineering, the biogeochemical barriers exert controlling effects on nonpoint pollution.

Mid-field water reservoirs also intercept chemical substances, immobilizing them in bottom deposits, where they are subjected to transformation by biogeochemical processes. The role of small mid-field ponds, neglected and often treated as wastelands, is particularly significant for more efficient use of fertilizers. They can serve as a tool for modification of matter cycling because the chemical compounds leached from fields could be returned to arable fields with sediment. Such forms of field fertilization were recommended long ago in Poland by General Dezydery Chłapowski in his book on agriculture, published in 1843.

The results of investigations concerning the functioning of landscapes led to the conclusion that a purposeful introduction of the biogeochemical barriers such as forest

PROSPECTS OF LANDSCAPE MANAGEMENT IN PROGRAMS OF SUSTAINABLE DEVELOPMENT

The majority of landscape studies deal with the description or modeling of patterns of change in land-use forms due to human actions (e.g., Latesteijn 1994, Lambin et al. 2000 and many others). Approaches concentrated on either a single process (e.g., deforestation, urbanization, desertification) or on one aspect of the driving processes (economic, political, environmental deterioration, etc.). One of the best reviews of the state of art in studies on land-use changes was provided by Turner et al. (1990). The change in human population during the last 300 years and its environmental impact on various ecosystems are well addressed in this study, which includes case studies of regional transformations. This analysis shows that human transformation of landscapes is caused by a complex interplay of behavioral, technological, social, and political factors affecting human demands for resources and living conditions. The mix of these factors has changed in time as well as in geographical scale, and changes are now beginning to proceed at such high rates that their impacts are being recognized in day-to-day observations of landscapes. Although various enacted policies have an important role in driving land-use changes, it is doubtful if the simple explanation for the observed changes can be found among characteristics of human population activities (Turner et al. 1990).

Energy sources at the disposal of human societies for performance of useful work can be partitioned in various ways, depending on the interplay of technical, societal, and political characteristics of communities. Those interactions evolved throughout history under the pressure of both internal and external factors. This process is illustrated well by the evolution of the CAP. The need to safeguard food supplies resulted in a policy protecting increased agricultural production, and for a period of about 20 years this driving mechanism of agricultural development determined its impacts on land-use changes in Western Europe. Environmental problems in the late 1980s and 1990s clearly demonstrated that ignoring the relationship of the human population to the environment was a mistake. Restriction of analysis to only the level of societal interactions therefore cannot lead to discovery of the driving mechanisms. According to the hierarchical systems theory, driving mechanisms can be discerned if one analyzes the total system into which the studied subsystem is embedded (O'Neill et al. 1986). Thus, not only should the socioeconomical processes be sought in an attempt to pinpoint the driving forces responsible for land-use changes, but environmental processes also should be involved in the analysis.

Feedback between land-use changes and processes of solar energy partition for various processes of matter cycling, climate formation, and biological organization is particularly useful for discerning control mechanisms of landscape changes. Landscape in this context is considered not as an aggregated set of landforms but rather

use increases. The possibility that human actions subjugating landscapes could simultaneously with alternations of habitats actually enhance processes that counterbalance their detrimental environmental effects is still poorly recognized. That situation persists despite the fact that the concept of sustainability is broadly discussed, and it is obvious that sustainability of the system could be maintained only if system-deteriorating processes are counterbalanced.

A short description of such possibilities suggested by modern landscape ecology studies is presented below. An increasing number of studies show that permanent vegetation strips can control migration of chemical compounds leached from cultivated fields with ground water (see published proceedings of conference on buffer zones edited by Haycock et al. 1997). Studies on riparian vegetation strips, shelterbelts, and meadow stretches have shown that when the ground- or surface water outflow from neighboring cultivated fields passes through those biochemical barriers, the concentrations of many chemical compounds decrease. Many studies show that nitrate concentrations in ground water from beneath all studied meadows and shelterbelts were significantly lower than those in ground water under adjoining fields. In some areas studied, the reduction in nitrate concentrations in ground water under the shelterbelts was 34-fold (from 37.6 mg $N-NO_3 \cdot dm^{-3}$ to 1.1 mg $N-NO_3 \cdot dm^{-3}$; Bartoszewicz and Ryszkowski 1996).

The studies of Kędziora et al. (1995) disclosed the complex control mechanisms exerted by the biogeochemical barriers. In studies on migration of phosphates across a riparian meadow 80 m wide, separating a cultivated field from the drainage channel, the input of $P-PO_4^{-3}$ with ground water was 5.1 g into the segment of meadow 1 m wide during the vegetation season (March 21–October 31). The output of $P-PO_4^{-3}$ to the channel amounted to 0.5 g. Thus, absorption by meadow during the plant growth season amounted to 4.6 g, that is, 90% of input. Total primary production (above- and belowground) was estimated for 1336 $g \cdot m^{-2}$ dry weight (d.w.) biomass. The estimated contents of P in plant tissue was 1.78 $mg \cdot g^{-1}$ d.w. Thus, the P uptake by plants was equal to 1.81 $g \cdot m^{-2}$. The grass band (80 m long and 1 m wide) absorbed 145 g of P (80 $m^2 \cdot 1.82$ $g \cdot m^{-2}$). Thus, the plan's uptake of P from ground water incoming from cultivated fields (4.6 g) constituted only 3% of the total amount of P absorbed by plants in the band, and the main source of P for plant growth was obtained from its recycling within the soil-plant system. The mechanisms of internal recycling within the biogeochemical barrier therefore play a very important role determining efficiency of control exerted by the biogeochemical barrier. The cases when the biogeochemmical barriers change from sink to source of chemical compounds (Ryszkowski et al. 1999) can be explained by inefficiency of the internal recycling and its decreased storing capacities, which depend to a large extent on biological processes. Further studies on recycling of elements in the biogeochemical barriers will provide guidelines for control efficiency improvement of diffuse pollution by various bio-

Table 2.1   Evapotranspiration (mm) during the Growing Season (March 21–October 31)
            under Warm and Dry Climatic Conditions (Mean Temperature 12.9°C, Precipitation
            234 mm) in Various Landscape Elements

Forms of Land Use	Cultivated Field C	Meadow M	Coniferous Shelterbelt A	C+10% of A	Field between Shelterbelts	C+20% of A	Field between Shelterbelts
Evapo-transpiration	362	437	558	376	355	390	348

Source: Modified after Ryszkowski and Kędziora 1995.

Table 2.2   Annual Rate of Ground Water Recharge (mm·y⁻¹) in
            Watershed Covered by Different Kinds of Plant Cover

Meteorological Conditions	Dry Year	Normal Year	Wet Year
Precipitation	627	749	936
Watershed covered by:			
Cultivated fields	108	233	351
Grasslands	0	155	271
Forests	0	149	181

Source: Modified after Werner et al. 1997.

such as air temperature, moisture, and wind speed are well known. Nevertheless,
feedback of those modifications on water cycling is not often used for conserving
water resources in the agricultural landscape. Thus, for example, shelterbelt planting
in a landscape consisting solely of cultivated fields, although increasing evapotranspiration rates of the whole area, protects fields against winds, and decreases their
evapotranspiration (Ryszkowski and Kędziora 1995). Thus, a shelterbelt can be
considered a landscape structure that redistributes evapotranspiration in the watershed. Areas under shelterbelts show higher evapotranspiration rates while cultivated
fields conserve some water.

The shelter effect is greater in dry and warm weather than in wet and cool
weather. In landscapes with 10% of the area under shelterbelts, the difference
between evapotranspiration rate from open cultivated fields and sheltered fields was
estimated to be 7 mm (362 mm to 355 mm), and when shelterbelts covered 20% of
the area this difference was 14 mm or 14 liters of water/m² (Ryszkowski and
Kędziora 1995; Table 2.1).

Studies in humid climatic conditions showed differences are not great; nevertheless, the conserved water in sheltered fields can delay wilting of cultivated plants.
The large differences in evapotranspiration rates between cultivated fields and afforestations result in substantial differences in the water runoff rates from watersheds

Changes in plant cover not only modify evapotranspiration rates and watershed water discharge rates but also influence vertical fluxes of energy leading to air heating (vertical fluxes of sensible energy). Cultivated fields convert much more solar radiation into air heating than do afforestations (Kędziora and Ryszkowski 1999). Through the change of sensible energy fluxes and evapotranspiration, the plant cover structure of a landscape may modify mesoscale air circulations, cloud formation, and precipitation. Stohlgren et al. (1998) indicated that in the plains of Colorado conversion of natural vegetation into cultivated fields and urban habitats influenced climate and vegetation in adjacent natural areas in the Rocky Mountains. That effect of landscape cover structure on climatic processes is especially interesting with regard to studies on the driving mechanisms of the global climatic change.

Many studies have shown that mosaic landscapes composed of cultivated fields and nonarable habitats house much richer and more diverse plant and animal communities than do uniform cultivated fields (see, e.g., Baudry 1989, Karg and Ryszkowski 1996). Those findings indicate that by structuring landscape with various refuge sites farmers could counteract biodiversity impoverishment caused by modern agricultural technologies.

There are also many studies indicating that mosaic landscapes efficiently counteract erosion processes (Pimentel 1993). Permanent plant stands reduce runoff to 90% or more and stimulate water infiltration into soil. The most efficient methods of erosion control consist of both within-farm and landscape measures. At the farm level, recommended measures include crop rotations, mulches, no-till, countour planting in hilly areas; at the landscape level, grass strips, hedges, field margins, and shelterbelts are proposed. These two approaches constitute the integrated soil conservation programs (Pimentel 1999).

The above discussion of options for counteractions to threats brought by intensive agriculture shows there is a growing body of ecological knowledge indicating that management of agricultural landscape for its structural diversity is becoming the important pillar of the sustainability of rural areas. Contemporary programs of environmental protection should not only aim at introducing environmentally friendly technologies of cultivation within the cultivated field or farm, important though that is; they should also be concerned with the challenge of increasing the resistance or resilience of the whole landscape against widely spreading threats. This challenge could be approached by stimulating natural processes underpinning the control of diffuse pollution, efficiency of water retention, wind and water erosion, heat balance changes, and maintenance of biodiversity. Landscape management that harnesses the natural processes of ecosystem and landscape functioning does not represent a refutation of modern means of agricultural production. Rather it represents a more rational strategy which seeks to meet the objectives of food production and of landscape protection and thus contributes to the sustainable development of

24   LANDSCAPE ECOLOGY IN AGROECOSYSTEMS MANAGEMENT

modified by actions at the farm level. Thus, for example, a farm constitutes the economical operational unit, while in regard to energy fluxes and matter cycling processes the landscape is a more autonomous unit (Ryszkowski 1998b). To discern the operational relationships between various levels of a hierarchical organization of the landscape constitutes the challenge for landscape ecologists (O'Neil et al. 1986).

Disentanglement of those relationships will also have great practical significance. The new CAP policy offers opportunities for integrating agricultural production with environment protection in the form of cross-compliance requirements. This approach requires farmers to observe a set of prespecified land-use conditions and practices in return for the receipt of government payments. These requirements concern introduction of more environmental friendly technologies to withdraw a land from production. Lack of landscape perspective in those requirements makes them less efficient. For example, there are no specific instructions indicating which parcels with respect to landscape functions should be withdrawn in the set-aside program. There is also a lack of regulation concerning landscape planning strategies. It seems that incorporating the landscape approach into proposed regulations of the new CAP policy will enhance its ecological effects. Landscape ecology, therefore, generates not only great scientific interest but also practical interest. Thus, landscape approach to the agriculture has become one of the pillars of sustainability.

REFERENCES

Balter M. 1998. Why settle down? The mystery of communities. *Science* 282:1442–1445.

Bartoszewicz A., Ryszkowski L. 1996. Influence of shelterbelts and meadows on the chemistry of ground water. In *Dynamics of an Agricultural Landscape*. Eds. L. Ryszkowski, N. R. French, A. Kędziora. Państwowe Wydawnictwo Rolnicze i Leśne. Poznań: 98–109.

Baudry J. 1989. Hedgerows and hedgerow networks as wildlife habitat in agricultural landscapes. In *Environmental Management in Agriculture*. Ed. J.R. Park. Belhaven Press. London: 111–124.

Boyden S. 1992. *Biohistory: The Interplay between Human Society and the Biosphere*. Parthenon Publishing Group. Paris. 265 pp.

Bülow-Olsen A. 1988. Disappearance of ponds and lakes in southern Jutland, Denmark 1954–1984. *Ecol. Bull.* (Copenhagen) 39: 180–182.

Chłapowski D. 1843. O rolnictwie. Druk Walentego Stefańskiego. Poznań. 164 pp.

Com, 1999. *Direction toward Sustainable Agriculture*. Commission of the European Communities. Brussels. 30 pp.

Denisiuk Z., Kalemba A., Zając T., Ostrowska A., Gawliński S., Sienkiewicz J., Rejman-Czajkowska M. 1992. *Integration between Agriculture and Nature Conservation in*

Gutry-Korycka M. 1978. Evapotranspiration in Poland (1931–1960) [in Polish]. *Przegląd Geograficzny* 23: 295–299.

Hannah L., Lohse D., Hutchinson C., Carr J., Lankerani A. 1994. A preliminary inventory of human disturbance of world ecosystems. *Ambio* 23: 246–250.

Haycock N.E., Burt T.P., Goulding K.W.T., Pinay G. (Eds). 1997. *Buffer Zones: Their Processes and Potential in Water Protection.* Quest Environmental. Harpenden, U.K. 326 pp.

IUCN. 1980. *The World Conservation Strategy.* IUCN-UNEP. Gland, Switzerland. 156 pp.

Kaniecki A. 1991. Drainage problems of Wielkopolska Plain during last 200 years and changes of water relationships [in Polish]. In *Ochrona i Racjonalne Wykorzystanie Zasobów Wodnych na Obszarach Rolniczych w Regionie Wielkopolski.* ODR Sielinko. Poznań: 73–80.

Kauppi L. 1990. Hydrology: water quality changes. In *Toward Ecological Sustainability in Europe.* Eds. A. M. Solomon and L. Kauppi. International Institute for Applied System Analysis. Laxenburg (Austria): 43–66.

Kleczkowski A.S., Mikulski Z. 1995. Water management forecast — State of resources (in Polish). In *Prognoza Ostrzegawcza Zmian Środowiskowych Warunków Życia Człowieka w Polsce na Początku XXI Wieku.* Ed. S. Kozłowski. Instytut Ekologii PAN. Oficyna Wydawnicza, Dziekanów Leśny: 35–46.

Kędziora A., Ryszkowski L. 1999. Does plant cover structure in rural areas modify climate change effects? *Geographia Polonica* 72: 65–87.

Kosturkiewicz A., Kędziora A. 1995. Problems of water management in agricultural areas [in Polish]. In *Zasady Ekopolityki w Rozwoju Obszarów Wiejskich.* Eds. L. Ryszkowski and S. Bałazy. Research Centre for Agricultural and Forest Environment. Poznań: 73–98.

Lambin E.F., Rounsevell M.D.A., Geist H.J. 2000. Are agricultural land-use models able to predict changes in land-use intensity? *Agric. Ecosys. Environ.* 82: 321–331.

Latesteijn van H.C. 1994. Ground of choice: a policy oriented survey of land use changes in the EC. In *Functional Appraisal of Agricultural Landscape in Europe.* Eds. L. Ryszkowski and S. Bałazy. Research Centre for Agricultural and Forest Environment. Poznań: 289–297.

Laude Y. 1996. *The Enlargement of the Common Agricultural Policy to the Ten Central and Eastern European Countries.* Club de Bruxelles. Brussels. 95 pp.

Lev-Yadun S., Gopher A., Abbo S. 2000. The cradle of agriculture. *Science* 288: 1602–1603.

Lubchenco J. 1998. Entering the century of the environment: a new social contract for science. *Science* 279: 491–497.

Medvedev V.W., Bulygin S.J. 1996. The experience of agricultural land conservation in the Ukraine. In *Landscape Diversity: A Chance for the Rural Community to Achieve a Sustainable Future.* Eds. L. Ryszkowski, G. Pearson and S. Bałazy. Research Centre for Agricultural and Forest Environment. Poznań: 176–179.

Miller G.H., Magee J.W., Johnson B.J., Fogel M.N., Spooner N.A., McCulloch M.T., Ayliffe L.K. 1999. Pleistocene extinction of *Genyornis newtoni*: human impact on Australian megafauna. *Science* 283: 205–208.

OECD. 1986. *Water Pollution by Fertilizers and Pesticides.* OECD. Paris. 144 pp.

O'Neill R.V., DeAnglis D.L., Waide J.B., Allen T.F.H. 1986. *A Hierarchical Concept of*

Pringle H. 1998. The slow birth of agriculture. *Science* 282: 1446–1450.

Reenberg A., Baudry J. 1999. Land-use and landscape changes — the challenge of comparative analysis of rural areas in France. In *Land-Use Changes and Their Environmental Impact in Rural Areas in Europe*. Eds. R. Krönert, J. Baudry, I. R. Bowler, A. Reenberg. The Parthenon Publishing Group. Paris: 23–41.

Ryszkowski L. 1994. Strategy for increasing countryside resistance to environmental threats. In *Functional Appraisal of Agricultural Landscape in Europe*. Eds. L. Ryszkowski and S. Bałazy. Research Centre for Agricultural and Forest Environment. Poznań: 9–18.

Ryszkowski L. 1998a. Ecological guidelines for the management of agricultural landscapes. In *Modern Trends in Ecology and Environment*. Ed. R.S. Ambasht. Backhuys Publishers. Leiden. 187–201.

Ryszkowski L. 1998b. Guidelines for environment protection against diffuse pollution (in Polish). In *Kształtowanie Środowiska Rolniczego na Przykładzie Parku Krajobrazowego im. Gen. D. Chłapowskiego*. Eds. L. Ryszkowski and S. Bałazy. Zakład Badań Środowiska Rolniczego i Leśnego PAN. Poznań: 81–88.

Ryszkowski L. 1998c. Farm in the agricultural landscape (in Polish). In *Dobre Praktyki w Produkcji Rolniczej*. Ed. I. Duer. Instytut Uprawy, Nawożenia i Gleboznawstwa. Puławy: 491–505.

Ryszkowski L. 2000. The coming change in the environmental protection paradigm. In *Implementing Ecological Integrity*. Eds. P. Crabbe, A. Holland, L. Ryszkowski and L. Westra. Kluwer Academic Publishers. Dordrecht: 37–56.

Ryszkowski L., Bałazy S. 1995. Strategy for environment and nature protection in rural areas (in Polish). In *Zasady Ekopolityki w Rozwoju Obszarów Wiejskich,* Eds. L. Ryszkowski and S. Bałazy. Zakład Badań Środowiska Rolniczego i Leśnego PAN, Poznań: 49–64.

Ryszkowski L., Karg J. 1992. Energy flow in rye grown in continuous and norfolk rotation cultures. *Acta Academiae Agriculturae AC Technicae Olstenensis. Agricultura* 55: 201–213.

Ryszkowski L., Bartoszewicz A., Kędziora A. 1999. Management of matter fluxes by biogeochemical barriers at the agricultural landscape level. *Landscape Ecol.* 14: 479–492.

Stanners D., Bourdeau P. 1995. *Europe's Environment*. European Environment Agency. Copenhagen. 676 pp.

Stern A. 1996. Environment. The mid-field review of the European Union's Fifth Action Programs. Club de Bruxelles. Brussels. 125 pp.

Stiner M.C., Munro N.D., Surovell T.A., Tchernow E., Bar-Yosef O. 1999. Paleolithic population growth pulses evidenced by small animal exploitation. *Science* 283: 190–193.

Stohlgren T.J., Chase T.N., Pilke R.A., Kittels G.F., Baron J.S. 1998. Evidence that local land use practices influence regional climate, vegetation and stream flow patterns in adjacent natural areas. *Global Change Biol.* 4: 495–504.

Szabolcs I. 1991. Salinization potential of European soils. In *Land Use Changes in Europe*. Eds. F.M. Brouwer, A.J. Thomas, M.I. Chadwick. Kluwer Academic Publishers. Dordrecht: 293–315.

Thenail C.B.J. 1996. Consequences on landscape pattern of within farm mechanisms and of land use changes. In *Land Use Changes in Europe and Its Ecological Consequences*.

DEVELOPMENT OF AGRICULTURE AND ITS IMPACT ON LANDSCAPE FUNCTIONS 27

Werner A., Eulenstein F., Schindler U., Müller L., Ryszkowski L., Kędziora A. 1997. *Ground-wasserneubildung und Landnutzung. Zeitschrift für Kulturtechnik und Landentwicklung* 38: 106–113.
Van Lynden G.W.J. 1955. *European Soil Resources.* Nature and Environment 71. Council of Europe Press. Strasbourg. 99 pp.
Vitousek P.M., Aber J.D., Howarth R.W., Likens G.E., Matson P.A., Schindler D.W., Schleisinger W.H., Tilman D.G. 1997a. Human alteration of the global nitrogen cycle: sources and consequences. *Ecol. Appl.* 7: 737–750.
Vitousek P.M., Mooney H.A., Lubchenco J., Melillo J.M. 1997b. Human domination of earth's ecosystems. *Science* 277: 494–499.
Voloshyn V., Lisovsky S., Gukalova I. Reshetnik V. 1999. Ecological-economic problems of land use in the Ukraine. In *Land-Use Changes and Their Environmental Impact in Rural Areas in Europe.* Eds. R. Krönert, J. Baudry, I.R. Bowler and A. Reenberg. The Parthenon Publishing Group. Paris: 221–234.

CHAPTER **3**

Mitigation of Radiation and Heat Balance Structure by Plant Cover Structure

Janusz Olejnik, Andrzej Kędziora, and Frank Eulenstein

CONTENTS

INTRODUCTION

The Earth's atmosphere is an extremely important medium that links the other elements of the environmental system. There are permanent fluxes of energy and matter exchange between the Earth's surface and its atmosphere. All processes occurring on the Earth are possible because of spatial differentiation of potential energy. The realization of each irreversible process is connected with entropy production — in other words, with the increase of "disorder" in a thermodynamic system. With regard to the Earth–atmosphere system, it means that if no energy flux flows into the system from outside, the temperature within the system will equalize and all processes will be stopped. Yet constant flux of solar energy acts as a brush; it

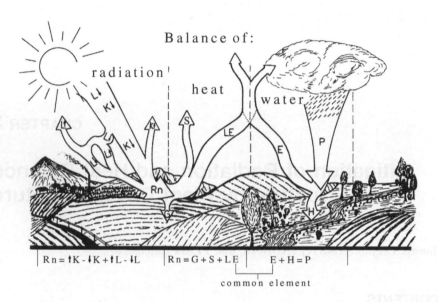

Figure 3.1   Schematic connection of radiation, heat, and water balances in agricultural landscape.

processes, which are essential for life on Earth, is possible. The most important of these are the matter-cycling processes, especially water cycling.

Due to these fluxes, one can consider the atmosphere a mirror, in which the present conditions of the Earth's surface environment are "reflected." On the other hand, the condition of the environment can be modified by the state of the atmosphere, both by short- and long-term interactions. Industrial and agricultural development in the 19th and 20th centuries led to unfavorable changes in the natural environment. At present, while these changes are commonly recognized on both the global and local scales, it is important to stop or at least minimize the effects of those processes that have negative effects on the environment. The crucial question is how to manage the landscape to create conditions that will enable us to return (if possible?) to ecological equilibrium (Stigliani et al. 1989, Freedmen 1992, Kędziora 1999). To answer that question, we first must learn more about interactions between the atmosphere and the Earth's surface.

One of the ways to achieve a better understanding of all the processes that occur in the natural environment is to measure or model them. Results of measurement of energy fluxes create the possibility of parameterization of these processes, consequently leading to development of these models, which can be applied to different

One way of describing environmental processes on different scales (from micro to global scale) is by the analysis of energy flux exchange between the Earth's surface and the atmosphere. The knowledge about these fluxes and their interactions provides information not only about energy exchange in the environment but also about water, carbon, and nutrient balances and can be helpful for their estimation when direct measurements over large areas are difficult or impossible. Therefore, the analysis of heat balance structure is one of the best descriptions of environmental conditions on both local and global scales (Paszyński 1972, Rosenberg 1974, Kim and Verma 1990, Olejnik 1996).

The driving force of energy in water, carbon, or nitrogen cycles and the interactions of these cycles provide a basic framework for understanding the potential feedback of the active surface and the atmosphere. Most of these relationships exert strong biological control on land-atmosphere interactions. Therefore, land-cover and land-use changes are driving variables for understanding energy exchange and, consequently, climate processes on both regional and global scales (Valentini et al. 1999).

The landscape is a very complex, open, dynamic system of relationships in which nonlinear processes and feedback mechanisms play an important role. Many of these feedback mechanisms depend on the process of energy and mass exchange between the atmosphere and the Earth's surface. The so-called surface processes are strongly connected with the transfer of water or water vapor from and into the atmosphere. Part of the water that reaches the Earth's surface returns to the atmosphere via evapotranspiration and part goes to sea by runoff. The available energy is the most important physical factor that limits the amount of evapotranspiration. On the other hand, the kinds of plants that cover the Earth's surface as well as the plant development stage are the most important biological factors. For model investigations on heat and water balance structure, both physical and biological factors have to be included in models for all scales — from field to global.

HEAT BALANCE COMPONENTS

The middle section of Figure 3.1 shows the heat balance of an active surface. In general, energy exchange processes are driven by solar radiation. The right part of Figure 3.1 shows the radiation balance. The short-wave radiation reaching the Earth's surface is called global radiation ($\downarrow$K). The short-wave radiation is partly absorbed by the active surface (soil, plant, water, etc.) but partly reflected ($\uparrow$K) from the active surface. The ratio of reflected to incoming global radiation is called albedo ($\alpha = \uparrow$K/$\updownarrow$K) and is one of the most important parameters characterizing the input of energy into agroecosystems, forests, deserts, etc. For example, in natural conditions, high albedo (e.g., 0.3) means that a substantial part of the short-wave radiation energy reaching the ecosystem is reflected and only 70% is absorbed by the active surface.

atmosphere and to space, but part returns to the Earth's surface ($\downarrow$L). If one adds all four fluxes — two incoming (short wave from the sun and long wave from the atmosphere back to the Earth's surface) and two outgoing (short wave reflected and long wave emitted by the Earth's surface) — one can construct the radiation balance of an active surface (Figure 3.1). The result of the balance of all these fluxes is the net radiation (Rn), and it denotes energy supply for all processes in the environment driven by natural energy, including photosynthetic processes, which utilize solar radiation for organic matter production. However, metabolic processes may be considered negligible relative to the magnitude of other energy fluxes, and, therefore, it is possible, following the energy conservation law, to write the energy (heat) balance equation as:

$$Rn = Le + S + G \qquad (3.1)$$

where Rn is net radiation, LE is latent heat flux (L is evaporative heat of water and E is evapotranspiration), S is sensible heat flux, and G is soil heat flux.

Evapotranspiration of the plant community (or its energy equivalent, latent heat flux), warming of the soil surface, and consequently, also deeper soil layer (due to conductivity) or convection heating of the atmosphere are well-known processes occurring in the environment. Rn sets the energy utilized by these processes. The partition of net radiation for these processes is described by the heat balance equation and depends on both physical and biological factors that determine environmental conditions: plant type, land-use structure, plant development stage, weather conditions, etc.

Through the latent heat flux (or evapotranspiration), there is a direct connection between heat and water balance of an investigated area. Figure 3.1 shows the connections between radiation, heat, and water balances. The links between these balances play a very important role for the complex description of energy-water conditions of the environment and are often used by investigators to calibrate or validate different models for estimating mass and energy exchange processes. At present, many modeling activities use these principles to calculate the mass and energy exchange for several land-use/cover types, from the patch (small area having the same characteristics of habitat) to regional or global scale. Nevertheless, since the concept is relatively new, it has not been tested widely against field measurements of mass and energy exchange, nor has this concept been tested over an assortment of vegetation types. Experimental measurements of biosphere-atmosphere mass and energy exchange have the additional advantage of providing a new synthesis view of ecosystem processes, which allows us to improve our conventional classification of vegetation types and forms, achieving a deeper insight into functional biodiversity at the landscape level (Kędziora and Olejnik 1996, Valentini et al. 1999, Kędziora and Ryszkowski 1999).

Recent improvements in the eddy covariance method, which provides a direct measure of biosphere-atmosphere mass and energy fluxes at the ecosystem level, were particularly important for studies (Ryszkowski and Kędziora 1993b, Tenhunen and Kabat 1999, Kędziora et al. 2000). In the past, most of our knowledge of carbon, water, or energy exchange in vegetation was based on measurements taken at the leaf level with enclosure chambers or portable cuvettes (Field et al. 1982). With the development of micrometeorological methods, a large change in scale of focus was achieved, moving from the leaf level, species/individual-dependent approach to an integration of overall ecosystem fluxes in a direct, nondestructive way. There are many micrometeorological methods based on physical principles, which allowed measurement of the turbulence components of heat balance (LE and S in Equation 3.1). According to the eddy covariance theory, the eddy flux of any scalar can be written as:

$$F_c = \overline{\rho_c \cdot w} \qquad (3.2)$$

where F_c is the flux density of scalar c, w is the vertical wind speed, and ρ_c is the density (or concentration) of the transported species. The overbar represents the mean of product over the sampling interval. The eddy covariance method technique, developed during the 1980s and the 1990s, needs very technologically advanced sensors, and therefore it is not so commonly used. To measure eddy fluxes with this technique, rapid response sensors are necessary. These sensors and incorporated data acquisition systems are able to measure and collect data with very high frequencies (10 Hz and more). To measure heat or mass fluxes, sonic anemometer and open- or close-path gas analyzers are used (Arya 1988).

Less technologically advanced sensors can be used for eddy flux measurements using gradient or heat balance methods. To determine the latent and sensible heat fluxes (LE and S in Equation 3.1) measurements of temperature and water vapor pressure in the air at several levels are necessary. For such measurements, adequate electrical psychrometers and cup anemometers can be used. It is known that latent (LE) and sensible (S) heat flux densities are proportional to water vapor and air temperature gradients, respectively (Monteith 1975), as follows:

$$LE = Kv\rho c_p \, \gamma^{-1} \frac{\partial e}{\partial z} \qquad (3.3)$$

$$S = K_H \rho c_p \frac{\partial T}{\partial z} \qquad (3.4)$$

where γ is the psychrometric constant equal to 0.65 hPa·K^{-1}, e is the water vapor

34   LANDSCAPE ECOLOGY IN AGROECOSYSTEMS MANAGEMENT

(Equation 3. 1) (for example, the so-called Bowen ratio method, Olejnik et al. 2001). Using this method, some results of heat balance components determination with respect to land-use conditions are discussed later in this chapter.

There are many models that describe explicitly the physical processes in the environment, but to use them on a large or global scale is very difficult because of computer power and input data limitations. Therefore, sometimes an empirical model can be used, and, after the modeling procedure, its results can be coupled with a larger or even global scale model (e.g., GCM) (Kundzewicz and Somlyódy 1993). Even empirical models do not explain different phenomena themselves, but the results from these models can be very helpful by quantitatively describing many environmental processes.

One of the considerable limitations for modeling is the data availability for the input data set. There are some quite precise models for heat or water balance components estimation (McNaughton and Spriggs 1985, Holtslag and De Bruin 1988, Beljaars and Holtslag 1991, Wegehenkel 2000), but their application for larger areas is very limited because the input data set needed for these models is very often inaccessible. In some cases it is impossible, by now, to collect the input data, such as moisture or physical properties of soils for a large area at satisfactory resolution. Therefore, it seems that the MBC model recently developed by the Agrometeorology Department of the Agricultural University of Poznań (Olejnik and Kędziora 1991, Olejnik 1996, Leśny 1998) is a compromise between the quality and resolution of conclusions and the possibilities of input data set collection and appropriate consideration of physical and biological processes included in the modeling procedure. The resolution of MBC (limited only by land-use data resolution) can be significantly higher than GCM's resolution, and the input data set is relatively easy to collect for larger areas. This model, or its earlier stage, has been successfully applied on different scales: on the field scale (Olejnik 1988), the catchment scale (Kędziora et al. 1989), the regional scale (Olejnik and Kędziora 1991), and the country scale (Ryszkowski et al. 1991, Ryszkowski and Kędziora 1993).

The MBC model is based on the empirical equation that links the Bowen ratio with meteorological parameters and plant development stage. In 1986 an agrometeorological index was proposed (Olejnik 1988), as follows:

$$W = \left[100\left(D\,v^{0.5}\right)^{\mathrm{atn}\,\Pi/2f}\right] \Big/ \left[T(s+0.4)\right] \tag{3.5}$$

where D is saturation vapor pressure deficit in [hPa], v is wind speed in $[\mathrm{m\cdot s^{-1}}]$, f is plant development stage (changing from 0 to 1), T is air temperature in [°C], and s is relative sunshine.

By definition, the Bowen ratio is equal to $\beta = S/LE$. Using the proposed index,

Finally, from the combination of the Bowen ratio and Equation 3.1, the latent and sensible heat fluxes can be estimated in the MBC model as follows:

$$LE = (Rn + G)/[1 + \beta(W)] \tag{3.7}$$

$$S = (Rn + G)/[1 + 1/\beta(W)] \tag{3.8}$$

Using the above equations and data about land-use structure, as well as plant cover development stage, the first version of MBC model for areal heat and water balance estimation was proposed (Kędziora et al. 1989, Olejnik and Kędziora 1991). Later in this chapter, some results of areal heat balance components estimation with respect to land-use conditions are discussed using the MBC model application.

DAILY VALUE OF HEAT BALANCE COMPONENTS FOR SELECTED ECOSYSTEMS

Using the above-described methodology, we can measure or model heat balance components (Equation 3.1). During the last decade, many measurements have been made at different sites from the climatic and plant cover viewpoints. This chapter discusses some results of heat balance daily dynamics on the basis of *in situ* investigations in different conditions carried out by the Agrometeorology Department of the Agricultural University in Poznań. The measurements were carried out in different climatic zones to study the impact of climatic conditions on heat balance components.

In general, the structure of heat balance is formed mainly by three factors:

- Amount of solar energy reaching the active surface depending on latitude and daily course of sunshine
- Physical features of the active surface, mainly richness and structure of plant cover
- Habitat moisture, which controls the intensity of evapotranspiration and thus the intensity of latent heat flux

An active surface is any surface through which energy and matter are passing and simultaneously changing their characteristics. For example, solar short-wave radiation is absorbed by the soil and re-emitted as long wave radiation. Similarly, water flowing in the soil as liquid changes its state to gaseous form by evaporation.

The typical daily courses of heat balance components on sunny days are presented in Figure 3.2 (Kędziora et al. 1997). In arid climatic conditions, when the soil surface is covered with very poor vegetation, the sensible heat flux constitutes

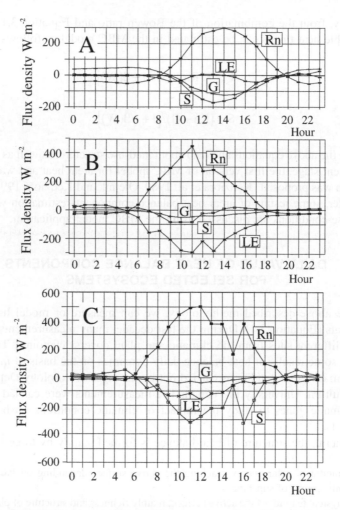

Figure 3.2 Typical 24-h courses of heat balance components in different climatic zones. A — arid zone (Kazakhstan), B — transition zone (Turew, Poland), C — humid zone (Cessieres, France).

climate conditions (Figure 3.2B) the LE/Rn ratio exceeded 90%. In the case of oceanic climate (Figure 3.2C) the LE/Rn ratio was almost 100% (Table 3.1).

An entirely different situation was observed in arid or semi-arid climatic conditions, when plant cover existed and water supply was unlimited. When a dry hot area surrounds an irrigated alfalfa field (Figure 3.3), very high latent heat flux was observed. The average daily net radiation on observational days was not higher than

Table 3.1 Heat Balance Components Measured at Different Sites Located in Different Climatic Zones

Site	Date	Heat Balance Components [Wm⁻²]				Ratios		
		Rn	LE	S	G	LE/Rn	S/Rn	S/LE
A	20.08.89	55	−59	−5	+9	−1.07	−0.09	0.08
	21.08.89	83	+2	−67	−18	+0.02	−0.81	−33.50
	25.08.89	70	−9	−44	−17	−0.13	−0.62	4.90
K	11.07.88	60	−38	−18	−4	−0.63	−0.30	0.47
	17.07.88	149	−133	−7	−9	−0.89	−0.05	0.05
T	13.07.83	149	−79	−75	5	−0.53	−0.50	0.95
	17.08.83	65	−25	−30	−11	−0.38	−0.46	1.20
	16.05.84	115	−127	37	−25	−1.10	0.32	−0.29
	30.06.85	126	−100	−12	−13	−0.79	−0.10	0.12
	09.08.87	106	−100	+2	−7	−0.94	0.02	−0.02
C	16.07.93	164	−152	−6	−6	−0.93	−0.04	0.04
	16.07.93	145	−56	−77	−12	−0.39	−0.53	1.37
Z	15.07.94	181	−178	+13	−16	−0.98	0.07	−0.07
	16.07.94	65	−129	+48	+16	−1.98	0.74	−0.37
	17.07.94	182	−209	+39	−12	−1.14	0.21	−0.18

In Alma-Ata (A) the climate is continental, arid, with low precipitation, high temperature in summer and very low temperature in winter, low air humidity, and a high percentage of sunshine and high global radiation.

In Kursk (K) the climate is less continental. It is humid, with moderate precipitation, temperature, sunshine and air humidity, and a relatively high global and net radiation.

In Turew and Müncheberg (M and T) the climate is transitional (between continental and oceanic), with moderate temperature, air humidity, and precipitation, and low sunshine and global and net radiation.

In Cessieres (symbol C) the climate is humid, with relatively high precipitation, and moderate sunshine and global and net radiation.

In Zaragoza (symbol Z) the climate is warm, semi-arid, with low precipitation and air humidity, high sunshine, and global and net radiation.

The names of these cities were used symbolically; the study sites were located near cities in agricultural landscapes.

ratio means that sensible heat flux is oriented toward the active surface, having a positive sign, while latent heat flux of evapotranspiration has a negative sign (water vapor is flowing from the surface into the atmosphere). Therefore, plant cover plays the most important role for the structure of heat balance, because it is the factor by which the evapotranspiration process controls the density of latent heat flux. This conclusion is supported by the case of heat balance structure for July 16 in Zaragoza (Figure 3.3B). The previous day was sunny, with very intense solar radiation and evapotranspiration. It was completely cloudy on July 16, but the evapotranspiration rate remained very high because of a deep temperature inversion over the alfalfa

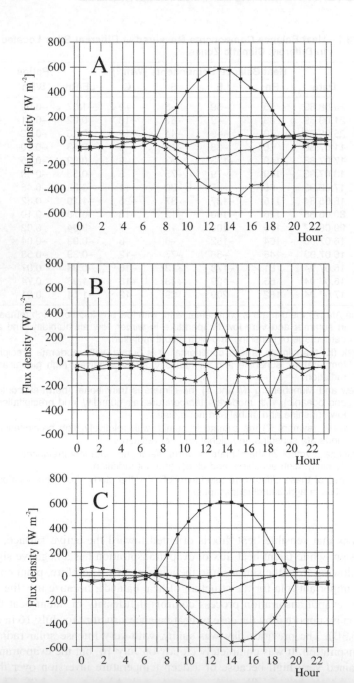

Table 3.2   Impact of Landscape Structure on Heat and Water Balance in the Growing
Season (March 21 to October 31) in Turew

Landscape	Rn	LE	S	ETP	ETR	ETR/ETP	ETR/OP
Grain crop monoculture	1542	1035	495	650	414	0.64	1.10
Grain crops with shelterbelts	1586	1078	496	586	431	0.76	1.15
Grain crops with windbreaks	1567	1010	546	581	404	0.86	1.08
Grain crops without shelterbelts under convection	1586	1258	315	898	503	0.56	1.34
Grain crops with shelterbelts under advection	1586	1181	412	592	464	0.78	1.24

Heat balance components are expressed in $W \cdot m^{-2}$ and water balance components are expressed in mm.

landscape structure is the main factor determining the structure of heat balance. If irrigated fields or tree stands are surrounded by the very hot and dry areas, strong spatial differentiation of air temperature is observed. It creates strong horizontal temperature gradients which are a driving force for energy flux flowing from hot air over intensely transpiring surface. Even in humid climate conditions, heat advection over fully developed plant cover causes excessive evapotranspiration. Creation of the proper landscape structure is the best way to counteract this unfavorable process. For example, introducing shelterbelts into a uniform cereal landscape can save as much as 40 mm water during the growing season under advection conditions (Table 3.2).

In humid climatic conditions during sunny days, the LE/Rn ratio also reached a very high value for cultivated fields (Figure 3.2B, Turew-alfalfa) amounting to –0.94 (Table 3.1). However, during cloudy days this ratio was not as high in a warm arid climate under irrigation conditions (Table 3.1).

Such an important influence on heat balance structure comes from only living plant cover. After maturity and death, plant cover plays a less important role in partition of net radiation into different fluxes. In Cessieres, France on July 16, 1993, heat balance observed for a living sugar beet field (Figure 3.4B) was quite different from that observed for a pea field after maturity (Figure 3.4A). For the sugar beet field, the LE/Rn ratio was 0.93 while for the pea field it was only 0.39. The Bowen ratios were 0.04 and 1.38, respectively (Table 3.1).

The above examples support the assumption that the impact of plant cover on heat balance structure is modified by habitat moisture conditions. This impact can easily be seen if longer periods of measurement of the heat balance components are analyzed. Comparison of average value of heat balance components for eight measurement periods in different agroecosystems are shown in Figure 3.5 (Olejnik et al. 2001). The measurement sites and time of investigation cover not only different plants and bare soil but also different weather conditions; measurements were carried

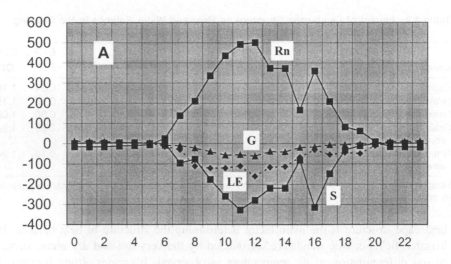

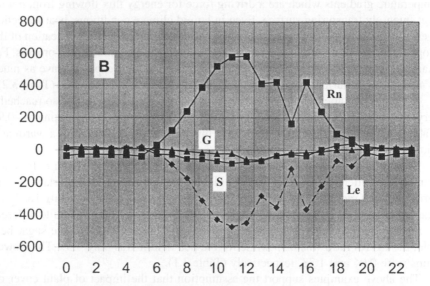

Figure 3.4   Daily course of heat balance components for peat field (A), and sugar beet field (B) in Cessieres, France on July 16.

September 24, 1997 for bare soil and was 232 Wm^{-2} on June 12, 1994 for oat; on these days latent heat flux varied from –10 Wm^{-2} to –203 Wm^{-2}, respectively.

The average values of the Bowen ratio varied from –0.04 for period P4 (sugar beet) to 1.0 for period P2 (bare soil) (Figure 3.5). The average value of Rn varied

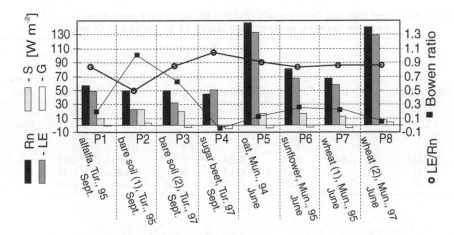

Figure 3.5 Comparison of averages of heat balance components and their ratios of eight several-day measurement periods (P1–P8).

flux (energy equivalent of evapotranspiration), varied from 0.49 to 1.04. The highest values of that ratio were observed for the fully developed plant community (sugar beat in September, or oat, wheat, and sunflower in their high plant development stage). On average, the α ratio (LE/Rn) is equal to 0.9 for well-developed plant cover, which means that almost 90% of available energy is utilized for latent heat flux. In cases of bare soil, that ratio is significantly lower. In the first period, P2, the α ratio is equal to 0.5 and rises to 0.78 for bare soil for period P3. In this case for bare soil, that ratio is relatively high because heavy rain occurred during the measurement period; consequently, the soil was very wet, and, although there were no plants, the vaporization was very intense and high latent heat flux was noted.

SEASONAL VALUE OF HEAT BALANCE COMPONENTS FOR SELECTED ECOSYSTEMS AND THE LANDSCAPE

To measure seasonal values of heat balance components, it is necessary to carry out long period measurements (over a growing season or whole year); one can use models that can estimate these components on the basis of meteorological and phenological data. The MBC model, described earlier, was used to estimate heat balance components during the growing season for some ecosystems in Turew (Wielkopolska region) in Poland.

During the growing season, the values of net radiation in the Turew ecosystems ranged from 1494 to 1883 MJ·m^{-2} (Table 3.3). The lowest net radiation was observed in the meadow ecosystem, while the highest was in water bodies and forest. Crops

**Table 3.3   Heat Balance Components (MJm^{-2}) for Different Ecosystems
(mean for growing season, from March 21 to the end of October)**

Component				Ecosystem			
				Cultivated Field			
	Water	Forest	Meadow	Winter Rape	Sugar Beets	Wheat	Bare Soil
Rn	1883	1730	1494	1551	1536	1536	1575
LE	1585	1572	1250	1163	1136	1090	866
S	120	121	215	327	339	385	651
G	179	37	29	61	61	61	47
LE/Rn	0.84	0.88	0.84	0.75	0.74	0.71	0.75
S/Rn	0.06	0.07	0.14	0.21	0.22	0.25	0.41
S/LE	0.08	0.08	0.17	0.28	0.36	0.45	0.75

(sensible heat flux, S) also ranged significantly for different ecosystems. For example, forest used nearly 5.5 times less energy for air heating than did the bare soil.

Energy used for soil heating (soil heat flux G) was the smallest part of net radiation and its average value during the growing season ranged from 29 MJ·m^{-2} for meadow to 61 MJ·m^{-2} for cultivated field. However, soil heat flux for bare soil during early spring could reach more than 300 joules per second per square meter, which is equal to the Rn value. The average value of soil heat flux, during the whole growing season, was small because during the first half of the growing season (March to the beginning of August) the soil warms and after this period the soil begins cooling. Consequently, at the beginning of the growing season soil can be considered an energy sink and after that an energy source — daily averages of soil heat flux change from positive to minus: at the beginning of the growing season it is the outgoing flux from the active surface (minus sign in Equation 3.1) and then from the beginning of August it is the incoming flux to the active surface (plus sign in Equation 3.1). Thus, although the seasonal average values of soil heat flux are rather small compared with other components of heat balance, at the beginning and end of the growing season the energy used for soil heating (spring) or cooling (autumn) can be high, equal to, or sometimes even exceeding, the net radiation value.

Various ecosystems use net radiation energy in different ways (Table 3.1). The energy used for evapotranspiration (latent heat flux, LE) ranges from 866 MJ·m^{-2} for bare soil to 1574 MJ·m^{-2} for forest and 1585 MJ·m^{-2} for water. Wheat has the lowest evapotranspiration value and water bodies have the highest (LE in Table 3.3) (Ryszkowski and Kędziora 1987, Ryszkowski and Kędziora 1993a).

The energy values used for evapotranspiration (latent heat flux, LE) also vary greatly for different ecosystems (Table 3.3). Forest uses about 40% more energy for evapotranspiration than does the wheat field, while the wheat field diverts approximately three times the energy to air heating than does the forest (Table 3.3). There

do, their leaves also have less stomata resistance. Second, shelterbelts also have greater canopy roughness than does wheat, which, together with higher wind speed in the shelterbelt canopy, results in more intense turbulent exchange over the shelterbelt (Ryszkowski and Kędziora 1987, Kędziora et al. 1989).

The proportion of energy used for air and soil heating also depends on plant cover. Soil and air heat fluxes depend mainly on temperature gradients existing in the air and soil strata near the active surface. These gradients depend on the amount of solar energy reaching the active surface, which in turn is related to the density and structure of the plant cover (Kędziora 1999). To illustrate the possibilities for mitigating heat and water balance structure of landscape under different weather conditions, the effect of introducing shelterbelts into agricultural landscape was evaluated (Ryszkowski and Kędziora 1987, 1995). Introducing shelterbelts into simplified landscapes is one of the best tools for managing heat balance in the landscape. Shelterbelts reduce wind speed and conserve the water supply of the field located between shelterbelts, but they increase little sensible heat flux. However, during strong advection of dry and warm air, irrigated fields can conserve during evapotranspiration as much as 10% more water in comparison with landscape without shelterbelts (Ryszkowski and Kędziora 1995).

Plant cover and landscape structure are the main factors determining the partitioning of net radiation into different internal energy processes of the ecosystem. Bare soil uses only 55% of available energy for evapotranspiration while forest uses 88% (Figure 3.6), but air is heated almost twice as much by bare soil (42% of net radiation) than by alfalfa field (only 20% of net radiation).

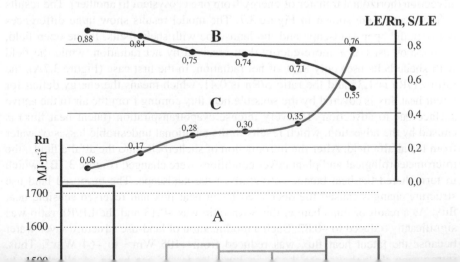

Again, using the MBC model, the heat balance components for different agricultural landscapes can be calculated. In the input data set of the MBC model, the structure of landscape can be modified and heat balance structure can be estimated for different (modeled or real) landscape structures. Such a procedure creates the possibility of analyzing the heat and water balance structure changes as a function of landscape structure before and after land-use changes. Analysis of the impact of the plant cover and land-use structure on differentiation of heat balance in agricultural landscape by the use of the MBC model shows that the more developed the landscape structure, the smaller the diversification of heat balance structure (Table 3.4). Three different landscape structures are considered. The first landscape, in the Turew region, consisted of 30% forest, 15% meadow, 10% row crops, and 45% cereal. The second was completely agricultural, a uniform landscape consisting of 20% row crops and 80% cereals. The third landscape was mosaic, consisting of 50% forest, 20% meadow, 10% row crops, and 20% cereals. The simplification of landscape structure caused a decrease in seasonal average value of latent heat flux density by 3.73 $W \cdot m^{-2}$ and an increase in sensible heat by 1.94 $W \cdot m^{-2}$ (Table 3.3, K2–K1). But changing landscape structure from simple (K2) to mosaic (K3) increased latent heat flux density by 6.2 $W \cdot m^{-2}$ and decreased sensible heat flux density by 3.31 $W \cdot m^{-2}$ (Table 3.4 K3–K2). So, the more complex landscape used more energy for evapotranspiration and less remained for air heating.

The other results of heat balance components modeling for open agricultural landscape compared with the same landscape with introduced shelterbelts are a good example of such analysis. For this calculation, the alfalfa fields together with bare soil in the neighborhood were chosen. Such a landscape structure creates energy advection (horizontal transfer of energy from one ecosystem to another). The results of the analysis are shown in Figure 3.7. The model results show huge differences between the open landscape and the landscape with shelterbelts. In the open field, evaporation uses 35% more energy than received by net radiation, while the field with shelterbelts uses only 82% of net radiation. In the first case (Figure 3.7A), the ratio LE/Rn is 1.35 and the ratio S/Rn is 0.41, which means the energy deficit for latent heat flux is covered by the sensible heat flux coming from the air to the active surface due to advection. The very intense evapotranspiration (latent heat flux) is caused by the advection, which results in the additional undesirable losses of water from the alfalfa field. After the introduction of shelterbelts into the alfalfa fields, the micrometeorological and plant cover conditions were changed (Figure 3.7B), which in turn altered the heat balance structure of the landscape. The proposed land use structure changes caused the decreased latent heat flux and reversed sensible heat flux. As a result of this change, the S/Rn ratio was –0.13 and the LE/Rn ratio was significantly reduced to –0.82. Such a modification of land-use structure saves water because the latent heat flux was reduced from –105 Wm^{-2} to –64 Wm^{-2}. Thus, introducing shelterbelts into the agricultural landscape can be an excellent way to

MITIGATION OF RADIATION AND HEAT BALANCE STRUCTURE   45

Table 3.4   Seasonal Average Values of Heat Balance Components under Present and Future Climatic Conditions and Different Landscape Structures

Parameter W·m⁻²	Landscape		
	K1	K2	K3
Latent heat flux density, LE, under present climatic conditions	49.93	46.20	52.40
Latent heat flux density, LE, under future climatic conditions	55.44	50.68	58.53
Sensible heat flux density, S, under present climatic conditions	17.85	19.79	16.48
Sensible heat flux density, S, under present climatic conditions	15.89	17.94	12.97

Change in Latent Heat Flux Density as a Result of Landscape Structure Change under Present Climatic Conditions

	K2-K1	K3-K1	K3-K2
	−3.73	2.47	6.20

Change in Sensible Heat Flux Density as a Result of Landscape Structure Change under Present Climatic Conditions

	K2-K1	K3-K1	K3-K2
	1.94	−1.37	−3.31

Change in Latent Heat Flux Density as a Result of Landscape Structure Change under Future Climatic Conditions

	K2-K1	K3-K1	K3-K2
	−4.76	3.09	7.85

Change in Sensible Heat Flux Density as a Result of Landscape Structure Change under Future Climatic Conditions

	K2-K1	K3-K1	K3-K2
	2.05	−2.92	−4.97

Change in Latent Heat Flux Density as a Result of Landscape Structure and Climatic Conditions Changes

	K1p-K1o	K2p-K1p	K3p-K1o
	5.51	−4.76	8.60

Change in Sensible Heat Flux Density as a Result of Landscape Structure and Climatic Conditions Changes

	K1p-K1o	K2p-K1p	K3p-K1o
	−1.96	2.05	−4.88

Landscape structure: K1 — 30% forests, 15% meadows, 10% row crops, 45% cereals
K2 — 20% row crops, 80% cereals
K3 — 50% forests, 20% meadows, 10% row crops, 20% cereals
o — present climatic conditions
p — future climatic conditions

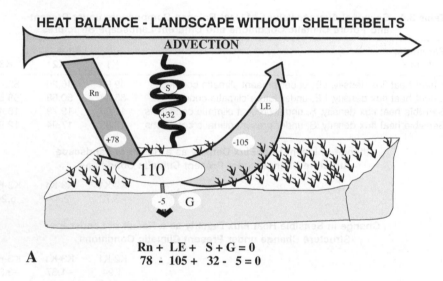

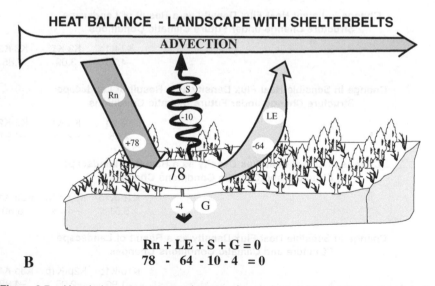

Figure 3.7 Heat balance component changes after shelterbelt introduction to the agricultural landscape in Wielkopolska region.

while in a wet year it was only 0.16, which means that the differentiation of temperature within landscape is much greater in a dry year than in a wet one and

Table 3.5 Latent (LE) and Sensible (S) Heat of Chosen Ecosystems in Growing Season (21.03 to 31.10) of Dry, Normal, and Wet Years in Turew and Their Diversification (Δ)

Ecosystem	LE MJ·m⁻²	Δ MJ·m⁻²	S MJ·m⁻²	Δ MJ·m⁻²	β S/LE	Δ MJ·m⁻²	α LE/Rn	Δ MJ·m⁻²
			Dry Year					
Meadow	1200		265		0.22		0.80	
Field	892	586	638	473	0.71	0.60	0.58	0.27
Field + Shelterbelts	998		538		0.54		0.63	
Forest	1478		165		0.11		0.85	
			Normal Year					
Meadow	1250		215		0.17		0.84	
Field	1035	487	495	374	0.48	0.40	0.67	0.21
Field + Shelterbelts	1078		458		0.42		0.68	
Forest	1522		121		0.08		0.88	
			Wet Year					
Meadow	1347		118		0.09		0.90	
Field	1242	298	288	185	0.23	0.16	0.81	0.09
Field + Shelterbelts	1300		236		0.18		0.82	
Forest	1540		103		0.07		0.89	

Table 3.6 Mean Weighted Values of Heat Balance Components and Their Ratios for Wielkopolska Region, 1956–1975

Period	Heat Balance Components W·m⁻²				Ratios			
	Rn	–LE	–S	–G	-LE/Rn	–S/Rn	–G/Rn	A/LE
Summer VI–VIII	108	77	28	3	0.71	0.26	0.03	0.39
Warm half-year IV–IX	93	63	25	5	0.68	0.27	0.05	0.41
Year I–XII	45	39	6	0	0.87	0.13	0	0.15

It is possible, using the MBC model, to estimate the heat balance components for longer periods and large areas (Ryszkowski and Kędziora 1995, Olejnik 1996). The calculations must be done in daily or monthly resolution for the whole period under consideration and for all land-use units that exist in the area under investigation. After that, the mean weighted values of all heat balance components can be calculated. Table 3.6 shows the results of such an estimation for the Wielkopolska region. The Rn flux density, average for the entire Wielkopolska region is equal to 108 W·m⁻² in summer, 93 W·m⁻² in a warm half-year, and 45 W·m⁻² in the whole year (Table 3.6). In summer, about 70% of this energy is used for evapotranspiration

The seasonal courses of heat balance components differ significantly when a different type of vegetation is taken into consideration (Figure 3.8.). Forest shows a regular course during the growing season and the course of latent heat (LE), sensible heat (S), and soil heat (G) never cross one another. Latent heat flux is greatest during the entire season, and sensible heat flux is greater than soil heat flux. Alfalfa shows a fluctuating type of seasonal course of heat balance components, the result of its two or three harvests during the season. Winter wheat shows intense growth in spring and early summer, while sugar beet grows mainly in summer and autumn. So, in the spring, winter crops have much higher values of latent heat than sensible heat, while row crops have much higher values of latent heat in summer (Figure 3.8). Seasonal courses of heat balance components are under significant influence from the landscape structure. Change in land use can essentially change the course of the fluxes (Figure 3.9A and 3.9B). Simplification of landscape structure (K2–K1 in Figure 3.9) decreases latent heat and increases sensible heat, while developing landscape structure causes the opposite changes. Changes in sensible heat are very small in spring when high moisture of the habitat ensures that simple landscape and composite have the same intense evaporation and the same amount of energy that can be used for air heating. In summer, however, the changes in sensible heat are much greater because the soil water supplies have been used and evapotranspiration from simple and composite landscapes differs significantly. The diversification of heat balance structure for the entire landscape is lower than for particular ecosystems (Table 3.7).

CHANGES IN HEAT BALANCE COMPONENTS UNDER CHANGED CLIMATIC CONDITIONS

The MBC model can be applied to the analysis of changes of heat and water balance components due to the global climatic change in a local-catchment scale with the use of quite high resolution (Olejnik 1996, Olejnik et al. 2000). Because the input data set consists partly of meteorological data, it is possible to run the model using new, anticipated climatological data. The results from global circulation models, GCM ($2 \times CO_2$), were used as new input data for the MBC model. The increase of temperature (by 4°C) and the increase of precipitation (by 0.4 mm/day) for this part of Europe were assumed (Hulme et al. 1990).

The results of this analysis are shown in Figure 3.10. The upper part shows the present conditions of heat and water balance of the Main River catchment in Germany. Scenario I (center section of figure) shows the heat and water balance conditions after climatic change (data from GCM). It is easy to see that the increase in temperature and precipitation will result in a very significant increase in runoff, from 235 mm at present to 306 mm by about 2050. For this scenario, current land use

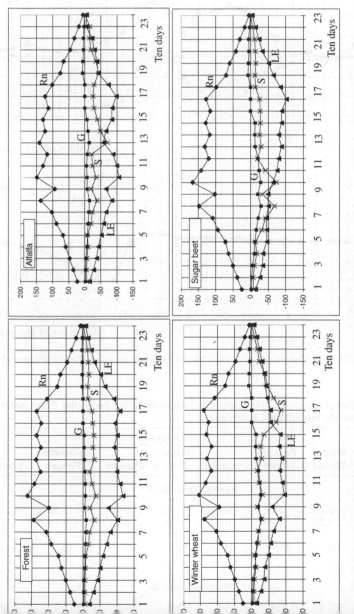

Seasonal courses of heat balance components for four chosen ecosystems within Turew. The first ten days are the first ten days of March.

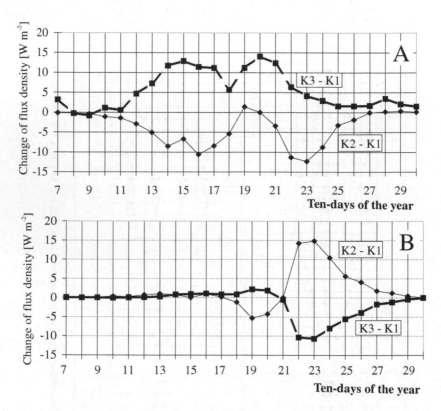

Figure 3.9   Change in latent (A) and sensible (B) heat flux as a result of landscape changes.

**Table 3.7   Diversification of Heat Balance Components for Particular Ecosystems
and for the Whole Landscape**

	Ecosystem						
Parameter	Winter Wheat	Spring Wheat	Potatoes	Sugar Beet	Meadow	Deciduous Forest	Landscape
Bowen Ratio S/LE							
Average	0.56	0.62	0.57	0.46	0.42	0.30	0.49
Standard deviation	0.34	0.28	0.27	0.25	0.16	0.12	0.14
Coefficient of variation	60	45	48	54	37	42	28
Sensible Heat, S							
Average	−28.4	−31.4	−29.4	−25.2	−24.4	−17.9	−26.1

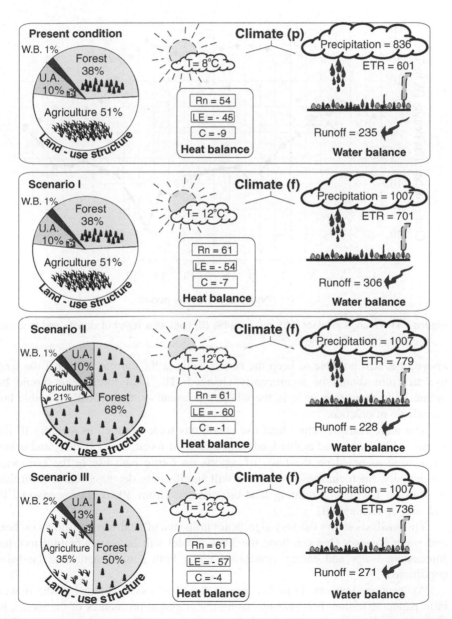

Figure 3.10 Mean annual values of meteorological data, heat balance, and water balance components in the Main River catchment in present climatic and land-use conditions as well as future conditions (scenarios I, II, and III). U.A. = urban areas; W.B. = water bodies.

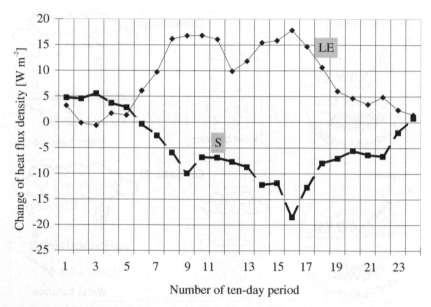

Figure 3.11 Change in latent (LE) and sensible (S) heat as a result of climate and land-use structure changes, K3p–K1o.

above, it is still possible to keep the runoff value at the current level, but the land use structure should be significantly changed. The total forest area should be increased to more than 68% in the whole catchment — theoretically possible but unrealistic in practice.

The more realistic future land use conditions were assumed in scenario III (in Figure 3.10). Decreased arable lands and increased forest, water bodies, and urban areas were assumed on the basis of reports on future land use in the European Community. Such changes in land use will significantly decrease runoff from that of scenario I without assuming land-use changes (from 306 mm of scenario I to 271 mm of scenario III).

This analysis shows the very significant influence of land-use conditions on heat and water balance. One can hope the MBC model will be a very helpful tool for forecasting water and energy conditions under both new climatic and land-use conditions.

As mentioned previously in this chapter, the MBC model is based on the Bowen ratio parameterization. It is possible, using the graphical procedures of the model, to show the areal distribution of the heat balance components. The use of such maps should be very helpful for land-use managers for making decisions about future land use. In the Main catchment there are areas where land use can be modified to increase the areal

There are no special limits in resolution increase for the MBC model. The only one is the resolution of the available land-use data. Because of this, we hope that the MBC model can find application in decision making for new land-use structure on the catchment but also on the local scale due to socioeconomic and climatic changes.

As was shown, the solar energy partition, the amount of energy fluxes, and the mass exchange between the Earth's surface and the atmosphere strongly depend on plant cover type and its condition. Habitat moisture and meteorological conditions also have very important influences on heat balance structure. Better understanding of all the processes of energy and water exchange in the environment, as well as of the factors that influence these processes, creates the possibility of modeling them. Using appropriate models, we can estimate the changes in heat and water balance structures with respect to land use or potential climatic changes. This gives us a new tool to predict the potential environmental changes due to expected climatic and land use changes. Knowledge of such potential changes gives us the opportunity to take action and to minimize the effects of negative environmental changes.

REFERENCES

Arya S.P. (1988) *Introduction to Micrometeorology*. Academic Press, San Diego, 305 pp.

Beljaars A.C., Holtslag A.A.M. (1991) Flux parameterization over land surfaces for atmospheric models. *J. Appl. Meteor.* 3: 327–341.

Field C., Berry J.A., Mooney H.A. (1982) A portable system for measuring carbon dioxide and water vapor exchange of leaves. *Plant Cell Env.* 5: 179–186.

Freedman B. (1992) *Environmental Ecology*. Academic Press, New York, 247 pp.

Gurney R.J. (1988) Satellite estimates of evaporation from vegetation. In *Climate — Vegetation Interactions*. Ed., Gurney R.J., Kluwer Academic Publishers, Dordrecht, 307 pp.

Holtslag A.A.M., De Bruin H.A.R. (1988) Applied modeling of the night time surface energy balance over land. *J. Appl. Meteor.* 27: 659–704.

Hulme M., Wigley T.M.L., Jones P.D. (1990) Limitations of climate scenarios for impact analysis. In *Landscape-Ecological Impact of Climatic Change*. Eds., Boer M., De Groot R. IOS Press, Amsterdam.

Kędziora A., Olejnik J., Kapuściński J. (1989) Impact of landscape structure on heat and water balance. *Ecol. Int. Bull.* 17: 1–17.

Kędziora A., Olejnik J. (1996). Heat balance structure in agroecosystems. In *Dynamics of an Agricultural Landscape*. Eds., Ryszkowski L., French, N., R. Kędziora A. PWRiL, Poznań, 45–64.

Kędziora A., Tuchołka S., Kapuściński J., Paszyński J., Leśny J., Olejnik J., Moczko J. (1997) Impact of plant cover on heat and water balance in agricultural landscapes located in humidity gradients. *Roczniki Akademii Rolniczej w Poznaniu*, 294, 271–301.

Kędziora A. (1999) *Foundation of Agrometeorology*. PWRiL, Poznań, 364 pp.

54   LANDSCAPE ECOLOGY IN AGROECOSYSTEMS MANAGEMENT

Kim J., Verma S. (1990) Components of surface energy balance in a temperate grassland ecosystem. *Boundary Layer Meteorol.* 51: 401–417.

Kirchner M. (1984) Influence of different land use on some parameters of the energy and water balance. *Progress. Biometeorol.* 3: 65–74.

Kundzewicz Z., Somlody L. (1993) Climatic change impact on water resources — a system view. International Institute for Applied Systems Analysis, working paper, 32 pp.

Leśny J. (1998) The influence of plant cover on heat balance structure. Ph.D. dissertation, Library of Agricultural University of Poznań, 150 pp.

McNaughton K.G., Spriggs W.T. (1985) A mixed-layer model for regional evaporation. *BLM* 34: 243–262.

Molga M. (1986) *Agricultural Meteorology.* PWRiL, Warszawa, 470 pp.

Monteith J.L. (1975) *Vegetation and the Atmosphere.* Academic Press, London, 278 pp.

Olejnik J., Kędziora A. (1991) A model for heat and water balance estimation and its application to land use and climate variation. *Earth Surface Processes Landforms* 16: 601–617.

Olejnik J. (1996) Catchment scale modeling of heat and water balance structure under present and future climatic conditions. [in Polish] *Roczniki Akademii Rolniczej w Poznaniu,* Z.268, 125 pp.

Olejnik J. (1988) The empirical method of estimating of mean daily and mean ten-day values of latent and sensible-heat near the ground. *J. Appl. Meteorol.* 12: 1358–1369.

Olejnik J., Eulenstein F., Willms M. (2000) Mesoscale heat and water balance structure under present and future climatic conditions — catchment case study in southern part of Germany. *Prace Geograficzne. Uniwersytetu Jagielońskiego,* Kraków, Z 107, 397–405.

Olejnik J., Eulenstein F., Kędziora A., Werner A. (2001) Evaluation of water balance model using data for bare soil and crop surfaces in Middle Europe. *Agric. Forest Meteorol.* 106, 105–116.

Paszyński J. (1972) Studies on the heat balance and on evapotranspiration. *Geographia Polonica*: 22.

Rosenberg N.J. (1974) *Microclimate: The Biological Environment.* Wiley, New York, 343 pp.

Ryszkowski L., Kędziora A. (1987) Impact of agricultural landscape structure on energy and water cycling. *Landscape Ecol.* 1: 85–94.

Ryszkowski L., Kędziora A., Olejnik J. (1991) Potential effects of climate and land use changes on the water balance structure in Poland. In *Land Use Changes in Europe.* Eds., Brouwer F.M., Thomas A.J., Chawick M.J. Kluwer Academic Publishers, Dordrecht.

Ryszkowski L., Kędziora A. (1993a) Agriculture and greenhouse effect [in Polish]. *Kosmos* 42: 123–149.

Ryszkowski L., Kędziora A. (1993b) Energy control of matter fluxes through land-water ecotones in an agricultural landscape. *Hydrobiologia* 251: 239–248.

Ryszkowski L., Kędziora A. (1995) Modification of the effects of global climate change by plant cover structure in an agricultural landscape. *Geographia Polonica* 65: 5–34.

Stigliani W.W., Brouwer F.M., Munn R.E., Shaw R.W., Antonovsky M. (1989) *Future Environments for Europe: Some Implications of Alternative Development Paths.* Interna-

MITIGATION OF RADIATION AND HEAT BALANCE STRUCTURE   55

Valentini R., Baldocci D.D., Tenhunen J.D. (1999) Ecological control on land-surface atmo-
    spheric interaction. In *Integrating Hydrology, Ecosystems Dynamics, and Bio-
    geochemistry in Complex Landscapes,* Eds., Tenhunen J.D. and Kabat P., John Wiley
    & Sons, Chichester, 177–197.
Wegehenkel M. (2000) Test of a modeling system for simulating water balances and plant
    growth using various different complex approaches. *Ecol. Modeling* 129(1): 39–64.

CHAPTER 4

Water Balance in Agricultural Landscape and Options for Its Management by Change in Plant Cover Structure of Landscape

Andrzej Kędziora and Janusz Olejnik

CONTENTS

58   LANDSCAPE ECOLOGY IN AGROECOSYSTEMS MANAGEMENT

INTRODUCTION

Owing to unusually strong hydrogen bonds between molecules, water is one of the most amazing substances in nature. Many of its properties are qualitatively different from those of other substances participating in processes important for biosphere functioning — for example, water has anomalous high temperature at melting and boiling points, one of the highest specific heat and latent heat of evaporation, the highest dielectric constant, and very high dipole momentum.

By determining the process of solar energy transformation into organic matter and thereby the conditions of plant growth and development, water determines the level of agricultural production. Thanks to its enormous thermal properties, water controls the thermal status of plants and allows the plant body to store a large amount of thermal energy, which buffers the plant against rapid changes in environmental temperature. Continuous sufficient flux of water flowing through the soil-plant-atmosphere system is indispensable for utilizing the potential for the ecosystem to achieve plant growth and high yields. Three scales of water cycle can be distinguished (Figure 4.1):

- Global hydrologic cycle (Figure 4.1C), which consists of water exchanged between oceans and continents through atmospheric circulation and river water flow
- Local hydrologic cycle (Figure 4.1B, marked by a dashed line), including water exchanged between the land and the atmosphere
- Micro-water cycle (Figure 4.1A) which occurs as water circulates between top soil layers and near-surface layers of the atmosphere within plant communities

The last cycle is very rarely considered, but its role in creating microclimatological conditions of agricultural landscapes is very important. In the presence of a dense

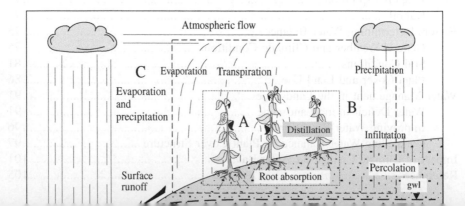

plant cover (for example, meadow or rapeseed field) water evaporating from the soil surface does not pass to the atmosphere but instead condenses on the bottoms of leaves, remaining within the plant cover, which explains why even during dry weather a very humid microclimate can exist inside plant cover.

One of the insufficiently recognized problems of formation of water balance is how the structure of plant cover in agricultural landscapes impacts the structure of water balance (Ryszkowski and Kędziora 1995, Kędziora 1999, Kędziora and Ryszkowski 1999, Valentini et al. 1999, Mills 2000). There are many studies on the impact of individual elements and characteristics of landscape on individual components of water balance. But, at the level of landscape, many interactions between processes in the landscape as well as between individual components of the landscape are observed. These phenomena are very poorly recognized because their final effects are not the simple sum of their individual effects (Caswell et al. 1972). The exact recognition of terrestrial hydrologic processes is very important for global circulation models (GCM) because the value of these models strongly depends on parameterization of surface processes of water transport and exchange between Earth and atmosphere (Thomas and Henderson-Sellers 1992, Viterbo and Illari 1994). Studies thus far show that the more developed a landscape structure is, the higher its resistance to many threats occurring in the environment. The evolution of nature brought about very high stability of the Earth's system, lasting until human civilization started.

The water cycling that stabilized during the long geological evolution has been disturbed by recent human action (Zektser and Loaiciga 1993). The environment is subject to very deep drought on the one hand and to flood on the other hand. These climatic disasters are becoming more frequent and less predictable. The global distribution of water resources is irregular. Very rarely is there enough precipitation to ensure soil water moisture favorable for plants during the whole growing season. In Poland and in most countries in Europe, water demands of plants in the growing season very often exceed available water supplies — the precipitation and water retained in the soil. During the summer months, evapotranspiration is higher than precipitation, leading to decreased soil moisture and lowering of the ground water table. Central Europe is rather poor in water resources and increased water demands from the human population and possible climate change brings new challenges in water management to support sustainable development of agriculture. The great challenge that faces humankind is to increase water supplies in the agricultural landscape. The average water deficit in the Wielkopolska region in Poland is equal to about 100 mm (100 l/m^2), that is, about 3 km^3 for the entire Wielkopolska region (total area of the Wielkopolska region is about 30,000 km^2). It is impossible to collect such a huge amount of water in artificial reservoirs. Thus, the technical efforts must be supported by the use of natural processes and mechanisms as well as by proper

Table 4.1   Water in the Hydrosphere

Water	Volume (thousands km³)	Percent of Total Volume	Percent of Fresh Water
Oceans	1,338,000.00	96.5	
Glaciers and snow cover	24,364.10	1.725	69.6
Ground water	23,400.00	1.69	30.1
Fresh water	10,530.00	0.76	
Salt water	12,870.00	0.93	
Lakes	176.40	0.013	0.26
Fresh water	91.00	0.007	
Salt water	85.40	0.006	
Soil water	16.50	0.0012	0.05
Atmospheric water	12.90	0.001	0.04
Wetlands	11.47	0.0008	0.03
Rivers	2.12	0.0002	0.006
Biological water	1.12	0.0001	0.003
Total	1,385,984.61	100.0	100.0
Fresh water	35,029.21	2.5	

Source: UNESCO 1978.

GENERAL WATER BALANCE

The structure of water balance depends mainly on precipitation and temperature. Total world water volume is nearly 1.4 billion km³, but 96.5% of it is gathered in oceans (Table 4.1). Fresh water constitutes only about 2.5%, more than two thirds of which is ice-bound. The most active part of the world's water is in the atmosphere and soil, constituting only 0.08% of fresh water and 0.002% of the total world water (Baumgartner and Reichel 1975, UNESCO 1978, Lwowich 1979). During a year, 577 km³ evaporates and falls as rain, which means that atmospheric water must circulate more than 40 times during a year because its total volume is equal to about 14 km³. Consequently, atmospheric water plays an important role in energy and mass transporting. The water balances of European countries vary considerably (Table 4.2). The lowest precipitation occurs in Poland, the Czech Republic, and Hungary (a little more than 600 mm). Because of its high evapotranspiration, Hungary's climatic water balance (precipitation minus evapotranspiration) is the lowest. The ratio of evapotranspiration to precipitation is also highest in Hungary (0.90) and very high in other central European countries (Figure 4.2). In Poland, especially in the Wielkopolska and Kujawy regions, the ratio of evapotranspiration to precipitation is also very high (Figure 4.3). In other European countries, including Spain, the ratio of evapotranspiration to precipitation (calculating for the whole country) does not exceed 0.70. The water supplies can be well characterized by water

WATER BALANCE IN AGRICULTURAL LANDSCAPE 61

Table 4.2 Water Balance of Select Countries in Europe

Country	Precipitation P	Evapotranspiration E	Runoff R = P — E	Runoff per capita (10³ m³)	E/P	R/P
Europe	733	415	318	5.11	0.57	0.43
Poland	604	424	180	1.72	0.70	0.30
Germany	725	430	295	1.4 (1.91)	0.59	0.41
Hungary	610	519	90	0.81 (3.81)	0.85	0.15
Czech Republic and Slovakia	735	442	293	1.9 (4.73)	0.60	0.40
Netherlands	676	427	249	0.78 (6.86)	0.63	0.37
Spain	636	380	255	3.88	0.60	0.40
France	965	541	424	4.57	0.56	0.44
Russia	620	410	210	6.23	0.66	0.34
Finland	549	234	315	22.5	0.43	0.57
Sweden	664	233	431	24.1	0.35	0.65
Norway	1343	182	1160	96.9	0.14	0.86

Figures in parentheses relate to the case when transit water is included, the Danube in Hungary, the Czech Republic, and Slovakia, and the Rhine in Germany and the Netherlands.

Source: Lwowicz 1979.

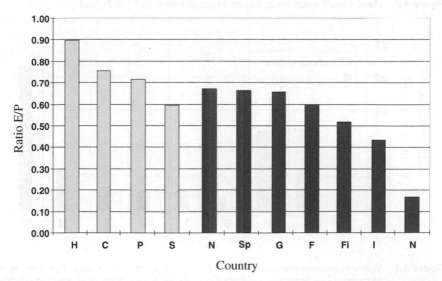

Figure 4.2 Ratio of real evapotranspiration to precipitation (E/P) for select countries in Europe. H — Hungary, C — Czech Republic, P — Poland, S — Slovakia, N — the Netherlands, Sp — Spain, G — Germany, F — France, Fi — Finland, I — Italy, N — Norway.

62   LANDSCAPE ECOLOGY IN AGROECOSYSTEMS MANAGEMENT

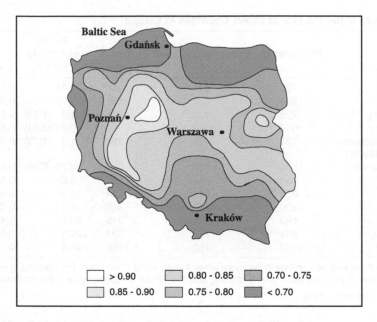

Figure 4.3   Ratio of real evapotranspiration to precipitation (E/P) in Poland.

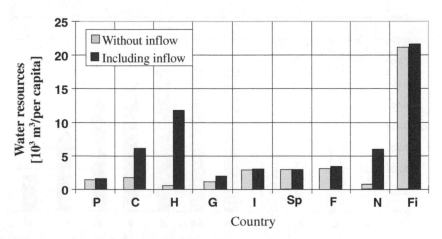

Figure 4.4   Water resources per capita [$10^3 \cdot m^3$]. P — Poland, C — Czech Republic, H — Hungary, G — Germany, I — Italy, Sp — Spain, F — France, N — the Netherlands, F — Finland.

increase of 10% in the area subject to drought will reduce the Warta River flow the
following year by 5.5 m³/s, that is, about 4% of the average flow in the years without

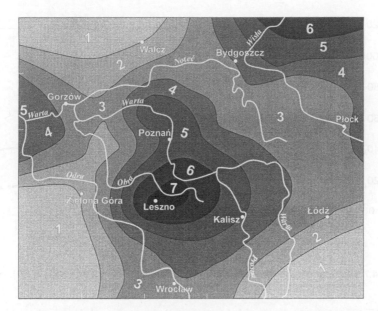

Figure 4.5   Map of drought risk in Wielkopolska. Class 1 — lowest risk, class 7 — highest risk.

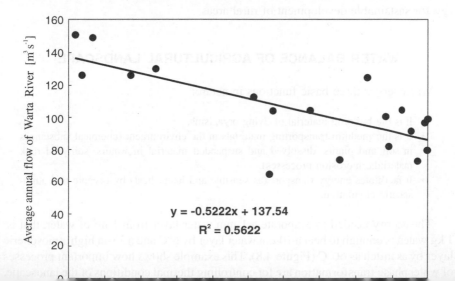

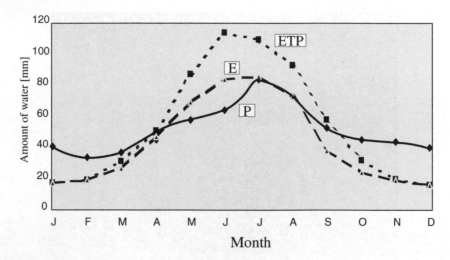

Figure 4.7   Annual course of precipitation (P), potential evapotranspiration (ETP), and real evapotranspiration (E), the Wielkopolska, 1951–1995.

All circumstances mentioned above show that water conditions in the agricultural landscape of Poland, as well as in all of Central Europe, require very wise and economical water management, which can be executed only when the factors determining components of water balance are well recognized, thus allowing scientists, decision makers, local government officials, and farmers to construct a proper strategy for sustainable development of rural areas.

WATER BALANCE OF AGRICULTURAL LANDSCAPE

Water serves three basic functions in nature:

- It is the building material of living organisms.
- It is the medium transporting materials in the environment (chemical substances in soil and plants, dissolved and suspended material in waters, soil, and rock materials in erosion processes).
- It facilitates energy transport (as sensible and latent heat) by oceanic and atmospheric circulation.

The energy needed to evaporate a 1-mm water layer from 1 m² of water, that is, 1 kg water, is enough to heat a 10-cm water layer by 6°C and a 33-m high atmospheric layer by as much as 60°C (Figure 4.8). This example shows how important processes of water phase transformation are for controlling thermal conditions of the landscape.

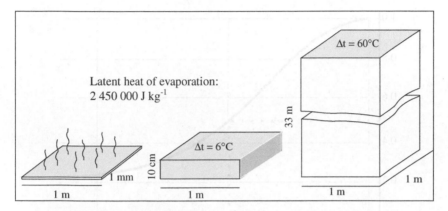

Figure 4.8   Effect of applying the same amount of energy for evaporation, water heating, and air heating.

The strong linkage between energy flow through the landscape and matter cycling within environment exists. The energy flux is the "driving force" for matter cycling. The maintenance of steady (within limits) flux of energy and matter is needed to ensure the stability of a system. The most important task is to ensure proper water conditions in the landscape because of the multifunctional role of water mentioned above. Any processes, natural or caused by human activity, that disturb the process of energy flow and water cycling could have substantial effects on landscape functioning and could create serious threats for sustainable development of the agricultural landscape.

The worsening of water conditions in rural areas has been observed for several decades. Increasing water deficits, decreasing soil retention ability in the face of growing water demands are the main threats to agricultural development in central Europe. The following causes of this situation must be taken into consideration:

- Changes of natural climatic conditions
- Changes in land use and landscape structure leading to simplification of landscape structure
- Human activity in water management incompatible with fundamental rules of energy flow and water cycling

The broad studies carried out during the second half of the 20th century showed that climatic conditions (precipitation and temperature) generally changed too little to cause the worsening water conditions in Poland (Lambor 1953, Pasławski 1992, Kędziora 1999).

However, an unfavorable phenomenon has been observed recently — the increas-

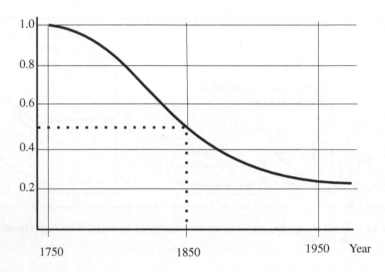

Figure 4.9   Change in ratio of meadows and pastures to arable lands in the Wielkopolska.

of water balance, as well as wind speed and water saturation deficit, did not change sufficiently to explain the worsening water condition (Gutry-Korycka 1978). Observations of ground water level show that hydrogeological conditions did not change significantly either (Wójcik 1998). A deep variation in the depth of ground water level occurred, but no trend has been observed in the agricultural landscape. Depletion of ground water level is observed only in the places where very deep transformations of land surface had occurred, for example, brown coal mines or gravel excavations.

The last millennium was a period of increasing transformation of the environment in central Europe. At the beginning of the period in the Wielkopolska region, the ground water level was about 1 m lower than it is today mainly because of high evapotranspiration of forests, which covered three quarters of the area (Czubiński 1947). Precipitation was the same as today (Kaniecki 1991). The rate of land transformation increased in the 15th century as colonization increased. At the end of the 14th century, forests covered more than 50% of the total country area, while arable land constituted only 18% of the total area. At the end of the 16th century, forested area decreased to 41%, to 31% at the end of the 18th century, and to 21% just before World War I (Miklaszewski 1928, Błaszczyk 1974). Cleared areas were converted to arable land. Also, pastures and meadows were very quickly converted to arable land. In 1750, the area of grassland was equal to arable land area, in 1850 it dropped to half that of arable land, and in 1950 the grassland area was five times smaller than the area of arable land (Figure 4.9). Decreasing water retention in the environment, accelerated runoff, and decreasing precipitation are the main negative results of land-

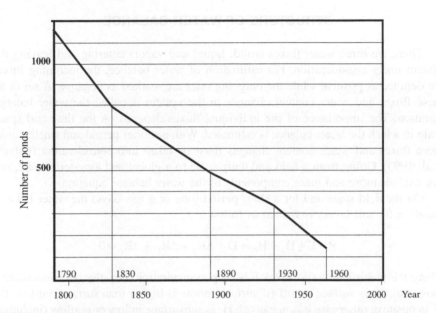

Figure 4.10 Disappearance of water millponds in the south Wielkopolska region.

in the Middle Ages, people have made many mistakes in water management. They began to regulate riverbanks, to straighten streams, and to drain wetlands (Kowalewski 1988, Mathias and Moyle 1992). These activities led to increased river current speed and cutting into the bed, as well as depleted water content in the environment, especially in soils. Many of these activities were done well from the engineering point of view but were completely wrong from the ecological point of view. They provided new land for agriculture, but they increased the amount of water quickly removed from the landscape, destroying many small ponds and degrading soil (Dembiński 1956, Kosturkiewicz and Kędziora 1995, Ryszkowski and Kędziora 1996a). Aridification of soil cover increases organic matter decomposition and decreases the soil's ability to retain water. The introduction of new agricultural technology, especially mechanization, accelerates the disappearance of many post-glacial midfield ponds, ditches, and other small meadow strips and wetlands. The use of electric mills instead of water mills almost totally removed small millponds (Gołaski 1988). Of 1208 water mills located in the Wielkopolska region in an area of 15,000 km² in 1790, only 70 remained in 1960 (Figure 4.10).

Thus, land-use changes and errors in water management must be regarded as the main causes of the present water conditions, which are unfavorable for agriculture. This unfortunate landscape management brought about simplified plant cover

STRUCTURE OF WATER BALANCE

There are three water fluxes (solid, liquid and vapor) entering and leaving the system under consideration. For estimation of water balance, the incoming fluxes are denoted as positive while the outgoing ones are marked as negative. A set of all these fluxes and water content changes in the system is called the water balance equation. The importance of the individual fluxes depends on the time and space scale in which the water balance is estimated. With a shorter period and smaller area, more fluxes and water content changes must be taken into consideration (Gilvear et al. 1993). Going from a field and daily scale to a global and long-term scale, one can exclude more and more components of the water balance equation.

On the field scale and for a short period (one or a few days) the water balance equation for soil layers is written as follows:

$$P + E + H_S + H_G + D + \Delta R_S + \Delta R_G + \Delta R_I = 0$$

where P is precipitation (positive), E is evapotranspiration (negative) or condensation (positive), H_S is surface runoff (if surface inflow is higher than surface outflow, the H_s is positive; otherwise it is negative), H_g is subsurface inflow or outflow (including lateral flow), D is percolation to the ground water (negative) or capillary upward flow (positive), ΔR_S is change of surface water retention, ΔR_G is change of soil water retention, and ΔR_I is change of plant cover water retention (change of interception).

Lengthening the time scale to a month or longer, we can neglect the change of plant cover retention, ΔR_I, and increasing the scale to a catchment, the water balance equation can be expressed as follows:

$$P + E + H_S + H_G + \Delta R_S + \Delta R_G = 0$$

Increasing the time scale to a decade or more (if neither turning to wetlands nor desertification is observed), we can neglect the change of water retention and write the equation of catchment water balance as follows:

$$P + E + H = 0$$

Finally, for the earth surface the water balance equation is the following:

$$P + E = 0$$

The structure of the catchment water balance depends mainly on:

- Variability and time distribution of precipitation, the parameter which is discrete

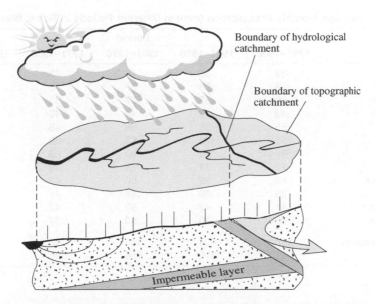

Figure 4.11 Incompatibility of hydrological and topographic catchment and its impact on ground water outflow.

The size of catchment has significant impact on the accuracy of the water balance estimation. In the case of small catchment, the incompatibility of topographic catchment and hydrological catchment can introduce an essential error in estimating water balance. If arrangement of the permeable and impermeable layers is such that a part of subsurface water runoff can flow out of catchment (Figure 4.11) the evapotranspiration calculated as the difference between precipitation and outflow measured at point A can be overestimated. As the part of catchment from which water flows increases, the error increases. This problem disappeared at the landscape level.

At the landscape level, the following four components of water balance must be taken into consideration: precipitation, evapotranspiration, runoff, and soil moisture changes. The last component disappears when a long period is analyzed. One must keep in mind that processes and fluxes important at a lower level of environmental organization form the higher system and can become less important at the level of this higher system, but they are also controlled by mechanisms occurring at this higher level of environmental organization (Tansley 1935, Allen and Starr 1982, O'Neill et al. 1986). For example, water vapor fluxes originating at the level of individual ecosystems depended on microclimatic conditions of the active surface to create the total water vapor flux outgoing from the landscape to the atmosphere. But they are controlled by meteorological conditions of the landscape, which deter-

Table 4.3   Average Monthly Precipitation (mm) in Different Periods in Turew, Wielkopolska

Month	Period				
	1881–1930	1921–1970	1951–1970	1971–1985	1881–1995
January	39	38	40	43	39
February	28	36	37	33	32
March	36	33	37	38	35
April	43	42	44	46	43
May	55	59	68	48	56
June	53	70	69	68	62
July	86	76	84	83	82
August	70	74	78	68	71
September	53	52	50	43	51
October	39	50	48	41	43
November	41	42	50	42	42
December	38	38	50	37	38
Growing season	410	434	453	410	410
Year	581	610	655	590	594

Precipitation

Precipitation is a vital water flux entering the landscape and largely determining the water balance structure. Annual distribution and intensity of precipitation are factors determining conditions of plant production and the quantity of annual runoff. Precipitation is the only component of water balance not under human control at the landscape level. All other components are subject to human activity. The average annual precipitation of the Wielkopolska region ranges from about 580 mm to 650 mm (Table 4.3). The average value of annual precipitation for a period of 100 years is 594 mm (Pasławski 1990). The average amount of rainfall in the growing season (from the third 10-day period of March to the end of October) varies between 400 and 450 mm. In comparison with other regions of Poland, the Wielkopolska region has one of the lowest amounts of precipitation. However, the distribution of precipitation over the entire year is favorable for agriculture. The amount of summer precipitation (May to August) is 271 mm, or 46% of annual precipitation.

Both 24-h and monthly rainfall distribution fit a gamma distribution, as can be seen in Figures 4.12 and 4.13. The density function of such a distribution, f(x), is given by the following equation (Kędziora 1996b):

$$f(x) = \frac{\lambda^k}{\Gamma(k)} \cdot x^{k-1} \cdot e^{-kx}$$

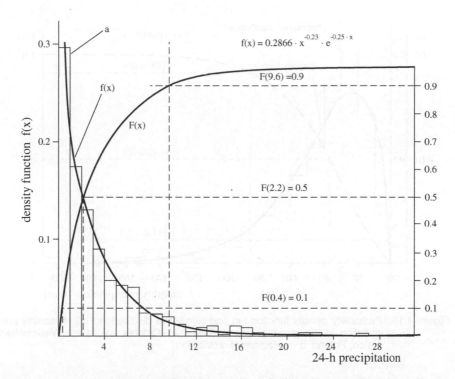

$$f(x) = 0.2866 \cdot x^{-0.23} \cdot e^{-0.25 \cdot x}$$

Figure 4.12   Probability density function f(x) and cumulative distribution F(x) of 24-h precipitation in the growing season (March 21–October 31) in the Wielkopolska region, Poland. a — diagram of distribution.

The median 24-h rainfall distribution is 2.2 mm, and there is only a 10% probability that the 24-h rainfall will exceed 9.6 mm or be less than 0.4 mm. The median of the average monthly rainfall distribution is 35 mm. There is a 10% probability that the monthly amount of rainfall will be less than 10 mm or more than 85 mm. Comparing the latter amount with the average monthly potential evapotranspiration (Table 4.4) shows that irrigation is necessary in the summer months in the Wielkopolska region. On average, for the growing seasons of 1978–1985, about 65% of the days were rainy days, while during the 1920–1970 period the average number of rainy days per month ranged from 10 in September to 14 in January. The mode of monthly rainfall distribution is 20 mm, and this means that in this region the most frequent monthly rainfall reaches 20 mm. However, a higher amount of rainfall occurs in the summer months, and a lower amount occurs in winter.

The structure of landscape has no direct impact on precipitation, but by influencing surface processes it can modify the local water cycling, which can indirectly affect precipitation. Even if landscape structure has no distinct impact on the amount of

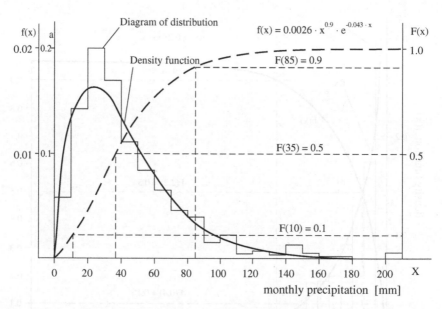

Figure 4.13   Probability density function f(x) and cumulative distribution F(x) of monthly precipitation in the growing season (March 21–October 31) in the Wielkopolska
region, Poland. a — diagram of distribution.

**Table 4.4   Average Monthly Potential and Real Evapotranspiration
in the Turew Landscape, Poland, 1951–1970**

Month	Precipitation P [mm]	Evapotranspiration [mm] Potential ETP	Real E	ETP/P	E/P
January	39	15	15	0.38	0.38
February	32	17	17	0.53	0.53
March	35	29	25	0.83	0.71
April	43	49	45	1.14	1.05
May	56	85	67	1.52	1.20
June	62	112	82	1.81	1.32
July	82	107	84	1.30	1.02
August	71	91	71	1.28	1.00
September	51	56	36	1.10	0.71
Oktober	43	30	22	0.70	0.51
November	42	18	17	0.43	0.40
December	38	14	14	0.37	0.37
JJA	215	310	237	1.44	1.10
Growing season	410	540	416	1.32	1.01
Year	594	623	495	1.05	0.83

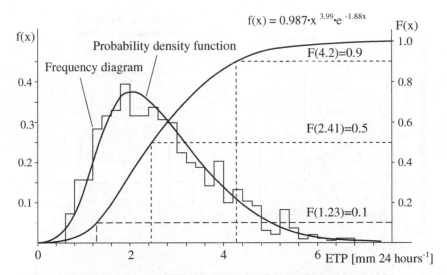

Figure 4.14   Probability density function f(x) and cumulative distribution F(x) of 24-h potential evapotranspiration in the growing season (March 21–October 31), Turew, Wielkopolska.

as well as variation of plant development stage, cause essential diversity of evapotranspiration in space and in time (Penman 1948, Kędziora 1999).

In the agricultural landscape of the Wielkopolska during three summer months (June, July, and August) 237 mm of water can be evaporated, during the warm half-year (April to September) 416 mm, and during the whole year 495 mm. The annual precipitation in the Wielkopolska region amounts to about 600 mm (Table 4.4), (Kędziora 1996b). During the warm period, potential evapotranspiration exceeds precipitation. Actual evapotranspiration exceeds precipitation considerably during May and June, but during April and July it exceeds precipitation only slightly.

The 24-h amount of potential evapotranspiration shows a gamma distribution (Figure 4.14). The average value for five growing seasons (1981 to 1985), from April to September, was 2.66 mm; however, the mode of this distribution was 2.2 mm. There is only a 10% probability that the 24-h potential evapotranspiration in the agricultural landscape of the Wielkopolska region will exceed a value of 4.2 mm, and a 10% probability that it will be lower than 1.2 mm.

During the monthly course of potential evapotranspiration, the maximum value is observed at the end of June and beginning of July when it reaches 110 mm per month, while the lowest value occurs in December or January and falls as low as 14 mm per month (Table 4.4).

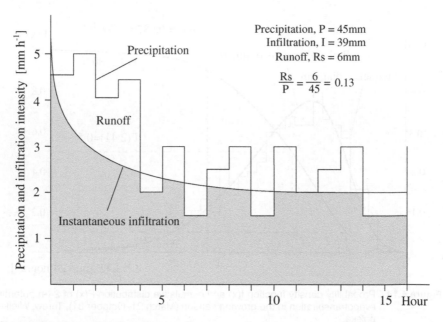

Figure 4.15 Formation of surface runoff.

the soil (Ben-Hur et al. 1995). One of the most important factors is the relation between infiltration rate and intensity of rainfall. Thus, in the case of low infiltration capacity, rainfall intensity exceeding the basic infiltration rate cannot infiltrate the soil surface, and it becomes wholly or partly surface runoff (Figure 4.15). In the case analyzed, the intensity of rainfall during a rainstorm lasting 10 h oscillated between 3 and 6 mm/h. It was higher than the infiltration rate, which changed from 4 mm/h (at the beginning of the rain) to 2.4 mm/h at the end (basic infiltration rate). As a result of such a relation, of the 42 mm of rainfall, only 31 mm of rain infiltrated the soil, and 11 mm formed the surface runoff.

Land use and plant cover structure are the other important factors in formation of runoff. The time lapse of landscape reaction on intensity of rainfall is greater in the presence of rich plant cover, and the maximum runoff is reduced in comparison with bare soil (Figure 4.16). On average in the Wielkopolska region, surface runoff accounts for approximately 13 to 20% of rainfall, but sometimes it can reach as much as 50 to 60% (Pasławski 1990). Such conditions are, of course, unfavorable for agriculture because of soil erosion, especially on sloping, light bare soil surfaces. That part of the rainwater not retained by the soil profile percolates through the soil, and leaches and dislocates a material within the soil profile to the ground water.

Thus, water from precipitation that runs over the soil surface or percolates through the soil profile plays a most important role in processes of transporting

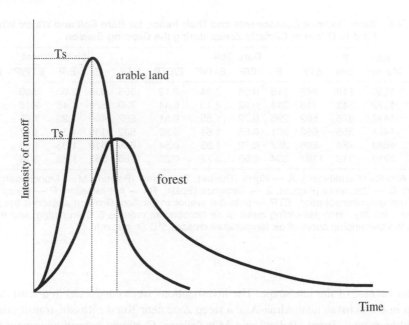

Figure 4.16   Intensity, timing, and time lapse (Ts) of surface runoff from arable land and from a forest.

FACTORS DETERMINING WATER BALANCE

Many factors determine the values of individual water fluxes and water balance components; they can be divided into three groups:

- General weather and climatic conditions
- Physical and hydraulic features of soil
- Plant covers characteristics

Many relations and much feedback exist among individual factors, factors and water balance components, and components themselves. As a result of all these mechanisms and interactions, water balance structure shows very high changeability, in both time and space (Kosturkiewicz et al. 1991, Kędziora 1994).

General Weather and Climatic Conditions

The total amount of water coming into the landscape as well as potential and

Table 4.5   Water Balance Components and Their Ratios for Bare Soil and Winter Wheat Field in Different Climatic Zones during the Growing Season

Site	Rn MJ·m⁻²	P mm	Bare Soil					Winter Wheat				
			ETP	E	E/P	ETP/P	E/ETP	ETP	E	E/P	ETP/P	E/ETP
A	1680	119	942	116	0.98	7.94	0.12	955	336	2.83	8.05	0.35
K	1572	342	718	314	0.92	2.10	0.44	730	506	1.48	2.13	0.69
T	1442	375	582	295	0.79	1.55	0.51	592	460	1.23	1.58	0.78
M	1461	355	582	301	0.85	1.64	0.52	592	466	1.31	1.67	0.79
C	1663	494	666	357	0.72	1.35	0.54	685	510	1.03	1.39	0.74
Z	2210	319	1187	304	0.95	3.72	0.26	1188	553	1.73	3.73	0.47

A — Alm-Ata (Kazakhstan), K — Kursk (Russia), T — Turew (Poland), M — Müncheberg (Germany), C — Cessieres (France), Z — Zaragoza (Spain). Rn — net radiation, P — precipitation, E — real evapotranspiration, ETP — potential evapotranspiration. Growing season is the period between the day when ascending curve of air temperature crosses 5°C (in spring) and the day when the descending curve of air temperature crosses 5°C (in autumn).

water balance of the landscape. The investigations were conducted in a semi-desert area in Kazakhstan near Alma-Ata, a steep zone near Kursk, Russia, transit climate conditions near Turew, Poland and Müncheberg, Germany, humid zone near Cessieres, France, and arid climatic zone near Zaragoza, Spain (Kędziora et al. 1994).

The solar energy flux is high in the arid climatic zone, but the very high surface temperature causes high-earth long wave radiation, and low concentrations of water vapor in the atmosphere cause low atmospheric reradiation toward the Earth's surface. Thus net radiation is not as high as in the Mediterranean climate but is higher than in humid climatic zones (Table 4.5). However, this net radiation with a very high-saturation water-vapor deficit causes very high potential evapotranspiration. Low precipitation and low soil water retention lead to low real evapotranspiration. In such conditions, the ratio of ETP/P (potential evapotranspiration to precipitation) is very high, the ratio of E/P (real evapotranspiration to precipitation) is also high, but ratio of E/ETP is small. In a transitional climatic zone or semi-arid zone, potential evapotranspiration is also high but can differ significantly mainly because of temperature and saturation vapor pressure deficit differences as well as length of the growing season. In the continental climate zone (Kursk), summer air temperature is higher than in the arid zone (Zaragoza) where spring and autumn months are much warmer. The growing season in Zaragoza also lasts the whole year, which is much longer than in Kursk where it lasts 7 months. As a result, net radiation in Zaragoza is 40% higher than in Kursk, and ETP is higher by about 25%. In humid climatic conditions (Turew, Müncheberg, Cessieres, Table 4.5) net radiation and potential evapotranspiration are lower but real evapotranspiration is on the same order, with the exception of Zaragoza. In all places studied, the ratio of E/P for wheat fields is

Table 4.6 Latent (LE) and Sensible (S) Heat of Selected Ecosystems in the Growing Season (21.03 to 31.10) of Dry, Normal, and Wet Years in Turew, as Well as Their Diversification (Δ)

Ecosystem	LE MJ·m⁻²	Δ MJ·m⁻²	S MJ·m⁻²	Δ MJ·m⁻²	β S/LE	Δ MJ·m⁻²	α LE/Rn	Δ MJ·m⁻²
				Dry Year				
Meadow	1200		265		0.22		0.80	
Field	892	586	638	473	0.71	0.60	0.58	0.27
Field + Shelterbelts	998		538		0.54		0.63	
Forest	1478		165		0.11		0.85	
Meadow	1250		215		0.17		0.84	
				Normal Year				
Field	1035	487	495	374	0.48	0.40	0.67	0.21
Field + Shelterbelts	1078		458		0.42		0.68	
Forest	1522		121		0.08		0.88	
Meadow	1347		118		0.09		0.90	
				Wet Year				
Field	1242	298	288	185	0.23	0.16	0.81	0.09
Field + Shelterbelts	1300		236		0.18		0.82	
Forest	1540		103		0.07		0.89	

On the other hand, evapotranspiration depends on humidity during any individual year (Rosenberg 1974). For example, in the Turew region, in the case of alfalfa, the ratio of real to potential evapotranspiration was 0.74 in the dry year of 1982 and 0.90 in the moderately moist year of 1983. In the moist year of 1984, this ratio was as much as 1.0.

The diversity of energy and water fluxes in an agricultural landscape is strongly influenced by general water conditions in any individual year (Table 4.6). In a dry year, the difference in total latent heat flux (energy used for evapotranspiration) between forest (using as much as 1478 MJ·m⁻² for evapotranspiration during the growing season) and field (using only 892 MJ·m⁻²) was very high, reaching as much as 586 MJ·m⁻². This amount of energy is enough to evaporate 235 mm of water. During a normal year, this difference is lower by about 100 MJ·m⁻², but during a wet year it reaches only half of the value of a dry year. The differences between latent heat flux (LE) of individual landscape elements in wet and dry years were as follows: 62 MJ·m⁻² for the forest, 147 MJ·m⁻² for the meadow, 350 MJ·m⁻² for the field, and 302 MJ·m⁻² for the field with shelterbelts. These examples prove the thesis

120 mm. Thus, higher diversification of landscape structure has higher stability of water cycling and water balance at the landscape level. Also the diversification of efficiency of the solar energy utilized for evapotranspiration is higher in a dry year than in a wet year. In a dry year, forest can use as much as 85% of net radiation for evapotranspiration while the field uses only 58% (Table 4.6). In a wet year, the efficiency of solar energy utilization by forest and field differs only by 9%. The increase of habitat moisture causes the increase of the ratio LE/Rn by 4% in the forest and by 23% in the field. Thus, richer plant cover means higher efficiency of energy utilization even when a water shortage occurs. In the humid climate of the Turew region, heat advection above the plant canopy can be observed quite often. In these cases, plants consume more energy for evapotranspiration than is absorbed as net radiation.

Weather conditions have a strong impact on the daily course of evapotranspiration (Figure 4.17). This impact is strongly linked with that of plant cover (discussed

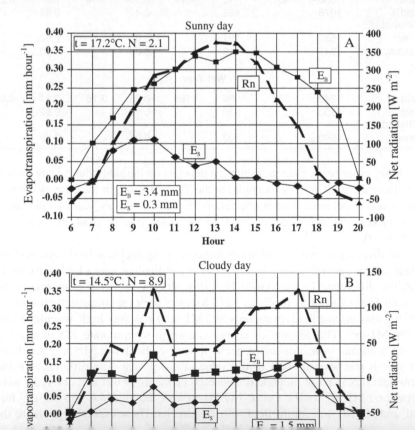

later in this chapter). During a sunny day (Figure 4.17) a daily course of plant cover evapotranspiration is regular, and its intensity can reach a level as high as 0.35 mm/h in the early afternoon hours. The course of evapotranspiration from bare soil or soil covered by nonactive plant detritus is quite different. The maximum is a few times lower, and it decreases before noon. This difference is because plants can use the water stored in topsoil as well as in the deeper layer of the soil profile. Thus, only solar energy input limits intensity of evapotranspiration. There is no limit in access to water supply. In the case of a field without plants, the quickly growing atmospheric water demands and solar energy input force intensive evaporation but only to the point when water stored in a thin soil surface layer has been evaporated. In humid climatic zones, water is usually stored in the thin soil layer during the nocturnal condensation process. When this water is exhausted, moisture of the soil surface layer decreases, causing the reduction of hydraulic conductivity in this layer, which finally leads to a decrease in or a halt to the evaporation process. In such conditions, condensation is usually observed in late afternoon or early evening. During a cloudy day (Figure 4.17) the daily course of evapotranspiration is irregular, maximum evapotranspiration intensity is low, and differences between a plant-covered field and field without plants are not significant.

The impact of solar energy flux on the intensity of evapotranspiration increases simultaneously with increasing plant development stage (Figure 4.18), (Kędziora et al. 2000). During the days with low solar flux (Rn < 40W·m^{-2}) the differences between evapotranspiration of a field covered by poorly developed plants (plant development stage <0.3) and one with well-developed plant cover (plant development

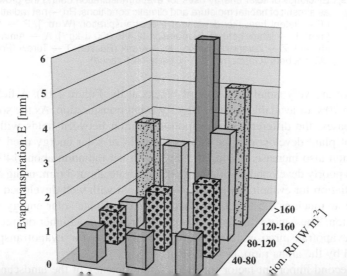

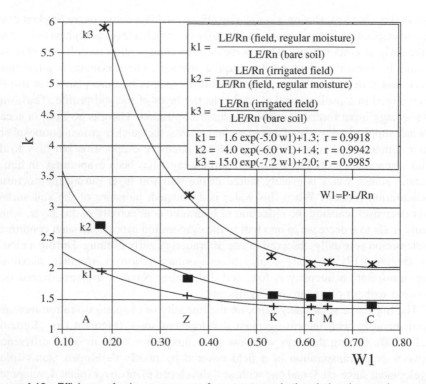

Figure 4.19   Efficiency of solar energy uses for evapotranspiration during the growing season as a result of habitat moisture and climatic conditions. Rn — net radiation [W·m⁻²], LE — latent heat flux density of evapotranspiration [W·m⁻²], P — precipitation [mm], L — latent heat of evaporation [2,448,000 J·kg⁻¹]. A — Alma-Ata (Kazakhstan), Z — Zaragoza (Spain), K — Kursk (Russia), T — Turew (Poland), M — Müncheberg (Germany), C — Cessieres (France).

stage >0.8) are very small (first row of blocks at the Figure 4.18). A field uses no more than 70% of available solar energy for evapotranspiration. As the solar energy flux increases, the differences in evapotranspiration between fields with different degrees of plant development as well as the ratio of solar energy used for evapotranspiration also increase. During the days with net radiation about 140 W·m⁻², a field with poorly developed plant cover can evaporate about 3 mm, using about 75% of net radiation for evapotranspiration, while a field with well-developed plants can evaporate as much as 4.5 to 5.0 mm, using total available solar energy for evapotranspiration. Thus, more developed plant cover results in a higher degree of energy use for evapotranspiration (efficiency of energy use for evapotranspiration is expressed by the alpha ratio, $\alpha = LE/Rn$).

The second important factor for increasing efficiency of the landscape's energy

of total precipitation (during a growing season) only 10% of net radiation is needed. This value changes from about 12% in an arid zone to 75% in a humid zone. The least efficient energy use is observed in the case of bare soil. In this case evaporation originates only from a thin surface layer, and it is quickly reduced when that layer dries, so intensity of this process is low. Higher efficiency will be observed for the field with plant cover under regular moisture conditions, but the highest efficiency will occur in the case of irrigated fields. The ratio between individual α ratios, denoted as the **k** ratio, is a measure of plant cover and habitat moisture impact on increasing of energy use efficiency of the fields and the same of the landscape. Thus, the ratio k_1 (Figure 4.19) shows how the ratio α of plant cover under regular moisture will increase in comparison with a bare field; however, the ratio k_2 shows how the ratio α increases when the field is irrigated.

The impact of plant cover or habitat moisture on efficiency of solar energy use is strongly affected by weather and climatic conditions, and the relationship between ratio **k** and climatic index W1 is nonlinear (Figure 4.19). For example, in the humid climate of Europe, fields with plant cover use about 40% more solar energy for evapotranspiration than bare soil does (k_1 equals 1.40) (Figure 4.19). In an arid zone this ratio is equal to 2 (fields with plant cover use solar energy for evapotranspiration two times more efficiently than bare soil does). The impact of habitat moisture on efficiency of solar energy use is even higher than the impact of plant cover. In humid climatic conditions energy use efficiency amounts to nearly 50% but in arid zones it is as high as 170% (k_2 is 1.5 and 2.7, respectively). Simultaneous impact of plant cover and irrigation on efficiency of solar energy use shows a synergistic character. In the humid climate condition of Europe, the ratio k_3 is equal to 2.0, which means total impact of plant cover and irrigation is equal to 100% (a little more than the sum of their individual impacts: 40% + 50%), but in arid climates this synergistic effect is as high as 500%, while the sum of separate effects of plant cover and irrigation is much lower (100% + 270%). Irrigated and well-developed plant cover use nearly the same or even more energy for evapotranspiration than that determined by value of net radiation, independent of general climatic conditions.

Thus, an agricultural landscape shows a much more stabilized efficiency of solar energy use for evapotranspiration than does any individual element making up that landscape. The importance of landscape structure for creating stable efficient solar energy use and for controlling the structure of water and heat balance is higher when moisture conditions are strained.

Soil Conditions

From the water-balance point of view, soil plays the role of water reservoir. The amount of water stored in the soil, its availability to plants, and its movement depend on the soil's structure and mineral composition as well as on its organic matter

82   LANDSCAPE ECOLOGY IN AGROECOSYSTEMS MANAGEMENT

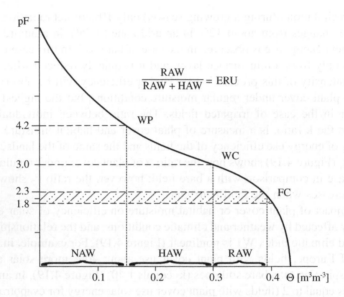

Figure 4.20   Soil water characteristics — pF curve. WP — wilting point, WC — critical moisture point (soil moisture level at which the capillary conductivity is vitally reduced), FC — field capacity, RAW — readily available water, HAW — hardly available water, NAW — nonavailable water, ERU — effective useful water retention.

water stored in the soil is available for plants nor does it all support intensive evapotranspiration. One must keep in mind that soil moisture, which determines the amount of water available for plants, as well as water potential, which expresses the work a plant must do to get water, are both factors. The best water characteristic of soil is the pF curve (Figure 4.20). The water retained in soil between about pF = 2.0 and 3.0 fills medium-sized capillaries. This water is readily available for plants (RAW) and ensures enough efficient capillary flow from the deeper soil layers toward the surface to support intensive evapotranspiration. When the soil moisture is running low (pF higher than 3.0), continuity of capillary water is broken, and evapotranspiration intensity decreases. The amount of water readily available for plants depends mainly on the density of medium-sized pores, which in turn depends on soil structure (Figure 4.21). The highest values of RAW occur in medium-textured, well-structured soils and can reach as much as 130 mm in a soil layer 1 m thick. It is enough to ensure evapotranspiration of a sugar beet field during one summer month. The fine-textured but poorly structured soil and coarse-textured soil have very low values of RAW, even below 30–40 mm in a 1-m soil layer (Figure 4.22). The capability for water movement in soil is best characterized by hydraulic conductivity, which in saturated zones mainly depends on soil structure and pore size distribution, but in unsaturated zones rapidly declines as soil moisture decreases (Figure 4.23). Thus

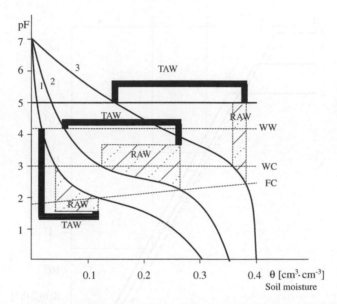

Figure 4.21 Total (TAW) and readily (RAW) available water for (1) coarse, (2) medium, and (3) fine textured soils.

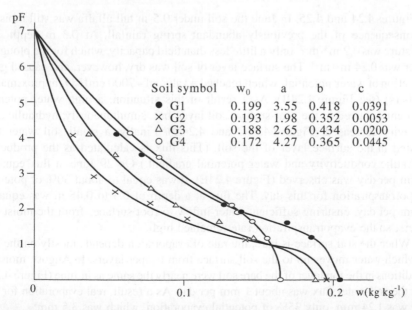

Soil symbol	w_0	a	b	c
● G1	0.199	3.55	0.418	0.0391
× G2	0.193	1.98	0.352	0.0053
△ G3	0.188	2.65	0.434	0.0200
○ G4	0.172	4.27	0.365	0.0445

Figure 4.22 Water retention curves of light soils of the Turew landscape, Wielkopolska, G1 —

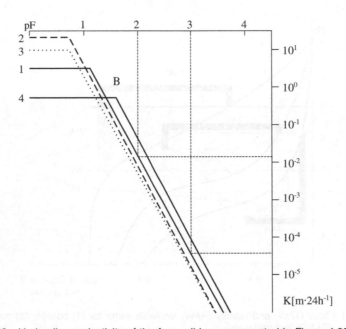

Figure 4.23 Hydraulic conductivity of the four soil layers presented in Figure 4.22.

in Figures 4.24 and 4.25. In June the soil under 0.5-m tall alfalfa was still moist as a consequence of the previously abundant spring rainfall. At 0.3 m depth, soil moisture was $0.2 \text{ m}^3 \cdot \text{m}^{-3}$, only a little less than field capacity, which for the ploughed layer was $0.24 \text{ m}^3 \cdot \text{m}^{-3}$. The surface layer of soil was dry, however, and caused great depletion of water potential, which reached a value of $-7000 \text{ cmH}_2\text{O}$ (approximately −7 bars) (see Figure 4.24B). As a result of this situation, a high water potential gradient occurred in the near-surface soil layer, but, simultaneously, hydraulic conductivity decreased (Figures 4.24C and 4.24D). Finally, a small soil water flux existed in the surface layer of the soil. (This flux is calculated as the product of hydraulic conductivity and water potential gradient.) On 29 June, a flux equal to 1 mm per day was observed (Figure 4.24E). It was equal to about 30% of potential evapotranspiration for this day. The flux at a depth of 0.3 to 0.45 m was equal to 3 mm per day, ensuring sufficient water inflow to root surfaces from the moist soil layers, so the evapotranspiration rate remained high.

When the soil surface is bare, the rate of evaporation depends mostly on the rate at which water moves up to the soil surface from deeper layers. In August, moisture conditions in the top layer of the bare soil were nearly the same as in June (Figure 4.25), and the soil water flux was about 1 mm per day. As a result, real evaporation for this day was 1.24 mm, only 35% of potential evaporation, which was 3.5 mm.

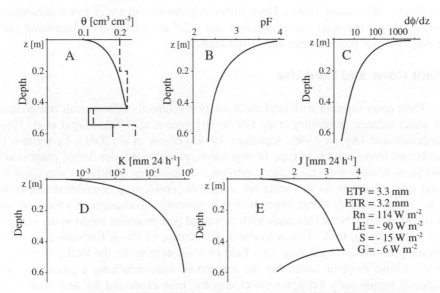

Figure 4.24 Water characteristics and water flux in the soil profile of an alfalfa field. A — soil moisture, B — soil water potential (expressed as pF), C — gradient of soil water potential, D — hydraulic conductivity (K), E — water flux (J), $\phi = 10^{pF}$, $J = K \cdot d\phi/dz$, Rn — net radiation, LE — latent heat of evapotranspiration, S — sensible heat, G — soil heat, ETP — potential evapotranspiration, ETR — real evapotranspiration.

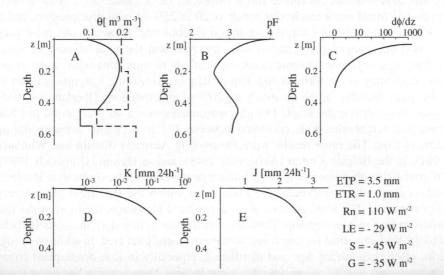

conditions (Richmond 1990). Thus, differentiation of soil cover in an agricultural landscape and the plant cover structure and relief are all factors determining variability of water balance structure in the landscape.

Plant Cover and Land Use

Plant cover together with land use directly or indirectly influence all components of water balance (Kędziora et al. 1987a, Gregoire et al. 1992, Magid et al. 1994, Kędziora and Olejnik 1996, Kędziora 1999, Olejnik et al. 2001). Ecosystem or landscape features such as type of vegetation, plant height, density of plant cover, and plant development stage affect individual components in different ways, but the final effect depends on the scale for which the problem is considered (Kędziora et al. 1992, Kędziora 1999). Precipitation is insensitive to changes of a forested area at the local scale, but it increases with increased percentage of forest at the regional scale (Bac 1968). In the Polish lowland, an increase of 1% in forestation increases annual precipitation by 5 mm. At a field or ecosystem scale, the total precipitation is insensitive to plant cover, but the amount of water reaching a ground surface depends significantly on plant cover density, best expressed by leaf area index (LAI — the ratio of the area of the top sides of all the leaves in the canopy projected onto a flat surface to the area of the surface under the canopy) (Neal et al. 1993). It is equal to the number of leaves that would be crossed by a vertical line passing through the canopy, on average. Water interception by plant cover changes from a fraction of a millimeter in the case of poor vegetation to as much as more than 10 mm in very dense, old spruce forest (Monteith 1975, Zinke 1967). Annual interception in forest areas reaches as much as 20 to 25% of total precipitation, and it is higher in a coniferous forest than in a deciduous one. However, one must keep in mind that gauged precipitation does not represent the total amount of water recharging ecosystems. In some cases, especially in mountainous areas, water entering ecosystems in the form of fog-drip or dew can constitute a significant part of total water income, up to as much as 20% of precipitation. Oberlander (1956) observed fog-drip under single Douglas spruce trees over a 40-year period in a San Francisco neighborhood. It oscillated between 175 to 425 mm annually during rainless days. The same results were observed in Australia (Costin and Wimbush 1961), in the Belgian Congo (Auberville 1949), and in Hawaii (Gingerich 1999). Of course this phenomenon is more intense on the edge of the forest than inside it, and, consequently, the structure of plant cover within the landscape plays a very important role. The more ecotons in an agricultural landscape, the more water this landscape can gain as fog-drip. Thus, it is much better to introduce many shelterbelts into the landscape than for one forest to have the same total area. In addition to fog-drip, other phenomena, dew and distillation, especially in low, dense plant cover, can sometimes constitute considerable water income. Dew occurs when water vapor

can reach as much as 100 mm. This phenomenon rarely occurs in forest or tree stands. Thus, the mosaic landscape demonstrates some advantages over the uniform one because all the above phenomena can occur (contrary to a uniform landscape) and total water income is higher, due to internal recycling, than in uniform landscape. This compensation mechanism plays a very important role in formation of water regime in an agricultural landscape.

Surface runoff is very sensitive to features and changes of plant cover in any scale. In forests or meadows, the topsoil is usually loose and overgrown by plants. High infiltration capacity of such soils and plant resistance to flowing water significantly reduces runoff. Especially in lowlands, runoff from such ecosystems is reduced to a minimum, excluding unusual rainstorms. In these areas, the subsurface outflow is quite high in comparison with surface runoff and is very stable in time. In contrast to this, in row crop fields or even in grain crop fields, and especially in the case of bare soil, the surface runoff is very high. These areas, because of compaction of the topsoil layer by machinery, very often have low infiltration rates, even in light soil. Moreover, the furrows in row crop fields and small grooves in grain crop fields, especially when they are oriented from the top to the foot of a hill, increase water outflow. Thus, surface runoff from the cultivated fields can be observed even in the case of mild precipitation, and it usually exceeds the subsurface runoff. In general, total runoff, as well as its partition into surface and subsurface outflow, depends on many factors, among which plant cover appears to be the most important. Consequently, in an agricultural landscape, the intensity, timing, and structure of runoff change as landscape diversification changes. In a mosaic landscape with a rich shelterbelt network, many meadow strips, and bushes, total runoff as well as the ratio of surface to subsurface outflow is low. As a result, the water outflow from such a landscape is slow and stable and lasts a long time, creating favorable conditions for rebuilding soil water supply on the one hand, and preventing the small streams from drying up on the other, which in turn allows survival of the many animals and plants living in or near those streams. Within uniform agricultural landscape, the runoff develops quickly, lasts a short time, is very intensive, and occurs mainly in the form of surface outflow (Figure 4.16). Such a situation runs counter to improving water conditions in the environment and may bring about serious soil erosion. Therefore, the more diverse a landscape structure is, the better will be its characteristics of runoff and hence the better the water conditions in the environment.

Evapotranspiration is the most sensitive component of water balance affected by plant cover. Evaporation from bare soil is determined by the moisture in the thin surface layer of soil, which in turn depends on hydraulic conductivity of the soil, determining the transport of water from deeper, moist soil layers or from ground water to the surface. But hydraulic conductivity in unsaturated zones strongly depends on water moisture. Thus, the evaporation process from bare soil leads to overdrying of the soil surface, and causes reduction

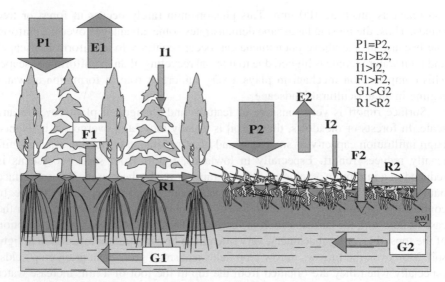

Figure 4.26   Differences between (1) forest and (2) crop field water balance components. P —
precipitation, E — evapotranspiration, I — interception, F — infiltration, R —
surface runoff, G — subsurface runoff.

are moist. This effect is more pronounced the taller the plant is and the deeper its
root zone. For example, the potential evapotranspiration over the forest canopy is
much higher than over the low plant canopy because the forest canopy, thanks to
its greater roughness, absorbs much more solar energy than does low plant canopy
and because the saturation water vapor pressure deficit and wind speed, as well as
water exchange coefficient, over the forest canopy are much greater than they are
over the short plant cover. The trees of shelterbelts function as the water pump,
intensifying the water cycling in an agricultural landscape (Figure 4.26).

When the daily course of potential or real evapotranspiration is examined, the vital
importance of the plant factor can be seen (Figure 4.27). The maximum hourly values
of real evaporation from bare soil (Figure 4.27B) are lower than the values of potential
evapotranspiration (Figure 4.27A), calculated by the Penman method. The hourly rate
of real evaporation reached 0.25 mm, while the rate of potential evapotranspiration was
as high as 0.43 mm. At the same time, over the plant community the real evapotranspiration rate (Figure 4.27D) was a little higher than the potential rate (Figure 4.27C),
reaching 0.38 mm/h. It should be noted that the potential evapotranspiration rate over
plants is lower than over bare soil because of the lower saturation deficit and air
temperature over the plant canopy as a result of cooling by intensive evapotranspiration.

When the seasonal course of evapotranspiration is considered (Figure 4.28) the
prevalence of real (A) over potential (B) evapotranspiration can be observed at the

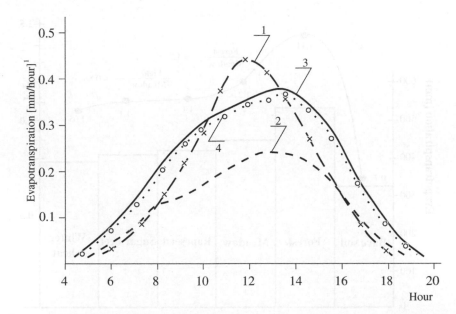

Figure 4.27 Daily courses of evapotranspiration from different surfaces. 1 — bare soil potential evaporation, 2 — bare soil real evaporation, 3 — potential evaportranspiration of alfalfa, 4 — real evapotranspiration of alfalfa.

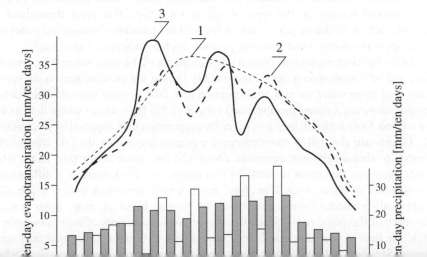

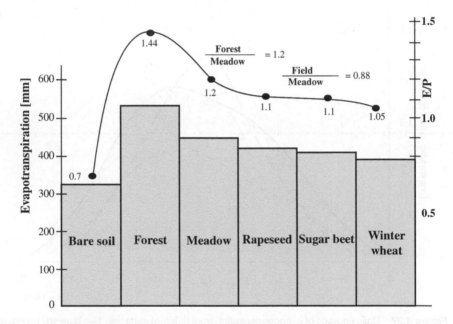

Figure 4.29   Evapotranspiration and ratio of evapotranspiration to precipitation (E/P) for main ecosystems of the Turew landscape during the growing season (March 21–October 31).

In the Turew region, no great differences were observed between evapotranspiration rates of different agricultural crops; however, some small differences do exist. They depend mainly on the type of vegetation. Crops that grow throughout the season, such as alfalfa or grass, show a higher total amount of transpired water than do crops with shorter plant growing periods, such as wheat or sugar beet.

In the Turew climatic conditions, a forest evaporates 20% more water than a meadow does and 30% more than a field (Figure 4.29). During the growing season, all ecosystems need more water for evapotranspiration than falls as precipitation. Water deficits (evapotranspiration minus precipitation) vary from 5% for a winter wheat field to 44% for a forest. Only bare soil uses less water for evaporation than supplied by precipitation.

Forests use about 40% more energy for evapotranspiration than do wheat fields. Forests or shelterbelts can evaporate about 170 mm more water than can a wheat field. There are two main reasons for this difference. First, there is a difference in the structure of plant cover. Trees have much longer roots than wheat, which allow them to absorb water from deeper layers of the soil. In effect, more water is within the reach of the trees' roots. Since trees have greater amounts of water available for their use than do cereals, tree leaves have smaller stomatal resistance than cereal leaves. Second, shelterbelts also have a greater canopy roughness than does wheat

reaching the soil. The latter is a function of the density and structure of the plant cover. Plant cover and habitat moisture are the main factors determining the partitioning of net radiation into different internal energy processes of the ecosystems, especially under convection conditions.

The direction of air advection is determined by wind direction, but the value of energy transported by advection depends on the temperature gradient, which is controlled by plant cover and moisture of the habitat. So, if a dry area surrounds a wet area covered by fully developed plants, advection brings the energy from the dry area to the wet field, enhancing the difference between latent heat flux of dry and wet fields. The diversification of water balance on the level of landscape is smaller than that on the level of the ecosystem. The more mosaic the structure of landscapes, the smaller the difference between water balance structure of the two compared landscapes.

WATER MANAGEMENT IN THE LANDSCAPE

For water management in the landscape to be effective it must be based on full recognition of the interrelations between energy flow and water cycling and must consider the following items:

- Water deficits or water surplus in the landscape
- Possibility of improving water retention
- Possibility of using natural mechanisms and processes to improve water use efficiency in the landscape

Water Deficit in the Landscape

As was mentioned in the section on water balance of the agriculture landscape, many water management errors and unfavorable land use practices brought about very strained water conditions in many regions around the world. An especially serious problem of worsening water conditions has been observed in the Polish lowlands, where water deficit is a crucial factor limiting agricultural production. Intensive evapotranspiration combined with rather low precipitation usually leads to water shortage, especially during the growing season.

However, the indices of evapotranspiration and precipitation do not sufficiently characterize water conditions in an agricultural landscape from the point of view of water demands and supplies. For this purpose, the interrelations between available energy and water also must be analyzed.

These interrelations can be characterized by the following two indices:

Table 4.7   Ten-Year Averaged Values of Water Balance Components in Period IV–IV
            for Seven Ecosystems in an Agricultural Landscape (D. Chłapowski's
            Landscape Park)

Ecosystems	Rn [W m⁻²]	LE [W m⁻²]	P [mm 24h⁻¹]	E [mm 24 h⁻¹]	P-E	RWDI	PWDI
Winter crop	93	65	2.38	2.30	0.08	0.97	1.38
Spring crop	94	62	2.38	2.20	0.18	0.92	1.40
Potatoes	94	62	2.38	2.20	0.18	0.92	1.39
Sugar beet	93	65	2.38	2.30	0.09	0.96	1.37
Perennial crop	92	68	2.38	2.38	0.00	1.00	1.37
Coniferous forest	103	83	2.38	2.93	−0.55	1.23	1.52
Deciduous forest	91	72	2.38	2.53	−0.14	1.06	1.34

Rn — net radiation [W m⁻²], LE — latent heat flux density [W m⁻²], P — precipitation [mm],
E — real evapotranspiration [mm], RWDI — real water demand index, PWDI — potential water
demand index.

Such an approach was used to characterize the water conditions within D. Chłapowski's Landscape Park, which is situated in the center of the Wielkopolska region and which is representative of the Polish lowlands.

During normal weather conditions, seasonal evapotranspiration of main crops in agricultural landscapes is only a little less than precipitation, for perennial crops it is equal to precipitation, and for forests it exceeds precipitation (Table 4.7 — RWDI). However, potential water demands (PWDI) are greater than precipitation for all ecosystems. These demands exceed precipitation by 30 to 40% which shows that soil water supplies collected during winter play a crucial role in satisfaction of plant cover water needs.

During the entire growing season (from March 21 to the end of October), evapotranspiration values varied from 350 mm for bare soil to 600 mm for moist forests. Crop fields showed values from 430 mm (winter wheat) to 470 mm (rapeseed). Meadow evapotranspiration was in the range of 500 mm. But, if only an individual crop growth period is considered, the differences are much greater. During the period of sugar beet growth (from June 10 to the end of October) precipitation normally amounts to 268 mm and can cover water needs (250 mm). But usually during this period soil water supplies are not readily available for plants. Thus, any disturbance of precipitation must unfavorably impact the growth and yield of this crop. Quite a different situation is observed with winter wheat. From the beginning of April to the end of July, winter wheat needs 295 mm of water. Precipitation during this period amounts to only 243 mm, but soil water supply in a 1-m layer is 40 to 80 mm. Consequently, the winter wheat crop rarely suffers from drought compared to sugar beet.

If seasonal courses of evapotranspiration and precipitation are analyzed in more detail, a short period with surplus precipitation is observed, especially during summer when rainstorms occur. This water surplus cannot infiltrate the soil and should be

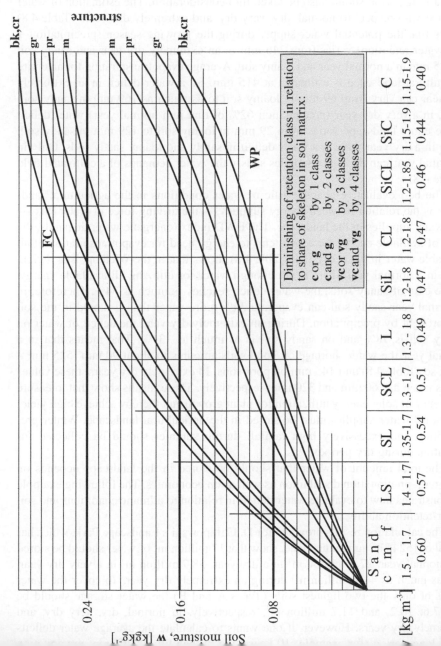

ependence of soil moisture on textural classification, bulk density, ard soil structure. TAW — total available water = FC – WP, RAW —
adily available water, FC — field capacity, WW — wilting point, c — cobble, g — gravely, bk — blocky, cr — crumb, pr — prismatic,
— massive.

over a long period should also be taken into consideration. The estimation of water balance with respect to normal, dry, very dry, and extremely dry years (Table 4.8) shows that the potential water supply during the growing season (precipitation + soil water retention) varies from 246 mm in an extremely dry year and sandy soil to 525 mm in a normal year and loamy soil. Average evapotranspiration for an entire agricultural landscape is estimated at 415 mm. So a water deficit in an extremely dry year can vary from 69 mm in loamy soil to as much as 169 mm in sandy soil. Also, in a very dry year (precipitation 62% of that in a normal year) water deficit of the entire landscape varies from 29 mm in loamy soil to 129 mm in sandy soil. Even in a dry year there is a water deficit in sandy and loamy sandy soils. Only in a normal year can the water supplies satisfy the water demands of a cultivated field (Table 4.8).

The best evaluative characteristic of water conditions in the agricultural landscape is the relation between energy supplies — the driving force of water flux — and water supplies in the landscape. The EWD index (energetic water demands, the ratio of available energy for ecosystems to energy need for evapotranspiration of all available water in the ecosystem, Table 4.8) shows that in the Wielkopolska region, even in a normal year, there is not enough water for utilizing all available energy. In the case of sandy soils, the water deficit reaches as much as 34%. Therefore, in a normal year sandy soil can evaporate 34% more water than is stored in the soil and brought by precipitation. During an extremely dry year, the deficit of water on loamy soil is 78% and on sandy soil is as much as 131%. That means that in a normal year the water shortage on good soils amounts to about 90 mm (525 mm × 0.17; see Table 4.8) and 145 mm on poor soils. In extremely dry years, these values are as much as 260 mm and 320 mm, respectively. These results show that to ensure efficiency of solar energy utilization for plant growth and thereby achieve high yield, additional water supplies must be stored in the agricultural landscape. Winter precipitation and excessively high rainfall during summer should be collected for irrigation during dry periods.

The total amount of water that should be stored in the landscape depends on long-term precipitation distribution as well as on economics. The following example demonstrates how to evaluate water deficit in relation to climatic conditions and soil water-retention ability.

The calculation results provided for D. Chłapowski's Landscape Park, which has a total of 15245 ha agricultural land, show that 17 million m^3 of water should be stored in a normal year, 27.4 million m^3 in a dry year, 38.7 million m^3 in a very dry year, and as much as 44.7 million m^3 during an extremely dry year. To cover the water needs of only the two lightest soils (Types A and B) the water storage should be 5.2; 7.6; 10.2, and 11.7 million m^3, respectively, in normal, dry, very dry, and extremely dry years. However, if one wants to calculate the average water deficits

Water Deficits in Arable Land during the Period IV-IX in D. Chłapowski's Landscape Park, Wielkopolska

Textural Group	Rn Area [ha]	E $W\,m^{-2}$	mm	Precipitation [mm] for Year Normal 100%	Dry 82%	Very Dry 62%	Extra Dry 50%	Water Retention TAW [mm]	RAW [mm]
sandy soils	552	88	415	365	299	226	186	60	35
loamy sandy soils	3166	90	415	365	299	226	186	80	45
sand loamy soils	7587	93	415	365	299	226	186	120	65
loamy soils	3940	95	415	365	299	226	186	160	80

	Precipitation + TAW in Year				Precipitation + TAW - Evapotranspiration			
	Normal	Dry	Very Dry	Extra Dry	Normal	Dry	Very Dry	Extra Dry
sandy soils	425	359	286	246	10	-56	-129	-169
loamy sandy soils	445	379	306	266	30	-36	-109	-149
sand loamy soils	485	419	346	306	70	4	-69	-109
loamy soils	525	459	386	346	110	44	-29	-69

	EWD in a Year			
	Normal	Dry	Very Dry	Extra Dry
sandy soils	1.34	1.58	1.99	2.31
loamy sandy soils	1.31	1.53	1.90	2.19
sand loamy soils	1.24	1.43	1.73	1.97
loamy soils	1.17	1.34	1.59	1.78

Total available water = FC - WP, RAW — readily available water = FC - WC, FC — field capacity, WP — wilting point, WC — moisture, EWD — energetic water demands.

· (number of seconds in the period)/[2448000·(precipitation + TAW)]

f calculation: the period IV-IX has 183 days; in a normal year, precipitation + TAW for soil A is equal to 425 mm. EWD = 400/2448000·425 = 1.337.

each year to ensure maximal utilization of landscape agropotential and a high yield from cultivated fields.

However, if one wants to ensure not maximal evapotranspiration in dry years but only evapotranspiration at the level of a normal year, the water reserves that must be collected can be substantially reduced. In a dry year, the deficits will occur only in soil types A and B, reaching 1.4 million m^3; in very dry and extremely dry years the deficits will be observed in every soil reaching 10.5 and 16.6 million m^3, respectively. Taking again into consideration the frequency of different hydrological years, the average annual water deficits will reach 3.1 million m^3 of water, but in a very dry year as much as 10 million m^3 of water should be available. The last two figures are very important for water management in the agricultural landscape. This calculation means that within D. Chłapowski's Landscape Park each year 3 million m^3 of water should be stored, and the total volume of water stored should be 10 million m^3.

Thus, in the case of potential as well as real water deficits, economically justified water needs are too great to be stored by technical means (artificial water reservoirs). So, any activities that lead to increased soil water retention as well as a natural means that increase small landscape retention are of fundamental importance for sustainable development of agriculture. This problem is discussed in more detail in the next section.

Improving Water Retention

The increase of water retention in the landscape may be obtained by increasing the quantity of water infiltrating the soil and by increasing soil capacity for water retention. The first goal can be achieved by keeping a high rate of soil infiltration and percolation in the topsoil layer. Usually, the infiltration rate is lowered either by natural processes like compacting of soil surface by rain or by tillage work that compacts the soil layer situated just beneath the ploughed layer. Applying no-tillage technology restores natural soil structure and channels made by roots are a very efficient means of water transmission in the soil. Reduction of surface runoff rate increases the time water stays in the landscape, thereby allowing more water to infiltrate the soil (Alconada et al. 1993). From this point of view, bare soil of the cultivated field is a landscape element that deteriorates water balance, while a forest is an element that improves it. So, once again, richer plant cover means a higher amount of water coming into the soil.

Although the infiltration capacity of the soil surface is very important for improved water storage, the structure of the soil is the most important factor (see earlier section on soil conditions). Well-structured soil has many pores of medium size that enable the soil to retain a maximal water amount (see Figures 4.21 and 4.30) (Ilnicki 1983, Cieśliński and Miatkowski 1989, Cieśliński 1997). The best way to achieve this goal is to increase the organic matter content of the soil. Organic

organic matter while the G2 layer is only 0.92%. The bulk densities of these soils are similar — 1588 kg·m⁻³ and 1638 kg·m⁻³, respectively. The pF curves (water retention curves) are described by the following equations (Kędziora 1984):

$$\text{G1: } pF = 3.55\left(\frac{0.316 - w}{2 + 0.062}\right)^{0.418}$$

$$\text{G2: } pF = 1.98\left(\frac{0.316 - w}{w + 0.087}\right)^{0.352}$$

Using these equations, one can calculate that the water retention at field capacity of these soils is 225 mm and 90 mm, respectively. Assuming, for simplification, a linear relation between organic matter content and water content, the rate of water retention increases 34 mm per 1% of organic matter in a 1-m soil layer, or 10 mm in a ploughed layer. But one must keep in mind that the soil reservoir is a reservoir, collecting water not only one time but automatically after any intensive rain. At the same time it also reduces surface runoff. Assuming increased soil water retention of only 20 mm, an additional 3 million m³ of water can be stored in the Landscape Park.

The second possibility of increasing landscape capacity for water storage is building new small modified ponds and restoring degraded ones (Kosturkiewicz 1985, Kosturkiewicz and Kędziora 1995, Ryszkowski and Kędziora 1996b). These small ponds play a dual role very important for water cycling: they collect excess spring and drainage water and also intensify water cycling. A small pond might evaporate 24 to 32% more water than can a reservoir 10 times bigger in the same area. Currently in the Wielkopolska region, there are two small mid-field ponds in every one hectare, but there could be as many as 100 per hectare. A small pond not only collects water in its bowl, but it also increases water retention in the soil within its neighborhood. The ratio of water stored in the soil to water stored in the pond depends on pond size; the smaller the pond, the higher the ratio (Figure 4.31).

The ratio of water retained in the soil surrounding the pond to the water stored in the pond itself is greatest for ponds located in light soils rather than in heavy soils; in light soils the range of increase in ground water table as a result of elevated water table in the pond is greater than in heavy soil. Increased soil moisture in the pond neighborhood has another positive effect; it reduces wind erosion of soil.

Thus, enriching landscape with mid-field ponds leads to increased water storage, improved soil water conditions, and intensified water cycling. As more energy is used for evapotranspiration less energy remains for air heating. So, improving water conditions in the agricultural landscape can improve the efficiency of solar energy use by that landscape (Tolk et al. 1998).

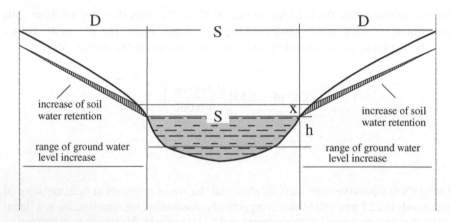

Figure 4.31   Increase of water retention in mid-field pond and in its surrounding soil as a result of elevation of water level in the pond. S — diameter of water surface in the pond, x — increase of water level, h — average depth of water pond, L — diameter of area impacted by increase of water level in the pond, D — distance of pond impact on ground water level. D = xz, z = distance of pond impact on ground water level when water level was elevated by 1 m.

and Kędziora 1987, Ryszkowski and Kędziora 1996a). Landscapes with a more mosaic structure have a higher degree of landscape resistance. The best way to improve landscape structure is the introduction of shelterbelts, strips of meadows, and hedges, rebuilding of damaged postglacial ponds, and maintenance of wetlands and riparian ecosystems. The saturation of landscapes by ecotons and biogeochemical barriers is the most efficient tool for controlling energy flow and matter cycling, which is also necessary for sustainable development of agriculture (Ryszkowski Kędziora 1987, Ryszkowski and Kędziora 1993, Kędziora et al. 1995, Ryszkowski et al. 1999).

A network of shelterbelts introduced into a uniform agricultural landscape changes substantially the microclimatic conditions in that landscape (Jansz 1959, Jaworski 1962). The reduction of wind speed in particular causes, in turn, changes in the intensity of other processes and phenomena (Figure 4.32). At the distance from a shelterbelt equal to 4 to 12 times longer than the height of its trees (100 to 300 m if the shelterbelt is 25 m high) the wind speed is reduced to 60% of that in an open landscape and potential evapotranspiration is reduced to 75%. Snow cover is thicker in landscapes with shelterbelts. Thus, the amount of water infiltrating the soil is higher in a mosaic landscape than in a uniform one. The role of shelterbelts is especially important when advection of warm air over the cultivated field is observed (Figure 4.33). In such situations, the additional flux of heat inflow over the plant canopy intensifies the process of evapotranspiration. In the Wielkopolska landscape under such conditions, the alfalfa field can evaporate as much as 834 mm during growing season. Of course, in such situations irrigation is needed because

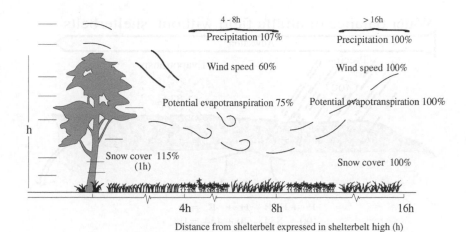

Figure 4.32 Impact of shelterbelts on microclimate in an agricultural landscape.

from the soil to the atmosphere and as a mechanical obstacle reducing wind speed in the entire landscape. So, on the one hand, shelterbelts have a positive impact on the water cycle and on the other hand they protect the cultivated field against the drying force of the atmosphere.

During the growing season, evapotranspiration of the uniform cereal landscape in Turew amounted to 414 mm, while the potential evapotranspiration was as high as 650 mm. By introducing shelterbelts into the landscape, one can reduce potential evapotranspiration from whole landscape to 586 mm, increasing simultaneously real evapotranspiration by 17 mm (Table 4.9). However, if one were to introduce only mechanical windbreaks, real evapotranspiration would decrease by 10 mm. Under advection conditions, the impact of shelterbelts is sufficient: the potential evapotranspiration is reduced by 35% and real evapotranspiration by 40 mm. This amount is equal to the typical irrigation dose and has essential economical importance. So, it is possible to control the water balance structure by adequate formation of landscape structure. Presently in Turew, with landscape consisting of 30% forest, 15% meadow, 10% row crops, and 45% cereals, the average value of latent heat flux density during the growing season (21.03–31.10) is equal to 49.93 W m^{-2} (landscape K1 in Table 4.10). Therefore, total evapotranspiration during this period equals 396 mm. Simplifying the landscape structure by removing forest and meadow reduces landscape evapotranspiration by 30 mm (landscape K2) and improving landscape structure by increasing forest and meadow increases evapotranspiration by 20 mm (landscape K3).

Generally speaking, the richer and more diversified the landscape structure, the better the water management in the landscape (Gilpin et al. 1992). Two guidelines

Water balance of alfalfa field without shelterbelts

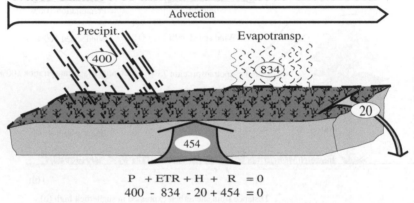

$$P + ETR + H + R = 0$$
$$400 - 834 - 20 + 454 = 0$$

Water balance of alfalfa field with shelterbelts

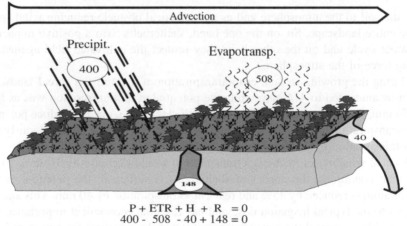

$$P + ETR + H + R = 0$$
$$400 - 508 - 40 + 148 = 0$$

Figure 4.33   Impact of shelterbelts on water balance of alfalfa field under advection.

Table 4.9   Water Balance Components of Different Agricultural Landscapes in the Turew Vicinity in the Growing Season (March 21–October 31)

Landscape	Rn MJ·m⁻²	LE MJ·m⁻²	ETP mm	E mm	E/ETP	E/P
Cereal monocultures	1542	1035	650	414	0.64	1.10
Cereal monocultures with a net of shelterbelts	1586	1078	586	431	0.76	1.15
Cereal monocultures with windbreaks	1567	1010	581	404	0.76	1.08

Table 4.10   Seasonal Values of Latent Heat Flux Density and Evapotranspiration under Different Landscape Structure

Parameter	Landscape		
	K1	K2	K3
Latent heat flux density, LE [W m^{-2}]	49.93	46.20	52.40
Evapotranspiration in growing season [mm]	396	366	416
Change of latent heat flux density, LE as a result of landscape structure change [W m^{-2}]	K2–K1 −3.73	K3–K1 2.47	K3–K2 6.20
Change of evapotranspiration as a result of landscape structure change [mm]	K2–K1 −30	K3–K1 20	K3–K2 50

K1 — 30% forest, 15% meadow, 10% row crop, 45% cereals;
K2 — 20% row crop, 80% cereals;
K3 — 50% forest, 20% meadow, 10% row crop, 20% cereals.

IMPACT OF CLIMATE CHANGE ON WATER BALANCE

Because of the strong link between energy flow and water cycling, the changes of atmospheric chemistry, plant cover, and land use observed recently, leading to transformation of the earth's energy conditions, have brought about changes in global, regional, and local hydrology (Ryszkowski et al. 1989, Valdés et al. 1994). The result of such changes are generally unknown because of the many feedback mechanisms between climate change and water in the environment (Ryszkowski et al. 1990, Ryszkowski and Kędziora 1995, Stohlgren 1998). One of the most complicated problems is that of the scale (Milly 1991). There have been many laboratory experiments on reactions of plants to enhanced CO_2 concentration, temperature increase, and soil and atmosphere moisture changes (Blasing 1985, Callaway and Currie 1985). But because of the many interactions between environmental factors, and the processes of competition and synergism, the results of these investigations cannot be uncritically applied to the landscape level (Ryszkowski and Kędziora 1999). Temperature and habitat moisture changes will affect the geographical range of many cultivated crops. Mean annual temperature increase of one degree in mid-latitude northern hemisphere countries would tend to move the thermal limit of cereal crops north by about 150 to 200 km, and raise the altitude limit by about 150 to 200 m (Parry and Swaminathan 1992). But in these countries where precipitation will not increase sufficiently, this positive effect of temperature increase might be reduced by water shortage caused by increased evapotranspiration. On the other hand, the increase of temperature mainly during the winter will unfavorably change the annual pattern of precipitation and evapotranspiration (Jaworski and Krupa-Marchlewska 1989). Duration of snow cover will be shortened or may disappear altogether, while evapotranspiration and surface runoff will increase and will change annual distribution (Vehviläinen and Huttunen 1997). For example, in the Wielko-

van Katwijk 1990). Many parts of the world can suffer from such a situation, including central and eastern Europe (Parry and Swaminathan 1992, Parry 1990). Other phenomena that can exert an influence on water conditions in agriculture are frequency of extreme events such as floods and droughts (Gleick 1992).

Increased global temperature will intensify global water circulation, which will in turn increase global precipitation (Jager 1988). But because of the very complex character of global air circulation, nobody is able to say where precipitation will increase and where it will decrease (IPCC 1996). Precipitation will probably continue to increase in regions where it is higher today and decrease in areas that are dry today. Because Poland and Central Europe are located in the transition climatic zone, it is very difficult to determine if precipitation will increase or decrease in this region. The crucial problem for the sustainable development of agricultural landscape is how global climate changes will alter the water balance structure of an agricultural landscape. The answer is even more difficult to determine if the impact of climate changes is linked to changes in land use (Brouwer and Chadwick 1991, Gottschalk and Krasovskaia 1997).

Taking into account different possibilities of temperature, precipitation, and forestation changes (Bach 1988), a few scenarios for Poland are possible (Table 4.11).

Table 4.11   Scenarios of Probable Climatic and Forestation Changes in Poland as a Result of Doubling of CO_2 in the Atmosphere

Scenario	Temperature Change	Precipitation Change	Forestation Change
0	Air temperature changes from 0 to −6°C in winter and from 16 to 19°C in summer	Lowest in winter, highest in summer; annual precipitation is equal to 700 mm	28% in country area
1	Increase in air temperature by 2°C in summer and −6°C in winter	Increase in precipitation by 0.6 mm/day in spring and by 0.4 mm/day in other period (average 23%)	Increase in forest area by 10% of present area
2	Increase in air temperature by 2°C in summer and −6°C in winter	Increase in precipitation by 0.6 mm/day in spring and by 0.4 mm/day in other period (average 23%)	Degradation in forest ecosystems in 50% of forest area
3	Increase in air temperature by 2°C in summer and −6°C in winter	Increase in precipitation by 0.6 mm/day in spring and by 0.4 mm/day in other period (average 23%)	Total degradation in forest
4	Increase in air temperature by 2°C in summer and −6°C in winter	Decrease in precipitation by 20%	Increase in forest area by 10% of present area
5	Increase in air temperature by 2°C in summer and −6°C in winter	Decrease in precipitation by 20%	Degradation in forest ecosystems on 50% of forest

The first scenario is the most optimistic: increased temperature (6°C in winter and 2°C in summer), increased precipitation (about 180 mm per year), and increased area of forest (10% of present area) will result in improvement of the water balance. During plant growth period in the Wielkopolska, evapotranspiration will increase by about 40 mm, but the increase of precipitation will be higher and water deficit will disappear. However, if evapotranspiration during the winter is higher than today, the increase in precipitation will not increase climatic water balance (P-E, Table 4.12). So the annual climatic water balance will increase only slightly (136 − 97 = 39 mm, Table 4.12, part for the whole year). But if the current process of forest degradation is continued in the future, and if the increase of precipitation is accomplished by the decrease of the energy used for evapotranspiration (impoverishment of the forested area), diminishing of forest function of the landscape will occur. This situation will lead to increased winter and annual climatic water balance, and it can give rise to serious threats to the agricultural landscape, especially water erosion.

In the case of scenario 4 (Table 4.12) — increased temperature and forested area, but decreased precipitation — new water conditions will cause a serious problem for the existence of agricultural ecosystems. Even with decreased forest evapotranspiration, due to forest degradation (scenario 5), land desiccation will be so high that plant cultivation without irrigation will be impossible. So in this case, the environment will be transformed into steppe, and grasslands will replace the forest ecosystems in the Wielkopolska.

In Central Europe, evapotranspiration during the plant growing period usually exceeds precipitation, so plants must use soil water supplies, thus depleting ground water level. Normally during the winter, snow precipitation rebuilds soil water supplies. But if in the future the temperature increases by 6°C, precipitation water will evaporate and the growing season will begin with small supplies of water, causing serious problems for agriculture development (Kędziora 1991, Kędziora 1996b). IIASA modeling results cite only Poland, England, and Spain as the European countries in which water scarcity will occur in the 21st century (Niemann 1994).

Generally, in the future, even with increased precipitation, the ratio of evapotranspiration to precipitation in summer months can increase. The wise strategy of water management is imperative in the face of increasing demands of irrigation water to sustain the development of agriculture. In such a situation, any activity in the frame of landscape structure leading to increased landscape water retention and increased efficiency of water use are crucial for future sustainable development of the agricultural landscape (Bouwer 1994).

Table 4.12 Water Balance (mm) in the Wielkopolska (W) and All of Poland (P), According to Scenarios in Table 4.11

Scenario	Region	Rn	LE/Rn	P	E	P-E	E/P	ETP
			Plant Growth Period (IV–IX)					
0	W	83	0.76	396	413	−17	1.04	492
	P	84	0.74	444	402	42	0.91	455
1	W	92	0.76	480	452	28	0.94	575
	P	92	0.75	528	445	83	0.84	578
2	W	85	0.73	480	400	80	0.83	523
	P	85	0.72	528	394	134	0.75	546
3	W	64	0.72	480	297	183	0.62	487
	P	65	0.70	528	297	231	0.56	462
4	W	92	0.76	324	452	−128	1.40	575
	P	92	0.75	365	445	−80	1.22	578
5	W	85	0.73	324	400	−76	1.23	523
	P	85	0.72	365	394	−29	1.08	546
6	W	64	0.72	324	297	27	0.92	487
	P	65	0.70	365	297	68	0.81	462
			Winter Period (X–III)					
0	W	4	4.29	222	110	112	0.50	108
	P	5	3.68	262	96	166	0.37	98
1	W	6	5.00	300	194	106	0.65	222
	P	6	5.56	340	188	152	0.55	208
2	W	4	7.00	300	181	119	0.60	210
	P	5	6.09	340	175	165	0.51	196
3	W	3	7.33	300	142	158	0.47	174
	P	4	6.60	340	135	205	0.40	162
4	W	6	5.00	186	194	−8	1.04	222
	P	6	5.56	217	188	29	0.87	208
5	W	4	7.00	186	181	5	0.97	210
	P	5	6.09	217	175	42	0.81	196
6	W	3	7.33	186	142	44	0.76	174
	P	4	6.60	217	135	82	0.62	162
			Whole Year					
0	W	43	0.93	612	515	97	0.84	600
	P	44	0.86	707	495	212	0.70	553
1	W	49	1.02	780	644	136	0.83	780
	P	49	1.00	869	629	240	0.72	745
2	W	45	1.00	780	580	200	0.74	744
	P	45	0.98	869	566	303	0.65	705
3	W	34	1.00	780	438	342	0.56	624
	P	34	0.98	869	429	440	0.49	592
4	W	49	1.02	504	644	−140	1.28	780
	P	49	1.00	580	629	−49	1.08	745

REFERENCES

Alconada M., Ansin O.E., Lavado R.S., Deregibus V.A., Rubido G., Gutiérrez Boem F.H. 1993. Effect of retention of run-off water and grazing on soil and on vegetation of temperate humid grassland. *Agric. Water Manage.*, 23: 233–246.

Allen T.F.H., Starr T.B. 1982. Hierarchy. *Perspectives for Ecological Complexity.* University of Chicago Press, Chicago, 310 pp.

Auberville A. 1949. *Climats, Forest et Desertification de l'Afrique Tropicale.* Soucte de Editious geographiques, Mantirue et Colonialos, Paris. 255 pp.

Bac S. 1968. The role of forest in water balance of Poland [in Polish]. *Folia Forestalia Polonica.* Seria A, 14: 5–65.

Bach W. 1988. Development of climatic scenarios: From General Circulation Models. In *The Impact of Climatic Variations on Agriculture.* Vol 1, 125–157. Eds. Parry M.L., Carter T.R., Konijn N.T. Kluwer Academic Publishers, Dordrecht.

Baumgartner A., Reichel E., 1975. *The World Water Balance.* R. Oldenbourg Verlag, Munich, 179 pp.

Ben-Hur M., Plaut Z.G., Levy J., Agassi M., Shainberg I. 1995. Surface runoff, uniformity of water distribution, and yield of peanut irrigated with a moving sprinkler system. *Agron. J.,* Vol. 87, No 4: 609–613.

Blasing T.J. 1985. Background: carbon cycle, climate, and vegetation response. In *Characterisation of Information Requirements for Studies of CO_2 Effects: Water Resources, Agriculture, Fisheries, Forest and Human Health.* p. 9–22. Ed. White M.R., Lawrence Berkeley Laboratory, Berkeley.

Błaszczyk M., 1974. Development of forestation in Wielkopolska [in Polish]. Kronika Wielkopolska, nr 3/4, Poznań.

Bouwer H. 1994. Irrigation and global water outlook. *Agric. Water Manage.* 25: 221–231.

Brouwer F.M., Chadwick M.J. 1991. Future land use patterns in Europe. In *Land Use Change in Europe.* Eds. Brouwer F.M., Thomas A.J., Chadwick M.J. 49–78. Kluwer Academic Publishers, Dordrecht.

Callaway J.M., Currie J.W. 1985. Water resources and changes in climate and vegetation. In *Characterisation of Information Requirements for Studies of CO_2 Effects: Water Resources, Agriculture, Fisheries, Forest and Human Health.* p. 23–67. Ed. White M.R., Lawrence Berkeley Laboratory, Berkeley.

Caswell H., Koenig H. E., Resh J. A, Ross Q. E. 1972. An introduction to system science for ecologists. *Systems Analysis and Simulation in Ecology. II.* Ed. Patten B.C. Academic Press, New York, 3–78.

Cieśliński Z., Miatkowski Z. 1989. Agromelioration as a factor of productivity increase and soil protection in Noteć valley [in Polish]. Conference "Noteć Valley Management." 172–188.

Cieśliński Z. 1997. *Agromelioration for Formation of Agricultural Environment* [in Polish]. Agricultural University Publishing, Poznań, Poland. 357 pp.

Costin A.B., Wimbush D.J. 1961. Studies in catchment hydrology in the Australian Alps. IV Interception by trees of rain, cloud and foe. Divn. Pl. Indust. Tech. Paper No. 16. CSIRO, Melbourne.

Gilpin M.G., Gall A.E., Woodruff D.S. 1992. Ecological dynamics and agricultural landscapes. *Agric. Ecosyst. Environ.* 42: 27–52.

Gilvear D.J., Andrews R., Tellam J.H., Lloyd J.W., Lerner D.N. 1993. Quantification of the water balance and hydrogeological processes in the vicinity of a small groundwater — fed wetland, East Anglia, U.K. *J. Hydrol.* 144: 311–334.

Gingerich S.B. 1999. Ground water occurrence and contribution to streamflow, Northeast Maui, Hawaii. U.S. Geological Survey Water Resources Report, 99-4090, 69 pp.

Gleick P., H. 1992. Effects of climate change on shared fresh water resources. In *Confronting of Climate Change. Risks, Implications and Responses.* Ed. I.M. Mintzer. Cambridge University Press, Cambridge, 127–140.

Gołaski J. 1988. *Distribution of Water Mills within Catchment of Warta, Brda and Barycz during 1790–1960 Period.* Agricultural University Publishing, Poznań. Parts I and II.

Gottschalk L., Krasovskaia I. 1997. Climate change and river runoff in Scandinavia, approaches and challenges. *Boreal Environ. Res.* 2: 145–162.

Gregoire F., Kędziora A., Kapuściński J., Paszyński J., Tuchołka S., 1992. Etude des change d'energie a la surface active. In *Contraintes Geoclimatiques et Environnement Tropical.* Publications de l'Association Internationale de Climatologie, 5: 425–430.

Gutry-Korycka M. 1978. Evapotranspiration in Poland (1931–1960). *Przegląd Geofizyczny,* t.XXIII (XXXI), z.4: 295–299.

Ilnicki P. 1983. Ecological guidelines for melioration [in Polish]. In *Ecological Agriculture, RCAFE,* PAN, Poznań, 152–175.

IPCC. 1996. Climate change 1995. Contribution of Working Group II to the Second Assessment Report on the Intergovernmental Panel on Climate Change. (Eds. R. Zinyowera M.C., R.H. Moss). Cambridge University Press, Cambridge, 878 pp.

Jager J. 1988. Development off climatic scenarios: B. Background to the instrumental record. In *The Impact of Climatic Variations on Agriculture.* 1: 125–157. Eds. Parry M.L., Carter T.R., Konijn N.T. Kluwer Academic Publishers, Dordrecht.

Jansz A. 1959. Impact of shelterbelt in Rogaczewo on microclimate of adjoining fields [in Polish]. *Roczniki Nauk Rolniczych,* ser. A, t. 79: 1091–1125.

Jaworski J. 1962. About purposefulness of shelterbelts introduction [in Polish]. *Postępy Nauk Rolniczych* nr. 4 (76): 35–44.

Jaworski J., Krupa-Marchlewska J. 1989. Evapotranspiration and soil moisture in the case of increasing of CO_2 concentration in the atmosphere [in Polish]. *Wiadomości Instytutu Meteorologii i Gospodarki Wodnej,* t. XII, z. 3–4: 45–55, Warszawa.

Kaniecki A. 1991. Problem of drainage of Wielkopolska Lowland during recent 200 years and changes of water conditions [in Polish]. Conference "Protection and rational use of water resources in agricultural areas of Wielkopolska region." Poznań, 73–80.

Kędziora A. 1984. *pF-Curve Estimation and Its Relation to the Other Physical Properties of Soil.* [in Polish]. Roczniki Akademii Rolniczej w Poznaniu, z. 144, 112 pp.

Kędziora A., Kapuściński J., Moczko J., Olejnik J., Karliński M. 1987a. Potential and real evapotranspiration of alfalfa field [in Polish]. *Zeszyty Problemowe Postępów Nauk Rolniczych,* 322: 83–104.

Kędziora A., Kapuściński J., Olejnik J., Moczko J., Karliński M. 1987b. 24-hour fluctuations of alfalfa field evapotranspiration [in Polish]. *Roczniki Akademii Rolniczej w Pozna-*

Kędziora A. 1991. Impact of global changes on water management [in Polish]. Conference "Protection and rational use of water resources in agricultural areas of Wielkopolska region." Poznań, 5–13.

Kędziora A. 1993. Climate and water conditions in environment of Wielkopolska region [in Polish]. *Kronika Wielkopolska,* Panstwowe Wydawnictwo Noukowe, 1993, 46–54.

Kędziora A. 1994. Energy and water fluxes in agricultural landscape. In *Functional Appraisal of Agricultural Landscape in Europe.* (EUROMAB and INTECOL Seminar). Eds. L. Ryszkowski and S. Bałazy, Research Center for Agricultural and Forest Environment, Polish Academy of Sciences, Poznań, 61–75.

Kędziora A., Kapuściński J., Olejnik J., Moczko J., Tuchołka S., Leśny J. 1994. Geographical variation of heat balance structure [in Polish], *Roczniki Akademii Rolniczej w Poznaniu,* 257: 175–194.

Kędziora A., Ryszkowski L., and Kundzewicz Z. 1995. Phosphate Transport and Retention in a Riparian Meadow — A Case Study. In *Phosphorus in the Global Environment,* Ed. H. Tiessen, John Wiley & Sons, Chichester, 229–234.

Kędziora A. 1996a. Impact of climate and land use changes on heat and water balance structure in an agricultural landscape. *Zeszyty Naukowe, Jagielonian University, Zesz.* 102: 55–69.

Kędziora A. 1996b. The hydrological cycle in agricultural landscape. In *Dynamics of an Agricultural Landscape.* Eds. Ryszkowski L., French N., R., Kędziora A. PWRiL, Poznań, 65–75.

Kędziora A. Olejnik J. 1996. Heat balance structure in agroecosystems. In *Dynamics of an Agricultural Landscape.* Eds. Ryszkowski L., French N., R., Kędziora A. PWRiL, Poznań, 45–64.

Kędziora A. 1999. *Foundation of Agrometeorology.* PWRiL, Poznań, 364 pp.

Kędziora A., Olejnik J., Tuchołka S., Chojnicki B. 1999. Impact of plant development stage on heat and water balance structure of cultivated field located in the vicinity of Cessieres (France) and in the Great Plain in Poland. In *Paysages agraires et environnement.* Ed. Wicherek S. CNRS Editions, Paris: 333–345.

Kędziora A., Ryszkowski L. 1999. Does plant cover structure in rural areas modify climate change effects? *Geographia Polonica,* 72(2): 65–87.

Kędziora A., Olejnik J., Tuchołka S., Leśny J. 2000. Evapotranspiration within Wielkopolska and Cessieres landscapes. In *L'eau de la cellule au paysage.* Ed. Wicherek S. Elsevier, Paris, 93–104.

Kirchner M. 1984. Influence of different land use on some parameters of the energy and water balance. *Prog. Biometeorol.* 3: 65–74.

Kleczkowski A. 1991. Threats and barriers to development of water management. In *Poland in the Face of Present Civilization Challenge* [in Polish]. Komitet Prognoz "Polska w XXI wieku" przy Prezydium PAN, Warszawa, 220–232.

Konopko S. 1985. Frequency of drought occurrence in Bydgoszcz region evaluated on the basis of long-term observations [in Polish]. *Wiadomości* IMUZ, t. XV, z. 4, 103–113.

Kosturkiewicz A. Kędziora A. 1995. Problems of water management in agricultural areas. In *Guidelines for Sustainable Development of Rural Areas* [in Polish]. Eds. Ryszkowski L., Bałazy S. Zalírad Badan Srodowish Rolniaeyi i Lesnego, Poznań, 73–98.

Kosturkiewicz A., 1989. Protection of water resources in agricultural landscape [in Polish].

Lambor J. 1953. Causes of drought increasing in our lands [in Polish]. Przegląd Meteorolog-
iczny i Hydrologiczny, Rocznik 1952, 3–4, 30.
Lwowicz M.I. 1979. *World Water Resources* [in Polish]. PWN, Warszawa, 438 pp.
Magid J., Christensen N., Skop E. 1994. Vegetation effects on soil solution composition and
evapotranspiration — potential impacts of set-aside policies. *Agric. Ecosyst. Environ.*
49: 267–278.
Manabe S., Wetherald R.T. 1986. Reduction in summer soil wetness induced by an increase
in atmospheric carbon dioxide, *Science* 232, 626–628.
Mathias M., Moyle E.P. 1992. Wetland and aquatic habitats. *Agric. Ecosys. Environ.* 42:
165–176.
McCulloch J.S.G., Robinson M. 1993. History of forest hydrology. *J. Hydrol.* 150: 189–216.
Miklaszewski J. 1928. *Forest and Forestry in Poland* [in Polish]. Zwiazek Zawodowy Lesni-
kow w Rzeczypospolity Polshing. Warszawa.
Milly P.C.D. 1991. Some current themes in physical hydrology of the land — atmosphere
interface. In *Hydrological Interactions Between Atmosphere, Soil and Vegetation,* Eds.
Kienitz G., P.C.D. Milly, M.Th. Van Genuchten, D. Rosbjerg, W.J. Shuttleworth).
IAHS, Wallingford, U.K. Publication No. 204: 3–10.
Mills G. 2000. Modeling the water budget of Ireland — evapotranspiration and soil moisture.
Irish Geography, 33(2): 99–116.
Montheith J.L. 1975. *Vegetation and the Atmosphere. Vol. 1, Principles.* Academic Press,
London, 278 pp.
Nash L.L., Gleick P.H. 1991. The sensitivity of streamflow in the Colorado Basin to climate
changes. *J. Hydrol.*, 125: 221–241.
Neal C., Robson A.J., Bhardwaj C.L., Conway T., Jeffery H.A., Neal M., Ryland G.P., Smith
C.J., Walls J. 1993. Relationships between precipitation, steamflow and throughfall
for a lowland beech plantation, Black Wood, Hampshire, southern England: findings
on interception at a forest edge and the effects of storm damage. *J. Hydrol.* 146:
221–233.
Niemann J. 1994. Water resources and climate change. Options. IIASA, 10–11.
Oberlander G., T. 1956. *Ecology,* 37: 851–852.
Olejnik J., Eulenstein F., Kędziora A., Werner, A. 2001. Evaluation of water balance model
using data for bare soil and crop surfaces in Middle Europe. *Agric. Forest Meteorol.*
106: 105–116.
O'Neill R.V. De Anglis D.L., Waide J.B., Allen T.F.H. 1986. *A Hierarchical Concept of
Ecosystems.* Princeton University Press, Princeton, 253 pp.
Parry M.L., Swaminathan M.S. 1992. Effect of climate change on food production. In *Con-
fronting of Climate Change. Risks, Implications and Responses.* Ed. Mintzer I. Cam-
bridge University Press, Cambridge, 113–125.
Parry M.L. 1990. *Climate Change and World Agriculture.* Earthscane, London, 165 pp.
Pasławski Z. 1990. Water balance of Wielkopolska region. In *Obieg wody i bariery bio-
geochemiczne w krajobrazie rolniczym* [in Polish]. Eds. Ryszkowski L., Marcinek J.,
Kędziora A. Wydawnictwo Naukowe, Poznań, 59–68.
Pasławski Z.1992. Hydrology and water resources of Warta catchment [in Polish]. Conference
"Protection and rational use of water resources in agricultural areas of Wielkopolska

Richmond G.S., Wang K.M., Stern W.R. 1990. Modifying a water balance program (WAT-BAL) to test the effectiveness of reclamation measures on degraded rangeland by chisel ploughing and waterponding. *Agric. Ecosys. Environ.* 32: 1–12.

Rosenberg N.I. 1974. *Microclimate: The Biological Environment.* Wiley & Sons, New York, 315 pp.

Ryszkowski L., Kędziora A. 1987. Impact of agricultural landscape structure on energy flow and matter cycling. *Landscape Ecol.* 1(2): 85–94.

Ryszkowski L., Kędziora A., Olejnik J. 1989. Critical ecological factors for a sustainable development of agriculture in Poland. Report on a contract study for the International Institute for Applied Systems Analysis. 29 pp. plus 24 tables.

Ryszkowski L., Kędziora A., Olejnik J. 1990. Potential effects of climate and land use changes on the water balance structure in Poland. In *Processes of Change Environmental Transformations and Future Patterns.* Kluwer Academic Publishers, Dordrecht. 253–274.

Ryszkowski L., Kędziora A. 1993. Energy control of matter fluxes through land-water ecotones in an agricultural landscape. *Hydrobiologia* 251: 239–248.

Ryszkowski L., Kędziora A. 1995. Modification of the effects of global climate change by plant cover structure in an agricultural landscape. *Geographia Polonica* 65: 5–34.

Ryszkowski L., Kędziora A. 1996a. Ecological guidelines for management of agricultural landscapes. In *Dynamics of an Agricultural Landscape.* Eds. Ryszkowski L., Frencz N.R., Kędziora A. PWRiL, Poznań, 213–223.

Ryszkowski L., Kędziora A. 1996b. Small retention in agricultural landscape. *Zeszyty Naukowe Agricultural University in Wrocław.* Nr. 289: 217–225.

Ryszkowski L., Bartoszewicz A., Kędziora A. 1999. Management of matter fluxes by bio-geochemical barriers at the agricultural landscape level. *Landscape Ecol.* 14(5): 479–492.

Stohlgren T.J., Chase T.N., Pilke R.A., Kittels G.F., Baron J.S. 1998. Evidence that local land use practices influence regional climate, vegetation and stream flow patterns in adjacent natural areas. *Global Change Biol.* 4: 495–504.

Tansley A.G. 1935. The use and abuse of vegetational concepts and terms. *Ecology* 16: 284–307.

Thomas G., Henderson-Sellers A. 1992. Global and continental water balance in a GCM. *Climate Change* 20: 251–276.

Tolk J.A., Howell T.A., Evett S.R. 1998. Evapotranspiration and yield of corn grown on three High Plains soils. *Agron. J.* 90(4): 447–454.

UNESCO. 1978. World water balance and water resources of the earth. *Studies and Reports in Hydrology* 25, Paris.

Valdés J.B., Seoane R.S., Gerald R.N. 1994. A methodology for evaluation of global warming impact on soil moisture and runoff. *J. Hydrol.* 161: 389–413.

Valentini R., Baldocci D.D., Tenhunen J.D., 1999: Ecological control on land-surface atmospheric interaction. In *Integrating Hydrology, Ecosystems Dynamics, and Biogeochemistry in Complex Landscapes,* Eds. J.D. Tenhunen and P. Kabat, John Wiley & Sons, Chichester, U.K., 177–197.

Vehviläinen B., Huttunen M. 1997. Climate change and water resources in Finland. *Boreal Environ. Res.* 2: 3–18.

Viterbo P., Illari L. 1994. The impact of changes in runoff formulation of a general circulation

110   LANDSCAPE ECOLOGY IN AGROECOSYSTEMS MANAGEMENT

Zektser I.S., Loaiciga H.A. 1993. Groundwater fluxes in global hydrological cycle: past,
    present and future. *J. Hydrol.* 144: 405–427.
Zinke P.J. 1967. Forest interception studies in the United States. In *Forest Hydrology.* W.A.,
    Sopper & H.W. Lull, Eds. 137–161. Pergamon Press, Oxford.

CHAPTER 5

Control of Diffuse Pollution by Mid-Field Shelterbelts and Meadow Strips

Lech Ryszkowski, Lech Szajdak, Alina Bartoszewicz, and Irena Życzyńska-Bałoniak

CONTENTS

INTRODUCTION

Water quality is one of the fundamental requisites for sustainable development of agriculture, and it constitutes the survival determinant of rich plant and animal assemblages. Interactions among physical, chemical, and biological processes characteristic of a watershed determine discharged water quality; alteration of any one of these processes will affect one or more water quality properties. This fact was recently learned by scientists and the public when growing problems of water

is, on the control of point sources of pollution by construction of water purification plants (Vollenweider 1968). Success was achieved in some reservoirs, such as Lake Constance, but eutrophication problems could not be totally eliminated, and, in addition, such problems started to appear even in water bodies located far away from point sources of pollution (Halberg 1989, Kauppi 1990).

As agricultural production intensified, land-use changes caused by agriculture became more apparent. Enlargement of farm sizes was linked to more efficient use of machines, which decreased costs of the cultivation of large fields not segmented by shelterbelts (mid-field rows of trees), open drainage ditches, and other obstacles to fast and powerful agricultural equipment. This trend of agricultural development resulted in homogenizing the countryside structure. For example, in France the average farm size increased from 19 to 28 ha in the period from 1970 to 1990. In the same time span, the average farm size in the U.K. increased from 54 to 68 ha, in West Germany from 13 to 18 ha, and in Belgium from 8 to 15 ha (Stanners and Bourdeau 1995). Consolidation and expansion of cultivated fields led to eradication of field margins, hedges, shelterbelts, small mid-field ponds or wetlands, and other nonproductive elements of the landscape. Thus, for example, 22% of hedgerows in the U.K. were eliminated by the mid-1980s (Mannion 1995). The disappearance rate of wetlands in the European Union, excluding Portugal, has amounted to 0.5% annually since 1973 (Baldock 1990). In Denmark, 27% of small water reservoirs disappeared from 1954 to 1984 (Bülow-Olsen 1988).

By intensifying production, farmers interfere with patterns of element cycling in landscapes using fertilizers and pesticides, and they are changing water regimes by drainage or irrigation. Feedback of the agricultural measures of production as well as induced changes in land use brought environmental problems, such as impoverishment of biological diversity or nonpoint (diffuse) water pollution. In the 1980s, it was recognized that control of point sources of pollution could not alone solve the problems of water quality. The water pollution, especially with nitrates, was detected in streams or lakes located far from urban or industrial point sources (Omernik et al. 1981, OECD 1986, Halberg 1989, Ryszkowski 1992). The diffuse water pollution problems were recognized worldwide in the 1990s.

Nonpoint water pollution is attributed to human-induced, above-natural-rate inputs of chemical compounds into subsurface and surface water reservoirs. At present, agriculture is undoubtedly the main reason for diffuse pollution problems (OECD 1986, Rekolainen 1989, Kauppi 1990, Ryszkowski 1992, Flaig and Mohr 1996, Johnsson and Hoffmann 1998). High concentrations of nitrates exceeding 50 mg per liter of soil solution were detected in Germany, northern France, eastern England, northwestern Spain, northern Italy, and Austria. Very high nitrate concentrations were detected in Denmark, the Netherlands, and Belgium (Stanners and Bourdeau 1995). So, at the beginning of the 1990s, it appeared that modern intensive

and Correll 1984, Pinay and Decamps 1988, Ryszkowski and Bartoszewicz 1989, Muscutt et al. 1993, Hillbricht-Ilkowska et al. 1995, and others). The majority of the studies concerned riparian plant buffer zones and their efficiency for the control of diffuse pollution. A thorough review of the riparian-strip functions for controlling diffuse pollution, both via surface and subsurface fluxes, published by Correll (1997) in proceedings of the 1996 buffer zone symposium, provides a review of studies on various aspects of diffuse pollution control (Haycock et al. 1997). A recent book edited by Thornton et al. (1999) addresses primarily the nonpoint pollution impacts on lakes and reservoirs, stressing the practical aspects of the control.

As stated above, most studies concerned protection of surface water reservoirs from diffuse pollution by riparian vegetation strips. Field studies have shown, for example, that nitrates are efficiently removed from shallow ground water passing through the root system of plants in a buffer zone. Mechanisms responsible for that process are still elusive (Correll 1997), but it is generally assumed that the following processes are important: ion exchange capacities of soil, plant uptake, and denitrification.

Long-term studies on the function of shelterbelts and stretches of meadows within the Turew agricultural landscape, carried out by the Research Centre for Agricultural and Forest Environment, Polish Academy of Sciences, provided information on control of diffuse pollution in upland parts of drainage areas, which enriched knowledge on control of nonpoint pollution outside riparian zones. Those studies also disclosed some mechanisms for a ground water pollution control, which can be useful for developing a strategy of water resource protection. Review of these studies will be used to evaluate the prospect for diffuse pollution control in agricultural landscapes.

ENVIRONMENT OF THE TUREW AGRICULTURAL LANDSCAPE

The Turew landscape (about 17000 ha) has been the object of long-term studies on agricultural landscape ecology (Ryszkowski et al. 1990, 1996), and detailed characteristics of climate, soils, hydrology, and land-use forms can be found in those publications. The landscape is identified by the adjacent village, Turew. The terrain consists of a rolling plain, made up of slightly undulating ground moraine. Differences in elevation do not exceed a few meters. In general, light soils are found on the higher parts of the landscape with favorable infiltration conditions (glossudalfs and hapludalfs). Endoaquolls and medisaprists occur in small depressions. The infiltration rates of upland soils range from a few to several $cm \cdot h^{-1}$ and can be classified as having moderate or moderately rapid infiltration rates. Thus, the water from rain or snow thaw can easily infiltrate beyond the depth of plant roots and then transport dissolved chemical compounds to ground water; however, in layers below

Table 5.1   Means of Physical and Chemical Characteristics of Hapludalfs and Glossudalfs

Soil Horizon	Thickness (cm)	Organic C (%)	N Total (T)	Contents of Clay Below 0.002 mm (%)	CEC mmol (+)·kg⁻¹	S mmol (+)·kg⁻¹	BS (S:CEC) (%)
ochric	30.8 ± 3.1	0.62 ± 0.14	0.075 ± 0.019	3.1 ± 1.2	49.2 ± 1.2	30.4	61.8
ric	26.9 ± 7.4	0.21 ± 0.12	0.025 ± 0.012	2.7 ± 0.9	34.8 ± 0.9	24.6	70.7
illic	37.8 ± 12.6	n.d.	n.d.	14.2 ± 3.7	95 ± 1.7	71.7	75.2
parent material	—	n.d.	n.d.	11.9 ± 3.3	81.2 ± 1.9	66.8	82.3

C — cation exchange capacity; S — sum of bases; BS — percent of saturation with bases

— not determined

Source: Bartoszewicz 2000.

CONTROL OF DIFFUSE POLLUTION   115

values of pH_{KCL}. The alkaline reaction is caused by the presence of calcium carbonates in the boulder loam.

The low values of cation exchange capacities of the soil, as well as small amounts of clay fractions and organic matter, indicate that with fast percolation of water there is intensive leaching of chemical solutes. Thus one can infer that sorption of ammonia ions as well as other cations is rather low in the upper horizons of soils located in the upland parts of the landscape and moderately low in deeper layers. The opposite situation is observed in endoaquolls and medisaprists situated in the depressions of the landscape. These soils are characterized by much higher contents of organic carbon (2.7 to 43.4%) and are poor or very poorly drained. Their adsorptive capacities for passing cations depends mainly on the content of organic matter because clay minerals are poorly represented. Pokojska (1988) has found positive correlation (r = 0.72) between values of cation exchange capacities and the percentage of organic carbon content in those soils. Thus endoaquolls and medisaprists of the Turew landscape have high potential to adsorb cations (Pokojska 1988, Marcinek and Komisarek 1990).

The area, from a Polish perspective, is warm, with an annual mean temperature of 8°C. Thermal conditions are favorable for vegetation growth. The growing season, with air temperatures above 5°C, lasts 225 days. On average, it begins March 21 and ends October 30. Mean corrected annual precipitation (1881–1985) amounts to 590 mm (uncorrected value to 527 mm). Although the amount of precipitation in the spring-summer period is more than twice that in winter, a water shortage often occurs in the summer. The annual evapotranspiration rate averages about 500 mm and runoff is 90 mm. Since a majority of the soils are characterized by high rates of infiltration, their water storage is not of great importance in dry summers. Water deficits are further intensified by drainage of a considerable part of the area.

The most advantageous component of the landscape is its shelterbelts (rows or clumps of trees), which were planted in Turew due to the initiative of Dezydery Chłapowski in the 1820s. In addition to shelterbelts, small afforestations are found in the landscape. Shelterbelts and afforestations cover 14% of the entire area and are composed of *Pinus sylvestris* (65.5% of total afforested area), *Quercus petraea* and *Q. robur* (14.5%), *Robinia pseudoaccacia* (5%), *Betula pendula* (4.3%) and others, totaling 24 tree species. But in shelterbelts oaks, false acacias, maples, lindens, larch, and poplars prevail. Oaks and larches have very deep root systems, while maples (especially sycamore maples) and lindens have moderately deep roots with broad root systems. The mix of the tree species creates a better screen to the seeping solutes in ground water than would a shelterbelt composed of one species (Prusinkiewicz et al. 1996). Cultivated fields cover 70% of the area. During the last 10 years, there has been a tendency for increased cereals (wheat, barley, rye, oats) in the crop rotation pattern, and it presently comprises 70% of arable land. Decreased

composed of lakes and mid-field ponds and channels, waterlogged areas, roads, and villages. The density of small reservoirs varies from 0.4 to 1.7/km^2. Mineral fertilization varied from 220 to 315 kg NKP/ha but since 1991 about a 40 to 50% decrease or greater in mineral fertilization was observed on some small farms because of the economic crisis associated with the change of the political system. Yields are high for cereals (rye, wheat, barley and oats), ranging from 3.2 t·ha^{-1} to about 4 t·ha^{-1}. The level of mechanization of field labor is high, amounting on the average to 1 tractor per 17 ha of cultivated fields.

In studies on the impact on ground water chemistry from the mid-field afforestations and shelterbelts or strips of meadows (called biogeochemical barriers), the dominant direction of subsurface water pathways was estimated by measurement of ground water table elevation in wells located in fields and adjoining shelterbelts, small forests, or meadows. The samples for nitrogen compound concentration measurements were collected from wells, drainage pipes, ditches, small ponds, and main drainage canals of the landscape over different periods but never during a time span shorter than 1 year.

Over the last 200 years, there were important habitat changes connected with land reclamation activities leading to drying of the area. The effects are observed not only in the drop in the ground water level but also in soil degradation caused by drainage. So, for example, fertile endoaquolls have been converted in many places into glossudalfs or hapludalfs with low carbon content. Thus, drying of the region is expressed in soil changes; although appearing slowly, the nature of the trend can be clearly recognized.

NITROGEN COMPOUNDS IN THE DRAINAGE SYSTEM OF THE TUREW LANDSCAPE

The Turew landscape is drained by a canal about 4 m wide with an average long-term water depth of 0.6 m. The annual mean concentrations of $N-NO_3^{-1}$ varied irregularly from 0.5 mg·dm^{-3} to 3.4 mg·dm^{-3}. Almost the same range of variation in $N-NH_4^+$ concentration was observed (Figure 5.1).

At the beginning of the 1990s, there was a decline in the use of fertilizers due to the economic crisis, amounting to a drop in application of 40 to 50%. But despite decreased input of fertilizers, the level of inorganic ion concentrations of nitrogen did not change, showing irregular cycles with a peak in 1993 and 1994, followed by a drop and then increasing since 1997 (Figure 5.1). Mean concentrations of the mineral forms of nitrogen in the canal water during the period 1973–1991, when higher doses of fertilizers were applied, were 1.40 mg·dm^{-3} for $N-NO_3^-$ and 1.70 mg·dm^{-3} for $N-NH_4^+$. In the period 1992–2000 when fertilizer use dramatically decreased, the mean concentration increased to 2.04 mg·dm^{-3} in the case of

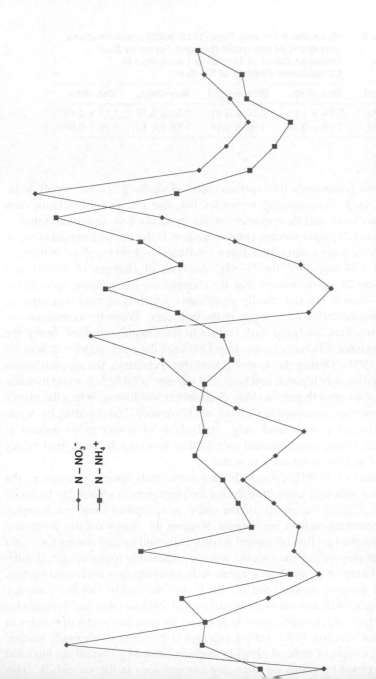

mean annual concentration in main drainage canal of the Turew landscape.

Table 5.2   Mean Monthly Long-Term (1973–2000) Concentrations (mg·dm⁻³) of Inorganic Nitrogen Forms in Main Drainage Canal of the Turew Landscape in Consecutive Periods of the Year

Period	Dec.–Feb.	March–April	May–Sept.	Oct.–Nov.
$N–NO_3^-$	2.79 ± 1.01	2.50 ± 0.71	0.88 ± 0.15	1.11 ± 0.48
$N–NH_4^+$	1.98 ± 0.30	1.92 ± 0.44	1.68 ± 0.19	2.10 ± 0.37

diverting nitrogen compounds into various routes of discharge (water runoff, volatilization), not only condition lag responses but also obscure the relationships between fertilizer input and their concentrations in water of the drainage system.

The long-term (28 years) average concentration of both mineral forms of nitrogen was the same in the main canal of the Turew landscape — 1.69 mg·dm⁻³ in the case of $N–NO_3^-$ and 1.74 mg·dm⁻³ for $N–NH_4^+$. Analysis of changes of $N–NO_3^-$ and $N–NH_4^+$ ions over 28 years showed that the changes are independent (correlation coefficient r = –0.06 is not statistically significant), which is another indication of the complex transformation of nitrogen in the landscape. When the concentrations of nitrogen forms were analyzed with respect to the monthly changes during the year, distinct seasonal differences were found between the cold and growth seasons (Bartoszewicz 1994). During the winter (December–February), the monthly mean nitrate concentration was highest, reaching 2.79 mg·dm⁻³ (Table 5.2) while its value during the full plant growth season (May–September) was lowest. When the plant's transpiration processes decreased in October and November (leaf shedding by deciduous trees, drying of grasses, and only small plants of winter crops present in cultivated fields), nitrate concentration increased so as to reach the highest values when biological activity is retarded in winter.

Concentrations of $N–NH_4^+$ cations did not show such distinct changes in the course of seasons although some drop during the plant growth season can be easily observed (Table 5.2). In the course of the entire year, nitrates show much higher variance of concentrations than ammonium. It seems the reason for this difference is connected with the fact that biological activity is in "full swing" during the warm season, although pinpointing the specific process responsible (plant uptake, denitrification, assimilatory or dissimilatory nitrate reduction) requires additional studies.

In the plant growing season (end of March until the end of October), average precipitation reaches 410 mm out of an annual total of 590 mm (Woś and Tamulewicz 1996). Despite high precipitation rates in summer, the concentrations of nitrates in water of the canal were low in this period, although $N–NO_3^-$ anions are easily leached from soil. Thus, effects of mineral nitrogen leaching caused by rainfall are modified by influences exerted by plants on migrating nitrogen ions in the watershed. (This

compounds. Analyses of mineral nitrogen distribution in the unsaturated zone below the cultivated field showed that in the spring large amounts are leached into the ground water. In ground water of some fields, high concentrations of $N-NO_3^-$, reaching 60 mg·dm^{-3}, can be found when fertilizers were applied during spring (Ryszkowski et al. 1997). But despite the fact that such situations occur in some fields, each spring the monthly concentrations of nitrates in the main canal draining the total area are very low, which again indicates strong modification effects of landscape structure on the control of diffuse pollution.

CONTROL OF MINERAL NITROGEN POLLUTION BY SHELTERBELTS AND MEADOWS

When ground water carrying nitrates is within direct and indirect (capillary ascension) reach of the root system, nitrate concentrations are substantially decreased. Knowing that the NO_3^- anion is practically not exchanged by soil colloids, these differences result mainly from the action of a complex set of biological factors involving the plant's uptake, denitrification processes, and release of gaseous products including NO, N_2O, and N_2. In addition, nitrates may undergo reduction to NH_4^+, which could be volatilized. The regulation of those processes under field conditions is poorly understood (Correll 1997).

The reduction of nitrates when ground water is seeping under shelterbelts, afforestations, or grasslands is pronounced, and under the Turew landscape conditions such reduction varied from 63 to 98% for shelterbelts and afforestations (Table 5.3). In the case of meadow strips, the reduction varied from 79 to 97%.

Table 5.3   Mean Concentrations of $N-NO_3^-$ (mg·dm^{-3}) in Ground Water under Cultivated Fields, Shelterbelts, Small Forests, and Meadows in the Turew Agricultural Landscape

Period of Sampling	Cultivated Field (a)	Shelterbelt or Forest Patch (b)	Meadow (b)	Reduction (a-b):a (%)	Reference
1982–1986	22.2	1.0	—	95	Bartoszewicz and Ryszkowski 1996
1982–1986	37.6	1.1	—	97	Bartoszewicz and Ryszkowski 1996
1972–1973	12.6	0.3	—	98	Margowski and Bartoszewicz 1976
1984–1986	33.1	8.1	—	75	Ryszkowski et al. 1997
1994	52.4	2.7	—	94	Ryszkowski et al. 1997
1995	13.1	4.9	—	63	Ryszkowski et al. 1996
1986–1989	48.3	—	6.5	87	Bartoszewicz 1990
1987–1989	15.9	—	0.7	95	Bartoszewicz 1990
1987–1991	13.1	—	2.8	79	Sznakowska and

Thus, both kinds of biogeochemical barriers (shelterbelts and meadow stretches) showed similar efficiency of nitrate reduction. Results are similar to the estimates gathered from various literature sources by Muscutt et al. (1993).

One can argue that changes in concentrations do not fully show the effects of nitrate limitation exerted by the biogeochemical barriers. The control effects depend on both the changes in concentrations and the rate of $N-NO_3^-$ flux through the barrier. Low concentration inflow can provide a large amount of chemicals if the rate of water flux is high. But the hydraulic conductivity of soils (up to $1.0 \ m \cdot day^{-1}$) as well as hydraulic gradients of ground water tables that determine flux are small in the landscape studied, which should render good approximations of nitrate fluxes by changes in their concentrations. This situation was confirmed by studies of the hydrology of water seeping under biogeochemical barriers (Ryszkowski et al. 1997). Estimated $N-NO_3^-$ ratios for output-input of annual flux estimates for birch mid-field forests amounted to 0.22, 0.25, and 0.28 in three consecutive years (Ryszkowski et al. 1997). The average for the 3 years was 0.25, which corresponds well with the estimate based on concentration changes, which was also 0.25.

In the case of the pine mid-field forest, both estimates also match very well (Ryszkowski et al. 1997). Thus, in an area where the slope of the ground water table is not too steep, the differences in concentrations of chemical compounds between an input and output characterize well the flux control efficiency of the barrier. Studies (Ryszkowski and Kędziora 1993) indicate that, as the steepness of slope increases, the shelterbelt and meadow are less efficient in regulating ground water flow and chemicals transported.

A great influence of plant cover structure on output of elements from watersheds was shown by Bartoszewicz (1994), and Bartoszewicz and Ryszkowski (1996). These studies were carried out in two small watersheds. The first was a uniform watershed (174 ha) covered 99% by cultivated fields and 1% by small afforestation. The second watershed (117 ha) was mosaic; cultivated fields made up 84% of the area, meadows 14%, and riparian afforestation 2%. During the 3-year period, the mean annual water output was $102.0 \ dm^3 \cdot m^{-2}$ from the uniform watershed and 70.2 $dm^3 \cdot m^{-2}$ from the mosaic watershed. The mean annual precipitation for both watersheds was the same, amounting to $514 \ dm^3 \cdot m^{-2}$, so the lower water runoff from the mosaic watershed was due to higher evapotranspiration rates characteristic of afforestations and grasslands (Ryszkowski and Kędziora 1987). This is clearly seen when water outputs are analyzed from both watersheds in summer (Table 5.4).

The water runoff from both watersheds during the hydrological years 1988/1989–1990/1991 differed by 32 mm on average. However, the water runoff during the winter half-years was almost the same from either watershed, whereas during the growing season water outputs from the uniform watershed (per unit of area) were three times higher than those from the mosaic watershed (Bartoszewicz 1994). Thus, shelterbelts and meadows making up 16% of the mosaic watershed

Table 5.4   Annual Mean Water Output (mm) and Nutrient Loss (g·m⁻²·year⁻¹) from Two Small Watersheds, Nov. 1988–Oct. 1991

Table 5.4   Annual Mean Water Output (mm) and Nutrient Loss ($g \cdot m^{-2} \cdot year^{-1}$) from Two Small Watersheds, Nov. 1988–Oct. 1991

Season	Precipitation (mm)	Uniform Watershed			Mosaic Watershed		
		Water Output (mm)	$N-NO_3^-$	$N-NH_4^+$	Water Output (mm)	$N-NO_3^-$	$N-NH_4^+$
Winter Nov.–April	220.7	60.8	12.3	3.0	56.8	0.90	0.95
Summer May–Oct.	292.9	41.2	4.0	1.1	13.4	0.05	0.25
Whole year	513.6	102.0	16.3	4.1	70.2	0.95	1.20

Source: Bartoszewicz 1994.

preponderance of nitrates over ammonium is clearly evidenced in water output from the uniform agricultural drainage area.

When the migration of mineral components from a mosaic watershed was analyzed, a low leaching rate of nitrogen constituents and a different ratio of nitrates to ammonia ions were observed. The annual leaching rates of mineral N from 1 ha of this watershed amounted to about 2 kg (ten times less than in the uniform watershed), and both ionic forms of N were represented by almost identical shares. Even more striking were the differences between the uniform arable watershed and the mosaic one with respect to seasonal variations in the migration of nitrogen. The majority of both nitrogen ion forms (86%) had leached from the mosaic watershed in the winter half-year, while during the plant growth period the leaching of either nitrogen form (particularly of nitrates) was negligible (Table 5.4).

The study of nitrogen leaching from the small watersheds with different plant cover structures supports the conclusion that shelterbelts, strips of meadows, and other biogeochemical barriers located both in upland and riparian parts of the Turew landscape effectively control the discharge of nitrogen from the drainage area. This conclusion explains the low concentrations of its mineral forms in the main canal (Figure 5.1). The smaller variability of ammonium cations in contrast to nitrates during the course of the year observed in the main Turew canal (Table 5.2) as well as the dissimilarity of these compound shares in discharge from the uniform and mosaic small watersheds (Table 5.4) indicate differentiated impacts of plant cover structures on dissemination of these inorganic forms of nitrogen in the landscape.

Contrary to nitrates, the concentrations of ammonium cations usually do not decrease when ground water is passing under shelterbelts or meadows. Comparing concentrations of $N-NH_4^+$ in ground water under cultivated fields and shelterbelts or grass strips one observes increased rather than decreased amounts of $N-NH_4^+$ (Table 5.5).

Because some ammonium ions incoming with ground water from fields are

Table 5.5   Mean Concentrations of $N-NH_4^+$ $(mg \cdot dm^{-3})$ in Ground Water under
             Cultivated Fields, Shelterbelts, Small Forests, and Meadows
             in the Turew Agricultural Landscape

Period of Sampling	Cultivated Field (a)	Shelterbelt or Forest Patch (b)	Meadow (b)	Reduction (a–b):a (%)	Reference
1982–1986	2.5	2.0	—	25	Bartoszewicz and Ryszkowski 1996
1982–1986	2.1	4.5	—	114[a]	Bartoszewicz and Ryszkowski 1996
1972–1973	1.4	2.7	—	–92	Margowski and Bartoszewicz 1976
1984–1986	1.7	1.7	—	0	Ryszkowski et al. 1997
1994	1.3	1.1		15	Ryszkowski et al. 1997
1986–1989	1.8	2.2		–22	Bartoszewicz 1990
1987–1989	1.8	2.1		–16	Bartoszewicz 1990
1987–1991	1.8	—	2.2	–22	Szpakowska and Życzyńska-Bałoniak 1994
1993	2.4	—	2.4	0	Ryszkowski et al. 1996a
1993	2.5	—	2.4	4	Ryszkowski et al. 1996a
1994	2.6	—	3.0	–15	Ryszkowski et al. 1996a
1994	3.9	—	7.1	–82	Ryszkowski et al. 1996a
1995	2.5	—	4.0	–60	Ryszkowski et al. 1996a

[a] Minus values mean increase of concentration under the biogeochemical barrier.

annual distribution are of considerable significance. In studies of mosaic and uniform small watersheds in the hydrological years of 1989/90 and 1990/91, during which the sums of precipitation were at the level of 550 mm·year^{-1}, the losses of nitrogen forms were much higher (in some instances twice as high) than in the hydrological year 1988/1989, during which the annual precipitation was 110 mm lower. In the case of nitrates, the pattern of annual precipitation distribution rates plays a significant role. When intensive rains occurred during the late autumn and winter of 1990/91 — i.e., at a time when appreciable amounts of nitrates were being released from the decomposing post-harvest plant remnants (Ryszkowski 1992) — the leaching of nitrates was 6 kg higher than in 1989/90 when precipitation was lower by 40 mm.

PROCESSING OF MINERAL NITROGEN
IN THE BIOGEOCHEMICAL BARRIERS

In order to study the distribution of mineral forms of nitrogen in the unsaturated zone of the soil profile, the method of moisture saturation extracts was used (Jackson 1964). According to this method, soil samples are treated with distilled water to

Table 5.6   Distribution of N–NO$_3^-$ and N–NH$_4^+$ (g·m^{-2}) in the Unsaturated Layers of Soil in the Cultivated Field and Adjoining Pine Afforestation

Sampling Term	Mineral Form of N	Cultivated Field Soil Layer Depth (cm)			Pine Afforestation Soil Layer Depth (cm)		
		0–80	81–150	0–150	0–80	81–150	0–150
April 1986	N–NO$_3^-$ (a)	13.9	7.9	21.8	5.2	1.1	6.3
	N–NH$_4^+$ (b)	2.3	4.9	7.2	6.7	3.3	10.0
	Sum	16.2	12.8	29.0	11.9	4.4	16.3
	a:b	6.0	1.6	3.0	0.8	0.3	0.6
September 1986	N–NO$_3^-$ (a)	3.7	1.2	4.9	0.3	0.2	0.5
	N–NH$_4^+$ (b)	1.6	0.9	2.5	2.1	0.5	2.6
	Sum	5.3	2.1	7.4	2.4	0.7	3.1
	a:b	2.3	1.3	1.9	0.1	0.4	0.2
April 1987	N–NO$_3^-$ (a)	9.7	8.6	18.3	2.8	1.0	3.8
	N–NH$_4^+$ (b)	1.7	0.3	2.0	3.9	0.8	4.7
	Sum	11.4	8.9	20.3	6.7	1.8	8.5
	a:b	5.7	28.6	9.1	0.7	1.2	0.8
May 1998	N–NO$_3^-$ (a)	0.4	4.5	4.9	0.7	2.9	3.6
	N–NH$_4^+$ (b)	2.2	1.1	3.3	2.4	1.7	4.1
	Sum	2.6	5.6	8.2	3.1	4.6	7.7
	a:b	0.2	4.0	1.5	0.3	1.7	0.8
September 1998	N–NO$_3^-$ (a)	5.5	1.9	7.4	1.3	0.7	2.0
	N–NH$_4^+$ (b)	1.9	1.2	3.1	3.3	1.6	4.9
	Sum	7.4	3.1	10.5	4.6	2.3	6.9
	a:b	2.9	1.6	2.4	0.4	0.4	0.4

Source: Modified after Bartoszewicz 2001a, Ryszkowski et al. 1997.

part of the watershed, soil samples were taken at different depths of soil profile during the 1986–1987 period (Ryszkowski et al. 1997). Almost 10 years later, again in the same place, soil samples were collected for moisture saturation extracts (Bartoszewicz 2001a). Comparing the distribution of inorganic nitrogen in soil profiles at various seasons in afforestation and adjoining fields, one can find higher concentrations of ammonium cations in the soil profile under afforestation than under the cultivated field. In the case of nitrates, the opposite situation was found (Table 5.6) as a result of more intensive nitrification processes due to the better soil aeration caused by tillage. The input of fertilizers resulted in higher concentrations of nitrogen mineral forms in the soil of the cultivated field. Applied nitrogen fertilizers consisted mainly of ammonium nitrate (NH$_4$NO$_3$), so the same amounts of both nitrogen forms were introduced into soil. Domination of nitrates indicated, therefore, intensive nitrification processes in the cultivated field soil due to aeration caused by tillage.

Comparisons of nitrogen ions concentrations at consecutive terms of sampling in April and September 1986 and April 1987, as well as those concentrations in May

Table 5.7 Distribution of N–NO$_3^-$ and N–NH$_4^+$ (g·m^{-2}) in the Unsaturated Layers of Soil under Cultivated Field and Newly Planted Shelterbelt

Sampling Term	Mineral Form of N	Cultivated Field Soil Layer Depth (cm)			New Shelterbelt Soil Layer Depth (cm)		
		0–80	81–150	0–150	0–80	81–150	0–150
October 1999	N–NO$_3^-$ (a)	4.8	17.8	22.6	0.4	0.6	1.0
	N–NH$_4^+$ (b)	1.7	0.4	2.1	1.0	0.6	1.6
	Sum	6.5	18.2	24.7	1.4	1.2	2.6
	a:b	2.8	44.5	10.7	0.4	1.0	0.6
May 2000	N–NO$_3^-$ (a)	4.0	11.5	15.5	1.1	0.1	1.2
	N–NH$_4^+$ (b)	1.2	0.4	1.6	1.7	0.3	2.0
	Sum	5.2	11.9	17.1	2.8	0.4	3.2
	a:b	3.3	28.7	9.6	0.6	0.3	0.6

Source: Modified after Bartoszewicz 2001a.

the distribution of nitrogen ions in the soil profile; in the majority of performed determinations, lower concentrations of nitrogen ions were observed in the soil strata below 80 cm depth (Table 5.6) as a result of ion uptake by root systems. Lack of direct measurements of water infiltration fluxes through soil profile obscures the precise estimates of nitrates and ammonium inputs into ground water from the unsaturated zone of soil with percolating water after precipitation events. Nevertheless, comparisons of nitrates and ammonium concentrations in ground water under fields and afforestations or meadows (Tables 5.3 and 5.5) clearly demonstrate that nitrates are reduced by those biogeochemical barriers while ammonium ions are not.

In order to evaluate the influence of shelterbelts on mineral nitrogen in the soil the special studies were done on the distribution of inorganic nitrogen in the soil profile 6 years after trees were planted on a cultivated field with hapludalf soils. In soil withdrawn from cultivation for 6 years, very low concentrations of mineral nitrogen were detected in comparison with an adjoining cultivated field (Table 5.7).

When inputs of fertilizers into soil under a growing shelterbelt were ceased, the amount of mineral nitrogen dramatically decreased almost 10 times in comparison with a field in October 1999 and 5 times in May 2000 when mineral nitrogen was regenerated due to greater decomposition rates brought by the higher temperatures of spring. The levels of mineral nitrogen in the soil under newly planted shelterbelts were lower than in soil under a 60-year-old afforestation planted also on the same hapludalf soils (compare Tables 5.6 and 5.7). This phenomenon is caused by the low level of litter and soil organic matter accumulation in a new shelterbelt due to short time lapse after tree planting. Decomposed organic matter is an important source of mineral nitrogen stocks in soil (this process is discussed with results of urease activity studies later in this chapter).

In the new shelterbelt, ammonium ions predominate over nitrates, which resem-

The other study of Bartoszewicz (2000) on the newly planted shelterbelt growing on endoaquolls showed that after 1 year of seedling growth, the ratio of $N-NO_3^-$ to $N-NH_4^+$ was already 0.8 (1.5 $g \cdot m^{-2}$ and 1.7 $g \cdot m^{-2}$, respectively) while in the same adjoining soil but under cultivation this ratio was 2.7 (3.5 $g \cdot m^{-2}$ and 1.3 $g \cdot m^{-2}$, respectively) in soil profiles to 150 cm of depth. This last result indicates that withdrawal of tillage activities alone has important bearing on decreased nitrification processes due to poorer soil aeration, although the levels of mineral nitrogen in organic-rich soil (endoaquolls) showed differences not as great between field and new shelterbelt (total mineral nitrogen in cultivated soil amounted to 4.8 $g \cdot m^{-2}$ and in 1 year the old shelterbelt was equal to 3.2 $g \cdot m^{-2}$).

Some ammonium ions are absorbed by roots as well as retained by the base-exchange complex, especially in the deeper strata of soil in the Turew landscape (see Table 5.1 for values of cation exchange capacities [CEC] and percent of satu-ration of sorption complex [BS]). The observed lack of decrease in $N-NH_4^+$ ions concentrations when ground water is passing through root systems of the bio-geochemical barriers (Table 5.5) should be related, therefore, to inputs of ammonium ions from decomposing organic matter.

Several biological processes could lead to production of $N-NH_3$. The first process is assimilatory nitrate reduction in which $N-NH_3$ is used for production of biomass, (proteins) which after mineralization could release ammonium ions. Assimilatory nitrate reduction takes place under oxygenic conditions. The second process is actually two processes: dissimilatory reduction of nitrates, which in denitrification releases gaseous forms of nitrogen, and in dissimilatory reduction of nitrate to ammonium releases $N-NH_4^+$ ions under anaerobic conditions (Tiedje et al. 1981). In addition, very small amounts of $N-NH_4^+$ can be exuded from tree roots as shown experimentally by Smith (1976) in the case of birch, beech, and maple trees.

In all plants, ammonia (NH_3) plays a key role in nitrogen assimilation because all nitrogen organic compounds are derived from ammonia assimilation regardless of the nutritional source of nitrogen to plants. Plant proteins and nucleic acids are built from low molecular organic compounds deriving nitrogen from the NH_3 form. Thus, nitrates absorbed by plants are converted by assimilatory nitrate reduction to ammonia, and in this form nitrogen is incorporated into the biomass. When plant tissues undergo decomposition, ammonia ions are released. This last process is controlled at the final stage by the enzyme urease, which is responsible for the conversion of urea nitrogen to ammonia nitrogen (Bremner and Mulvaney 1978). Urease activity analysis therefore monitors the release of $N-NH_3$ from decomposing organic compounds in the soil, which in soil solution appears as $N-NH_4^+$.

In studies reported here, urease activity was measured by the Hoffman and Teicher method described and calibrated by Szajdak and Matuszewska (2000). The urease activity was measured in the upper layer of soil (0–20 cm of depth) in the

Table 5.8 Urease Activity (UA; µg urea hydrolyzed·g⁻¹soil·h⁻¹), Organic Nitrogen (ON; mg·kg⁻¹) in Soils of Various Ecosystems in the Turew Landscape

| | Shelterbelts | | | | Fields Adjoining to Shelterbelts | | | |
| | 7-Year-Old | | 140-Year-Old | | 7-Year-Old | | 140-Year-Old | |
Date	UA	ON	UA	ON	UA	ON	UA	ON
March 12	4.35	590.5	16.88	1634.3	6.45	354.8	4.53	587.5
April 7	4.81	591.8	14.50	1677.3	5.98	350.6	4.38	567.8
May 8	4.56	609.1	18.96	1416.5	7.25	355.1	3.94	533.3
June 5	4.25	527.3	5.32	2541.4	5.27	497.1	2.50	789.1
July 9	6.20	602.7	4.88	5644.9	5.30	421.2	2.67	576.7
August 20	9.17	571.4	3.58	3400.7	8.40	472.4	6.41	566.1
September 14	7.76	538.4	7.94	2138.0	5.30	506.4	7.92	460.3
October 12	5.07	657.3	4.05	3133.0	3.40	488.9	2.63	521.1
November 14	4.35	549.1	2.15	2320.7	3.07	501.2	1.93	506.9
Mean	5.61	581.9	8.69	2656.3	5.60	438.6	4.10	567.6

According to the review by Bremner and Mulvaney (1978), urease activity is positively related to organic matter content due to microbial and plant metabolism, and its activity is high in soils under dense vegetation. Clay content to some extent protects urease against decomposition. Mineral fertilizers (e.g., ammonium nitrate) and soil oxygenation have no effects; slight effects are exerted by levels of soil moisture, but a rapid sequence of drying and rewetting of soil decreases its activity.

Fluctuations of urease activity are characteristic both for shelterbelts and cultivated fields (Table 5.8). Despite detected variability, the average values of urease activities in soil of the 7-year-old shelterbelt and adjoining cultivated field as well as a field adjacent to the 140-year-old shelterbelt were similar. The rate of urea [$CO(NH_2)_2$] hydrolysis into CO_2 and NH_3 brought by catalytic activity of urease depends on its concentration. Assuming that organic nitrogen contents in soil (estimated as the difference between nitrogen estimated by the Kjeldahl method [without reduction of nitrates] and $N-NH_4^+$), may be used as an index of urea concentrations, the similarity of urease activity in these three ecosystems can be explained by the same levels of substrates available for decomposition (Table 5.8). Soil under the 7-year-old growing shelterbelt did not store enough organic nitrogen, part of which could undego decomposition and provide significantly higher levels of urea concentration. But during the 140 years since the trees were planted on hapludalf soils, the organic matter accumulated in soil and the average contents of organic nitrogen are almost fivefold higher than in the adjacent field. In response to this increase the amounts of hydrolyzed urea almost doubled (Table 5.8). Thus, in shelterbelts not only is conversion of nitrates into ammonium ions by assimilatory nitrate reduction observed, but release of $N-NH_3$ during decomposition of biomass is also observed.

Because of the organic nitrogen accumulation during the growth of the shelter-

8.69 = 305. Thus, in the older shelterbelt more organic nitrogen is not decomposed into urea but stored in resistant-to-decomposition form. Nevertheless, when soil sorption capacities as well as nutritional demands of plants are met, then higher rates of leaching ammonium ion can be expected in older shelterbelts rather than in young ones, where low amounts of organic nitrogen are accumulated.

Interplay of these processes explains the results presented in the Table 5.5, showing the efficacy of biogeochemical barriers for control of ammonium ions seeping with ground water through the root systems of plants. In 9 of 13 analyzed cases, no decreases were detected when output-to-input $N-NH_4^+$ concentrations were analyzed. Efficient uptake of nitrates by plants in biogeochemical barriers and their processing by biota into $N-NH_4^+$ also explain lower variation of ammonium ions than nitrates in the main drainage canal during the course of the year (Table 5.2).

How much $N-NH_4^+$ is leached into ground water depends on the complicated interplay of several processes. Released $N-NH_4^+$ ions from decomposing organic matter could again be taken up by organisms for production of their biomass, converted to nitrates in nitrification processes, withdrawn from soil solution by sorption complex, incorporated into stored inert organic nitrogen, or leached into ground water. Moreover, ammonia (NH_3) could be volatilized into air. Because intense NH_3 volatilization appears from soil solutions when its reaction is above 7 pH, one can assume that this is not a significant form of ammonia loss from the Turew landscape soils, but drying of soils and high air temperature could have some effects on this process (Freney et al. 1981, Harper et al. 1996). The losses of NH_3 by volatilization may be reduced to some extent by the repeated absorption of released ammonia by plants when this gas is still within a vegetation stand (Harper et al. 1995). Due to those various recycling processes, the ammonia emissions from forests are small (Longford and Fehsenfeld 1992). All those processes are influenced by physical and chemical factors, but the central role is played by the biological processes of assimilation, decomposition, nitrification, and denitrification.

In the young and old shelterbelts and adjoining fields studied, CO_2 and N_2O evolution were measured. Carbon dioxide and nitrous oxide concentrations in gas samples were determined with Varian GC 3800, equipped with an electron capture at an operating temperature of 340°C and with a thermal conductivity detector operating at a temperature of 200°C. A Porapak 1.8-m Q 80/100 column (Alltech Associates, Inc., Deerfield, IL) at 50°C was used to separate N_2O, and at 74°C to separate CO_2 (Cabrera et al. 1993).

Evolution of CO_2 from soil can be used as an index of general biological activity of the cultivated fields and shelterbelts studied. The highest mean of CO_2 evolution was found from the soil of the 140-year-old shelterbelt while the two fields and the 7-year-old shelterbelt were characterized by similar (not statistically significant) rates of CO_2 evolution (Table 5.9). More than fourfold higher organic nitrogen

Table 5.9 CO_2 Evolution ($\mu g \cdot m^{-2} \cdot h^{-1}$) from Soil in Shelterbelts of Various Ages and Adjoining Cultivated Field

Date	Shelterbelts		Fields Adjoining Shelterbelts	
	7-Year-Old	140-Year-Old	7-Year-Old	140-Year-Old
March 12	82	119	82	79
April 7	83	120	82	78
May 8	84	141	87	81
June 5	86	146	91	100
July 9	88	167	115	129
August 20	87	159	107	99
September 14	83	155	87	83
October 12	82	123	83	80
November 14	79	119	85	68
Mean	83.8	138.8	91	88.6

With increasing age of shelterbelt, more and more nitrogen is stored in the resistant-to-decomposition organic compounds. The accumulation of soil organic matter under shelterbelts is the main mechanism of long-term withdrawal of various elements from the dynamic cycle of transformations in an ecosystem. Absorbed by plants or microbes, nutrients become built into plant biomass; then after the decay of biomass some of them are stored in humus. Estimates of the withdrawal rates of nitrogen into humus provided the figure of 1.7 $g \cdot m^{-2} \cdot year^{-1}$ in the 140-year-old shelterbelt and 1.4 $g \cdot m^{-2} \cdot year^{-1}$ for pine afforestation planted on hapludalf soils (Ryszkowski et al. 1997). Some nitrogen is also withdrawn from circulation for a considerable time in woody parts of trees. In the case of the pine afforestation this form of nitrogen storage attains a value of 0.0022 g $N \cdot m^{-2} \cdot year^{-1}$ (Ryszkowski et al. 1997) and is much lower than incorporation of nitrogen into humus. The highest level of organic nitrogen in the soil of the old shelterbelt was also correlated with very clear seasonal changes in the CO_2 evolution, reaching the highest values in the warmest months (July and August) of the year. The most intensive evolution of the nitrous oxide from soil was also observed in the summer (Table 5.10).

Table 5.10 N_2O Evolution (N $\mu g \cdot m^{-2} \cdot h^{-1}$) from Soil in Shelterbelts of Various Ages and Adjoining Cultivated Fields

Date	Shelterbelts		Fields Adjoining Shelterbelts	
	7-Year-Old	140-Year-Old	7-Year-Old	140-Year-Old
March 12	80	120	200	250
April 7	80	110	240	230
May 8	80	130	220	250
June 5	120	130	370	270
July 9	120	180	320	280
August 20	110	170	250	270

From the soil of cultivated fields, where nitrates prevail over ammonium ions (Tables 5.3, 5.6, and 5.7), much higher rates of N_2O evolution were observed than from the soil of shelterbelts where ammonium dominates or makes a substantial contribution to the pool of mineral nitrogen. This result confirms estimates of Robertson et al. (2000) indicating that N_2O fluxes are much higher from soil under annual crops than from poplar cultivation. With higher amounts of nitrates in soil, rates of N_2O evolution are also greater, though this relation is not linear and strongly depends on distribution of anaerobic sites in the soil. In the newly planted shelterbelt, where shortage of nitrates is apparent (1.0 to 1.2 $g \cdot m^2$ in 150-cm-deep soil stratum, Table 5.7) the evolution of N_2O was the lowest.

Converting the annual rates of nitrogen storage in humus and wood production into hourly intervals for the sake of comparison with N_2O evolution, 160 μg $N \cdot m^{-2} \cdot h^{-1}$ is found for pine afforestation and 190 μg $N \cdot m^{-2} \cdot h^{-1}$ for the 140-year-old shelterbelt. These rates are very similar to those for N_2O evolution in the old shelterbelt (Table 5.10). The rate of the nitrogen storage in wood is lower and after conversion is 22 μg $N \cdot m^{-2} \cdot h^{-1}$. One can infer that nitrogen storage in shelterbelt humus is greater than its storage in wood, and that both long-term storage mechanisms operate at the same level such as the release of nitrogen in N_2O evolution.

The intensive conversion of nitrate inputs to ammonium ions through assimilatory reduction and then decomposition of organic nitrogen, releasing ammonium, as well as the low intensity of nitrification processes are the reasons for lower rates of N_2O release into atmosphere in shelterbelts than in cultivated fields. Nitrification was not studied in the Turew landscape, but many researchers found low rates of nitrification in the terrestrial ecosystems with an unfertilized permanent vegetation (Verstraete 1981). The conversion of $N-NO_3^-$ into $N-NH_4^+$ constitutes, as one can hypothesize, a nitrogen-saving mechanism in shelterbelts or meadows, preventing nitrogen loss by nitrate leaching and N_2O evolution. Those nitrogen-saving mechanisms are evidenced by lower nitrate concentrations in ground water of the biogeochemical barriers than under fields (Table 5.3) and lower rates of N_2O evolution in the shelterbelts than in cultivated fields (Table 5.10) and generally low concentrations of ammonium ions both in ground water under cultivated fields and biogeochemical barriers (Table 5.5). The lower rates of N_2O evolution were directly measured. Considering operation of nitrogen-saving mechanisms in shelterbelts and meadows, the question still remains of how intensive is the release of N_2 in denitrification processes. In addition, field measurements of NH_3 volatilization should show if, as the literature suggests, very low rates of NH_3 loss in the Turew upland soils represent the real situation.

LANDSCAPE MANAGEMENT GUIDELINES FOR EFFICIENT

Nitrate input with precipitation in the Turew landscape, which makes up quite a substantial source,* as well as nitrate water runoff from cultivated fields are effectively controlled by the network of biogeochemical barriers that results in low mineral nitrogen concentrations in the main drainage canal. The mean annual mineral nitrogen ($N-NO_3^-$ plus $N-NH_4^+$) varied over a 28-year period within 1.5 $mg \cdot dm^{-3}$ to 6 $mg \cdot dm^{-3}$ (Figure 5.1). The landscape mechanisms of control kept nitrogen concentrations at this rather stable level despite dramatic decrease in fertilizer use, changes in crop rotation patterns, and various weather conditions. From the landscape management point of view, an interesting question is what should be the size of the area that should be covered by biogeochemical barriers to efficiently control diffuse pollution. The answer will be provided in three steps that reflect varying degrees of complexity. The first step concerns the evaluation of the effective width of the biogeochemical barrier. The second step is the functional relationship between buffer zone area and output of nitrates from the watersheds. The third step deals with the management of nitrogen storing capacities in shelterbelts.

Using the model of solar energy partitioning for various components of the heat balance of a large area incorporating meteorological characteristics and parameterization of plant cover structure developed by Olejnik (1988), Kędziora et al. (1989), Olejnik and Kędziora (1991), Ryszkowski and Kędziora (1993), Ryszkowski et al. (1997) estimated evapotranspiration rates of birch and pine afforestations under field conditions during the plant growth seasons for 10-day intervals. Estimates of evapotranspiration were then corrected for values of evaporation in order to obtain only plant stand transpiration estimates. Additionally, inputs of ground water from field into birch or pine afforestations through the 1 m × 2 m phreatic plane were estimated by special hydrological studies (Ryszkowski et al. 1997). Estimates of transpiration rates per square meter and the amount of ground water discharged from field into forest by under-surface flux were used to calculate the length of land band, under trees of 1-m width, necessary to transpire the incoming water. Thus, the subsurface water input was divided by the amount of water transpired by plants from 1 square meter. The length of that band was assumed to constitute an approximation of shelterbelt width necessary to perform effective removal of incoming mineral forms of nitrogen. Those calculations rely on the assumption that uptake of nitrates and ammonium ions by plants is mainly determined by water mass uptake for transpiration and that effects of the absorption by diffusion processes are small.

In these calculations describing a selective performance of plants for uptake, various mineral forms of nitrogen found in some studies (see, for example, Kirkby 1981, Prusinkiewicz et al. 1996) were not taken into account. 1996 among others). Regarding selective uptake of nitrogen mineral forms, Prusinkiewicz et al. (1996) show that in periods of high water consumption the differences between the uptake of NH_4^+ ions and NO_3^- ions diminish, while at low water consumption increased

inaccuracies, this proposed method of nutrient uptake estimation of an intact forest stand provides only approximations. Width of effective biogeochemical barriers during the plant growing season was estimated to vary from 5 m to 25 m in birch afforestation. The season-long mean was equal to 10.4 m. In pine afforestation, the estimates of effective width of biogeochemical barrier varied from 2.4 m to about 10 m, with an average value of 5.8 m. The differences between means were statistically significant, but the reason for it was probably not differentiated plant uptake by coniferous and deciduous trees but rather the poor stand of trees in the birch afforestation (Ryszkowski et al. 1997). One can assume, therefore, that the effective width of shelterbelt under Turew landscape conditions is about 6 to 8 m. This estimate was obtained for small slopes of ground water table and thus for slow ground water fluxes. For greater fluxes (e.g., higher water table slopes), the water will be passing faster and effectiveness of a given width of shelterbelt will be smaller. Ryszkowski and Kędziora (1993) show that the uptake of ground water by a 10-m wide shelterbelt or meadow can be so great on a warm and sunny day, having net radiation of 100 $W \cdot m^{-2}$, that flux of ground water is reduced almost completely if the ground water table steepness is 0.01. Under the same meteorological conditions but with the ground water table steepness of 0.04, the water flux is reduced by only 33%. These studies also show that shelterbelts have greater impact on ground water fluxes than meadow strips of the same width (Ryszkowski and Kędziora 1993) which explains the findings of Haycock and Pinay (1993) that grass riparian zones were less effective in control of nitrate pollution than were poplar riparian strips. The estimates of buffer barrier width for control of nitrates found by other scientists range from 5 m to 30 m. Individual estimates were as follows: 5 m (Cooper 1990), 8 m (Haycock and Burt 1991), 16 m (Jacobs and Gilliam 1985), 19 m (Peterjohn and Correll 1984), and 30 m (Pinay and Decamps 1988).

Despite different interpretations of these estimates presented by various scientists, results are quite similar. In view of results obtained in the Turew landscape, efficient removal of nitrates from ground water is related to their uptake by plants with transpired water but not mainly with denitrification processes as claimed by many authors studying riparian vegetation strips.

To disclose the relationship between an area of watershed covered by biogeochemical barriers and the output of nitrates, the studies were done on $N-NO_3^-$ concentration in the water discharged in ditches from six small watersheds situated in the studied landscape (Ryszkowski 2000). The studies were carried out from November 1995 until December 1996. The area of the watershed varied from 75 to 216 ha. Cultivated fields have hapludalf and glossudalf soils. The watersheds varied with respect to contribution of arable fields which ranged from 99 to 52% of total area. Meadows, shelterbelts, and small forests represented the perennial vegetation. Each watershed was drained by ditch from which water samples were taken every 2 weeks.

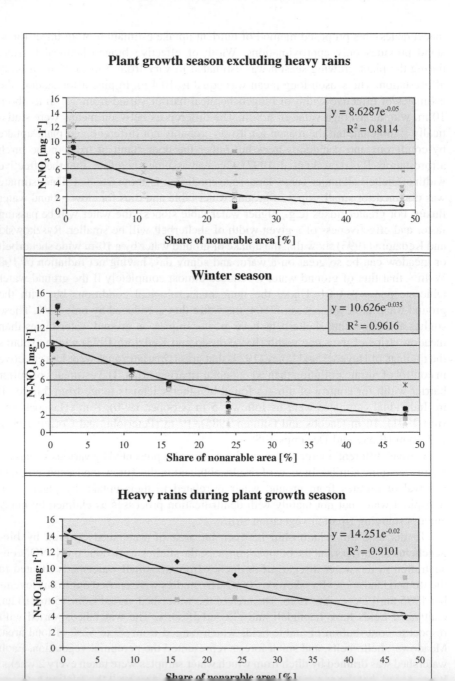

rains amounted to 106.4 mm, and at the beginning of July there was a second heavy rainfall of 111.1 mm. Measurements of $N-NO_3^-$ concentrations in ditch water runoff associated with these precipitation events were analyzed separately because of the appearance of intense surface flows. The exponent of the equation characterizing the relationship between the share of buffer zones in the total area of watershed and $N-NO_3^-$ concentration was higher for the plant growing season than for winter (Figure 5.2). This result can again be interpreted as caused by transpiration activity of plants — transpiring plants take up nutrients from soil solutions. Thus, greater area coverage by perennial plants with high transpiration rates shows larger uptake of nutrients and greater purification effect exerted by plants on diffuse pollution by nitrates. The lowest exponent value was found when heavy rains occurred. It seems that concentrated surface runoff caused by heavy rains masks to some extent the effects of biogeochemical barriers. Nevertheless, one can find a correlation between $N-NO_3$ concentrations and buffer zone areas even in the case of heavy rainfall (Figure 5.2).

The finding that the relationship between the share of biogeochemical barrier areas and nitrate concentration shows exponential character has an important bearing on landscape management practice. The small area under planted shelterbelts or grass strips shows unproportionally high control effects compared with those gained when it is extended to a larger size. That implication is true, of course, when biogeochemical barriers are strategically distributed with regard to directions of water runoff from the watershed. Permanent vegetation located in one patch will not exert such a control effect as would a network of shelterbelts and grass strips of the same area size. The network of biogeochemical barriers enables, therefore, optimization of the area withdrawn from agricultural production with arable land with respect to the economic issues of farming. In other words, the knowledge of the exponential effect exerted by the area under shelterbelts on water cleansing can help make economic decisions about the area of land that should be withdrawn from agricultural production for introduction of permanent vegetation strips.

Uptake of both nitrogen mineral forms by plants and the subsequent formation of NH_4^+ ions during biomass decomposition play a crucial role in regulation of nitrogen cycling in an ecosystem. One can therefore hypothesize that when NH_4^+ released during decomposition of organic nitrogen overcomes its storage in long-term withdrawal processes (incorporation into humus or in woody tissues) as well as its maintenance in dynamic recycling between various biota (plants, microbes, animals), the biogeochemical barrier turns from N sink to its source in the landscape. It can be presumed, therefore, that when accumulated plant litter is rapidly decomposed because of favorable conditions, substantial amounts of nitrogen are leached through the soil profile, and its concentration will increase in ground water under the biogeochemical barrier. This situation was documented in studies

Table 5.11   Nitrogen Loss from Litter in Park Afforestation Located on Hapludalf Soils

Period	Decomposition Rate k of Litter/day ($W_t = W_o e^{-kt}$)	Average Mass of Litter ($g \cdot m^{-2}$)	Loss of Nitrogen ($mg \cdot dm^{-2} \cdot 24h$)
April 23–June 18, 1996	0.0032	483.1	43.6
July 15–September, 1996	0.0042	312.2	27.0
November 28, 1996– March 27, 1997	0.0034	641.0	31.2
May 27–July 22, 1997	0.0046	464.0	94.1
September 15– November 10, 1997	0.0038	285.8	53.9

Source: Z. Bernacki (unpublished data).

is located on hapludalf soils. The park is kept as a nature reserve and no management practices are carried out. Along the direction of ground water flux from the cultivated field, through the park and then to the pond, piezometers were installed for sampling ground water chemistry. It was dry in 1996, and the decomposition rate of litter was low, especially low in autumn and winter, which resulted in accumulation of plant litter. After the fall of leaves, the average mass of litter reached 641 $g \cdot m^{-2}$ (Table 5.11). There were heavy rains in the summer of 1997; in June precipitation was 67.7 mm, in July 169.3 mm, and in August 139.4 mm, and some flooding was experienced. The total amount of nitrogen released from litter in plant growth in the period March 21–October 30, 1997 was estimated at 15.33 $g \cdot m^{-2}$ or 153 kg $N \cdot ha^{-1}$, which is equal to a very high dose of nitrogen application in fertilizer. The decomposition rate was estimated by exposition of known mass of litter in mesh bags, and the rate was calculated assuming exponential decay — W_t mass of litter after t days and W_o mass at the time of exposition, t time in days.

Studies on the contents of nitrogen mineral forms in the total unsaturated zone of hapludalf soils showed a constant decrease from May 26, 1997 to March 30, 1998 from 12.5 g $N \cdot m^{-2}$ to 1.4 g $N \cdot m^{-2}$. The share of $N–NO_3^-$ in total mineral form appearing in the unsaturated zone was 54% in May and 66% in March. But nitrification transformations of mineral nitrogen resulted in a preponderance of $N–NO_3^-$ concentrations in ground water under hapludalf soils (Table 5.12), although in the unsaturated soil zone only a slight dominance of $N–NO_3^-$ was detected.

The dominance of $N–NO_3^-$ over NH_4^+ ions in ground water increased with the distance from the edge of the park indicating intensive processes of nitrification of large amounts of nitrogen released in decomposition processes, although some contribution of $N–NO_3^-$ from rain was possible. Intensive nitrification was also confirmed by the decrease in organic nitrogen. However, the most important result of this study is the finding that high inputs of mineral nitrogen from decomposing

Table 5.12 Annual Mean Concentrations of N–NO₃⁻, N–NH₄⁺ and Organic Nitrogen (mg·dm⁻³) in Ground Water Seeping through Old Afforestations Calculated from Monthly Samples in 1997

Form of Nitrogen	Location of Piezometers in Distance from the Field-Park Boundary (m)		
	0	16.5	62
N–NO₃⁻	5.3	6.4	14.2
N–NH₄⁺	3.1	3.1	2.9
Total mineral nitrogen	8.4	9.5	17.1
Dissolved organic nitrogen	7.4	5.6	4.5

Results of long-term studies in the Turew landscape recommend the following guidelines for control of diffuse pollution (Ryszkowski 1998):

- The ground water table should be within the direct or indirect (capillary ascension) reach of plant root systems.
- Better control efficiency is obtained when the shelterbelt is composed of multiple tree species rather than one species.
- With less slope of the ground water table and hydraulic conductivity a narrower shelterbelt can be installed.
- Accumulated litter in shelterbelts should be removed.
- A network of shelterbelts is recommended.
- Other biogeochemical barriers, such as stretches of meadows, small mid-field wetlands or ponds, riparian buffer strips, and diversified crop rotation patterns all perform functions of diffuse pollution control.

PROSPECTS FOR CONTROL OF THE DIFFUSION POLLUTION THROUGH MANAGEMENT OF LANDSCAPE STRUCTURE

Diffuse pollution was recognized recently as a worldwide threat, and the need for its control is recognized in an increasing number of countries. For example, the European Union, revising its environment policy in 1996, stressed the need for a fundamental shift in its priorities from environmental protection legislation to using new policy instruments for pursuing environmental problems. One of the principal issues addressed is a clear need to tackle growing problems of ground water pollution, especially nitrate pollution of water reservoirs both underground and on the surface. Despite much environmental legislation, for example, the nitrogen directive of 1992, the success of implementation thus far is low, as 87% of agricultural area in Europe has a nitrate concentration above the guideline level of 25 mg·dm⁻³ and

Various environmental policies have recommended efforts to control diffuse pollution that focus on two directions of activity. The first direction concentrates on introduction of environmentally friendly cultivation methods. Those methods should be implemented at the farm level and, for example, consist of diversified crop rotation patterns, including use of after- or before-crops (catch crops), which grow when the main crop is absent in late autumn or early spring, to reduce nitrate leaching. Other proposed methods aim at conserving soil organic matter, application of fertilizers in quantities adjusted to cultivars' needs, integrated methods of pests and pathogens control, among others. The objective is to maintain high storage capacities of soil and to preserve or improve its physical, chemical, and biological properties in cultivated fields. But water cycling processes occur on a spatial scale much larger than the farm and waterborne migration of various chemical compounds is not restricted to one field. The meteorological conditions influencing nitrogen cycling also depend on processes occurring on regional or global scales. These constraints to efficient environmental threat control within the farm can be overcome by managing the landscape structure so as to increase its capacities for long-term nitrogen storage, recycling between ecosystems, or gas release in the form of N_2. The second direction of pollution control activity consists, therefore, in introduction or manipulation of agricultural nonproductive elements of landscape to increase their protection functions. But care should be taken with regard to effects obtained. The conversion of ground water nitrogen into N_2O fluxes could increase the concentration of greenhouse gases and therefore only changes the kind of environmental threat, namely, from water pollution to climate-warming phenomenon.

A majority of projects for controlling diffuse pollution direct attention to implementation of riparian vegetation strips (Haycock et al. 1997). The studies carried out in the Turew landscape have shown that the upland parts of the drainage area if diversified by a network of shelterbelts or meadows can also effectively cleanse ground water of nitrogen pollution caused by modern agriculture.

Although there is a consensus that vegetation strips or patches within the landscape show controlling effects on water quality, the reasons for that phenomenon are elusive. The emission of gaseous nitrogen forms in the denitrification processes is claimed to be the primary mechanism of nitrates control in the riparian buffer zones (Gilliam et al. 1986, Correll 1997, Haycock et al. 1997a, Lowrance et al. 1995). The nitrogen emissions during the denitrification processes can exceed 100 kg $N \cdot ha^{-1} year^{-1}$ (Groffman 1997) or approximately 1 mg $N \cdot m^{-2} \cdot h^{-1}$. This general estimate includes both N_2 and N_2O emissions, so the comparison with nitrous oxide evolution in the Turew landscape is very rough. The average N_2O evolution rate in shelterbelts during plant growth (Table 5.10) in the Turew landscape is about ten times lower than the general approximation of $N_2 + N_2O$ flux from the riparian buffer zones provided by Groffman (1997). Rudaz et al. (1999) studying grasslands found that the $N_2:N_2O$ ratio fluctu-

2 µg $N \cdot m^{-2} \cdot h^{-1}$ to 475 µg $N \cdot m^{-2} \cdot h^{-1}$. Kessel et al. (1993) estimated $N–N_2O$ emission rates in cultivated fields ranging from 46 µg $N \cdot m^{-2} \cdot h^{-1}$ to 904 µg $N \cdot m^{-2} \cdot h^{-1}$ during the vegetation season. In this last study, the N_2 release was almost equal to N_2O emissions. The emission rates estimated for cultivated fields in the Turew landscape were intermediary values (Table 5.10) with regard to the range of estimates reported by Eichner (1990) and Kessel et al. (1993).

Groffman and Tiedje (1989) estimated the mean annual nitrogen loss to denitrification to be 0.6 kg $N \cdot h^{-1}$ $year^{-1}$ for a well-drained, sandy forest soil and 0.8 kg $N \cdot h^{-1}$ $year^{-1}$ for somewhat poorly drained, sandy forest soils. These estimates are much lower than the approximation provided by Groffman (1997) for riparian buffer zones cited above. After conversion to units enabling comparison with estimates of nitrogen fluxes found in the Turew landscape, the Groffman and Tiedje (1989) figures are equal to 6.8 µg $N \cdot m^{-2} \cdot h^{-1}$ for the well-drained soil and 9 µg for the less-drained soil under forest. These rates of N emission are very low in comparison with N_2O release rates in shelterbelts. The reason for that discrepancy is unknown.

The important result of our studies is the indication that shelterbelts are characterized by lower rates of N_2O evolution than are cultivated fields. A shelterbelt's efficient control of the nitrates inputs with ground water from a cultivated field (Tables 5.3 and 5.4) is not tied to increased evolution of nitrous oxide, which would be connected with stimulation of greenhouse gases. In comparison to cultivated fields, the mean rates of N_2O evolution from soil of shelterbelts constituted only from 40 to 56% (Table 5.10). Although the rates of dinitrogen emissions were not estimated, shelterbelts do not clean water of nitrates at the expense of climate-warming stimulation.

The rapid uptake of nitrates by plants and microbes increases the infrasystem nitrogen cycling, which builds storage capacities of shelterbelts. As was shown, the long-term storage of nitrogen in humus and tree wood is higher than is the release of nitrogen as nitrous oxide (210 µg $N \cdot m^{-2} \cdot h^{-1}$ and 97–142 µg $N \cdot m^{-2} \cdot h^{-1}$, respectively). That process is caused by slow accumulation of humus, which will probably stop in a very old shelterbelt such as occurs in old natural forest (Hedin et al. 1995).

When decomposition of accumulated litter is very high, the old-growth shelterbelt turns from a sink to a source with respect to nitrogen cycling in the landscape (Table 5.12). Results of studies on afforestation in the Turew landscape support the N saturation hypothesis postulated by Omernik et al. (1981) and Aber et al. (1991). Thus, removal of litter is a very important condition for management of shelterbelts in order to implement successful control of diffuse pollution. The conversion of nitrates to ammonium due to plant and microbe metabolic processes could to some extent control nitrogen leaching. The amounts of NH_4^+ fixed by the soil's cation exchange capacity prevents leaching of this nitrogen form. Nevertheless, precipitation events introducing salt solution (e.g., potassium) wash out some amounts of fixed ammonium. The contents of fixed NH_4^+ change after rains, therefore, but the

The sharp decreases of $N-NO_3^-$ concentrations in ground water passing under the biogeochemical barriers (Table 5.3) indicate that biotic demands for nitrates are high relative to supplies. Nitrogen incorporated into biomass undergoes mobilization and repeated immobilization processes linked to biological interactions between plants, animals, and microbes. Efficiency of infrasystem recycling determines dynamic storing capacities of the system, and that volume is increased with the diversity of interacting organisms. The decomposition of urea may be used as the index of the dynamic internal recycling of nitrogen. For the sake of comparison with other fluxes presented in this analysis of the biogeochemical barrier functions, the estimates of urease activity (Table 5.8) were converted into approximation of ammonia production per 1 square meter. The conversions were made using the soil bulk density estimates as well as other correction factors such as those that 1 μg of urea hydrolyzed release 0.47 μg $N-NH_3 \cdot g^{-1}$soil$\cdot h^{-1}$ (Zantua and Bremner 1977), and those connected with volumes of the sampled air. All approximations of NH_3 production rates in urea decomposition were converted to the entire season means of $N-NH_3$ produced per hour in a soil block 1 m^2 in area and 0.3 m deep. In the 7- and 140-year-old shelterbelts studied, the average rate of NH_3 production was 1.11 g $N-NH_3 \cdot m^{-2} \cdot h^{-1}$ and 1.64 g $N-NH_3 \cdot m^{-2} \cdot h^{-1}$, while in adjoining fields the production of ammonia assumed values of 1.3 g $N-NH_3 \cdot m^{-2} \cdot h^{-1}$ and 0.83 g $N-NH_3 \cdot m^{-2} \cdot h^{-1}$. Estimates of the season mean-production rates of ammonia were obtained by analysis of soil samples under laboratory conditions and did not reflect real soil temperatures; standard quantities of reagents were also used. Nevertheless, such approximation illustrates well the magnitude of the transformation studied. The rates of $N-NH_3$ production in decomposition processes of urea were several thousand higher than rates of nitrogen emission in the form of N_2O or nitrogen long-term storage in humus. Such a great discrepancy can be interpreted as a result of fast nitrogen recycling in the dynamic system of interactions (Ryszkowski et al. 1997, Aber 1999).

The large pool of internally recycling nitrogen, which is evidenced by very high NH_3 production rates, is the result of interactions among a diversified set of components. Elimination of any one functional component — for example, plants — will limit assimilatory reduction of nitrates, which in turn will decrease the volume of recycling nitrogen. The internal nitrogen recycling, which is sensitive to a set of control factors, is poorly known. The complex interactions among plants, animals, and microbes modified by physical and chemical factors should be understood before management strategies to prevent nitrogen saturation of the biogeochemical barrier could be proposed. Such knowledge may also be useful in the purposeful steering of nitrogen fluxes for different output routes from the biogeochemical barrier. Any imbalance between components of the internal dynamic system of interactions results in promoting output products such as gaseous nitrogen forms and leaching of mineral or organic forms, as well as storing capacities of the system. At present, the best

CONTROL OF DIFFUSE POLLUTION   139

REFERENCES

Aber D. J. 1999. Can we close the water/carbon/nitrogen budget for complex landscapes. In *Integrating Hydrology, Ecosystem Dynamics, and Biogeochemistry in Complex Landscapes* (Eds. J. D. Tenhumen, P. Kabat). John Wiley & Sons, Chichester: 313–333.

Aber J. D., Melillo J. M., Nadelhoffer K. J., Pastor J., Boone R. 1991. Factors controlling nitrogen cycling and nitrogen saturation in northern temperate forest ecosystems. *Ecol. Appl.* 1: 303–315.

Baldock D. 1990. *Agriculture and Habitat Loss in Europe*. The Institute for European Environmental Policy, London, 60 pp.

Bartoszewicz A. 1990. Chemical compounds in groundwater of agricultural watershed under soil climatological conditions of Kościan Plain (in Polish). In *Obieg wody i bariery biogeochemiczne w krajobrazie rolniczym* (Eds. L. Ryszkowski, J. Marcinek and A. Kędziora). Adam Mickiewicz University Press, Poznań: 127–142.

Bartoszewicz A. 1994. The chemical compounds in surface waters of agricultural catchments. *Roczniki Akademii Rolniczej w Poznaniu* 250: 5–68.

Bartoszewicz A. 2000. Effect of the change of soil utilisation on the concentration of nitrogen mineral forms in soil and ground waters. *Polish J. Soil Sci.* 33/2: 13–20.

Bartoszewicz A. 2001. Content and quantitative relationships of exchangeable and water soluble forms of Ca, Mg, K and Na in arable lessive soils. *Roczniki Akademii Rolniczej w Poznaniu* 61: 3–21.

Bartoszewicz A. Mid-field shelterbelts as a way of confining nitrogen migration in soils and waters of rural areas. *Acta Agrophysica,* in press.

Bartoszewicz A., Ryszkowski L. 1996. Influence of shelterbelts and meadows on the chemistry of ground water. In *Dynamics of an Agricultural Landscape* (Eds. L. Ryszkowski, N. French and A. Kędziora). Państwowe Wydawnictwo Rolnicze i Leśne, Poznań: 98–100.

Bremner J. M., Mulvaney R. L. 1978. Urease activity in soils. In *Soil Enzymes* (Ed. R. G. Burns). Academic Press, London: 149–196.

Bülow-Olsen A. 1988. Disappearance of ponds and lakes in southern Jutland, Denmark 1954–1984. *Ecol. Bull.* (Copenhagen) 39: 180–182.

Cabrera M. L., Chiang S. C., Merka W. C., Thompson S. A., Pancorbo O. C. 1993. Nitrogen transformation in surface-applied poultry litter: effect of physical characteristics. *Soil Sci. Soc. Am. J.* 57: 1519–1523.

Com 1999, 22 final. 1999. Direction towards sustainable agriculture. Commission of the European Communities, Brussels, 30 pp.

Cooper A. B. 1990. Nitrate depletion in the riparian zone and stream channel of a small headwater catchment. *Hydrobiologia* 202: 13–26.

Correll D. L. 1997. Buffer zones and water quality protection: general principles. In *Buffer Zones: Their Processes and Potential in Water Protection* (Eds. N. E. Haycock, T. P. Burt, K. W. T. Goulding, G. Pinay). Quest Environmental, Harpenden, U.K.: 7–20.

Eichner M. J. 1990. Nitrous oxide emissions from fertilized soils: summary of available data. *J. Environ. Quality* 19: 272–280.

European Environment Agency. 1998. *Europe's Environment: The Second Assessment.*

FSA. 1997. Farm Service Agency Handbook 2 — CRP (Revision 3). USDA — Farm Service
    Agency. P.O. Box 2415. Washington, D.C. 2013–2415.
Gilliam J. W., Skaggs R. W., Doty C. W. 1986. Controlled agricultural drainage: an alternative
    to riparian vegetation. In *Watershed Research Perspectives* (Ed. D. L. Correll). Smith-
    sonian Press, Washington, D.C.: 225–243.
Groffman P. M., Tiedje J. M. 1989. Denitrification in north temperate forest soils: relationships
    between denitrification and environmental factors at the landscape scale. *Soil Biol.
    and Biochem.* 21: 621–626.
Groffman P. M. 1997. Contaminant effects on microbial functions in riparian zones. In *Buffer
    Zones: Their Processes and Potential in Water Protection* (Eds. N. E. Haycock, T. P.
    Burt, K. W. T. Goulding, G. Pinay). Quest Environmental, Harpenden, U.K.: 83–92.
Halberg G. R. 1989. Nitrate in ground water in United States. In *Nitrogen Management and
    Groundwater Protection* (Ed. R. F. Follett). Elsevier, Amsterdam: 35–74.
Harper L. A., Hendrix P. F., Longdale G. W., Coleman D. C. 1995. Clover management to
    provide optimum nitrogen and soil water conservation. *Crop Sci.* 35 (1): 176–182.
Harper L. A., Bussink D. W., Mear H. G., Corre W. J. 1996. Ammonia transport in a temperate
    grassland: I. Seasonal transport in relation to soil fertility and crop management.
    Agron. J. 88(4): 614–621.
Haycock N. E., Burt T. P. 1991. The sensitivity of rivers to nitrate leaching: The effectiveness
    of near-stream land as a nutrient retention zone. In *Landscape Sensitivity*. Wiley &
    Sons, London: 261–272.
Haycock N. E., Pinay G. 1993. Groundwater nitrate dynamics in grass and poplar vegetated
    riparian buffer strips during the winter. *J. Environ. Qual.* 22: 273–278.
Haycock N. E., Burt T. P., Gouding K. W. T., Pinay G. (Eds). 1997. *Buffer Zones: Their
    Processes and Potential in Water Protection*. Quest Environmental. Harpenden, U.K.:
    326 pp.
Haycock N. E., Pinay G., Burt T. P., Gouding K. W. T. 1997a. Buffer zones: current concerns
    and future directions. In *Buffer Zones: Their Processes and Potential in Water Pro-
    tection*. Eds. N. E. Haycock, T. P. Burt, K. W. T. Goulding, Pinay G. Quest Environ-
    mental, Harpenden, U.K.: 305–312.
Hedin L. O., Armesto J. J., Johnson A. H. 1995. Patterns of nutrient loss from unpolluted,
    old-growth temperate forests: evaluation of biogeochemical theory. *Ecology* 76 (2):
    493–509.
Hillbricht-Ilkowska A., Ryszkowski L., Sharpley A. N. 1995. Phosphorus transfers and land-
    scape structure: riparian sites and diversified land use patterns. In *Phosphorus in the
    Global Environment* (Ed. H. Tissen). John Wiley & Sons, Chichester: 201–228.
Jackson M. L. 1964. *Soil Chemical Analysis*. Prentice-Hall, Englewood Cliffs, NJ, 498 pp.
Jacobs T. C., Gilliam J. W. 1985. Riparian losses of nitrate from agricultural drainage waters.
    J. Environ. Qual. 14: 472–478.
Johnsson H., Hoffmann M. 1998. Nitrogen leaching from agricultural land in Sweden —
    standard rates and gross loads in 1985 and 1994. *Ambio* 27: 481–488.
Kauppi L. 1990. Hydrology: water quality changes. In *Toward Ecological Sustainability in
    Europe* (Eds A. M. Solomon, L. Kauppi). International Institute for Applied System
    Analysis, Laxenburg (Austria): 43–66.

Longford A. O., Fehsenfeld F. C. 1992. The role of natural vegetation as a source or sink for atmospheric ammonia. *Science* (Washington, D.C.) 255: 581–583.

Lowrance R. 1997. The potential role of riparian forests as buffer zones. In *Buffer Zones: Their Processes and Potential in Water Protection* (Eds. N. E. Haycock, T. P. Burt, K. W. T. Goulding, G. Pinay). Quest Environmental, Harpenden, U.K.: 128–133.

Lowrance R., Todd R., Asmussen L. 1983. Waterborne nutrient budgets for the riparian zone of an agricultural watershed. *Agric. Ecosys. Environ.* 10: 371–384.

Lowrance R., Vellidis G., Hubbard R. K. 1995. Denitrification in a restored riparian forest wetland. *J. Environ. Qual.* 24: 808–815.

Mannion A. M. 1995. *Agriculture and Environmental Change.* Wiley, Chichester, 405 pp.

Marcinek J., Komisarek J. 1990. Cation exchange capacity and travel time of solutes from the soil surface down to the ground water table (in Polish). Poznańskie Towarzystwo Przyjaciół Nauk. *Prace Komisji Nauk Rolniczych i Komisji Nauk Leśnych* 69: 71–85.

Margowski Z., Bartoszewicz A. 1976. Influence of agricultural use on the chemical composition of ground water. *Polish Ecol. Stud.* 2: 15–21.

Muscutt A. D., Harris G. L., Bailey S. W., Davies D. B. 1993. Buffer zones to improve water quality: a review of their potential use in U.K. agriculture. *Agric. Ecosys. Environ.* 45: 59–77.

OECD. 1986. *Water Pollution by Fertilizers and Pesticides.* OECD Publications, Paris, 144 pp.

Olejnik J. 1988. The empirical method of estimating mean daily and mean ten-days values of latent and sensible-heat fluxes near the ground. *J. Appl. Meteorol.* 27: 1358–1369.

Olejnik J., Kędziora A. 1991. A model for heat and water balance estimation and its application to land use and climate variation. *Earth Surface Processes and Landforms* 16: 601–617.

Omernik J. M., Abernathy A. R., Male L. M. 1981. Stream nutrient levels and proximity of agricultural and forest land to streams: some relationships. *J. Soil Water Conservation* 36: 227–231.

Pauliukevicius G. 1981. *Ecological Role of the Forest Stand on the Lake Slopes.* Pergale, Vilnius, 191 pp.

Peterjohn W. T., Correll D. L. 1984. Nutrient dynamics in agricultural watersheds: observations on the role of a riparian forest. *Ecology* 65: 1466–1475.

Pinay G., Decamps H. 1988. The role of riparian woods in regulating nitrogen fluxes between the alluvial aquifer and surface waters: a conceptual model. *Regulated Rivers* 2: 507–516.

Pokojska U. 1988. Potential possibilities of nutrient retention by soils of mid-field shelterbelts and meadows in agricultural landscapes. *Roczniki Gleboznawcze* 39: 51–61.

Prusinkiewicz Z., Pokojska U., Józefkowicz-Kotlarz J., Kwiatkowska A. 1996. Studies on the functioning of biogeochemical barriers. In *Dynamics of an Agricultural Landscape* (Eds. L. Ryszkowski, N. French, A. Kędziora). Państwowe Wydawnictwo Rolnicze i Leśne, Poznań: 110–119.

Rekolainen S. 1989. Phosphorus and nitrogen load from forest and agricultural areas in Finland. *Aqua Fennica* 19: 95–107.

Robertson C. P., Paul E. A., Harwood R. R. 2000. Greenhouse gases in intensive agriculture: contributions of individual gases to the radiative forcing of the atmosphere. *Science*

Ryszkowski L. 1998. Recommendations for environment protection against diffuse pollution (in Polish). In Kształtowanie środowiska rolniczego na przykładzie Parku Krajobrazowego im. gen. D. Chłapowskiego (Eds. L. Ryszkowski, S. Bałazy). Research Centre for Agricultural and Forest Environment, Poznań: 81–88.

Ryszkowski L. 2000. Protection of water quality against nitrate pollution in rural areas. In *L'eau, de la cellule au paysage* (Ed. S. Wicherek). Elsevier, Paris: 171–183.

Ryszkowski L., Bartoszewicz A. 1989. Impact of agricultural landscape structure on cycling of inorganic nutrients. In *Ecology of Arable Land* (Eds. M. Clarholm, L. Bergström). Kluwer Academic Publishers, Dordrecht: 241–246.

Ryszkowski L., Bartoszewicz A., Kędziora A. 1997. The potential role of mid-field forests as buffer zones. In *Buffer Zones: Their Processes and Potential in Water Protection* (Eds. N. E. Haycock, T. P. Burt, K. W. T. Goulding, G. Pinay). Quest Environmental, Harpenden, U.K.: 171–191.

Ryszkowski L., A. Kędziora. 1987. Impact of agricultural landscape structure on energy flow and water cycling. *Landscape Ecol.* 1: 85–94.

Ryszkowski L., Kędziora A. 1993. Energy control of matter fluxes through land-water ecotones in an agricultural landscape. *Hydrobiologia* 251: 239–248.

Ryszkowski L., Marcinek J., Kędziora A. (Eds.). 1990. *Water Cycling and Biogeochemical Barriers in Agricultural Landscape* [in Polish]. Adam Mickiewicz University Publications, Poznań, 187 pp.

Ryszkowski L., French N. R., Kędziora A. (Eds.) 1996. *Dynamics of an Agricultural Landscape.* Państwowe wydawnictwo Rolnicze i Leśne. Poznań. 223 pp.

Ryszkowski L., Życzyńska-Bałoniak L., Szpakowska B. 1996a. The influence of biogeochemical barriers on the expansion of non-point pollution [in Polish]. In *Oczyszczalnie Hydrobotaniczne* [Constructed wetlands for wastewater treatment]. (Ed. R. Błażejewski). Sorbus Publisher, Poznań: 147–156.

Smith W. H. 1976. Character and significance of forest tree root exudates. *Ecology* 57: 324–331.

Stanners D., Bourdeau P. 1995. *Europe's Environment.* European Environment Agency, Copenhagen, 676 pp.

Stern A. 1996. *Environment. New European Policies for 1996–2000.* Club de Bruxelles, Bruxelles, 125 pp.

Szajdak L., Matuszewska T. 2000. Reaction of woods in changes of nitrogen in two kinds of soil. *Polish J. Soil Sci.* 33/1: 9–17.

Szpakowska B., Życzyńska-Bałoniak I. 1994. Groundwater movement of mineral elements in an agricultural area of Southern-Western Poland. In *Functional Appraisal of Agricultural Landscape in Europe.* (Eds. L. Ryszkowski and S. Bałazy). Research Centre for Agricultural and Forest Environment, Poznań: 95–104.

Thornton J. A., Rast W., Holland M. M., Jolankai G., Ryding S. O. 1999. *Assessment and Control of Non-point Source Pollution of Aquatic Ecosystems.* Parthenon Publishing Group, New York, 466 pp.

Tiedje J. M., Sorensen J., Chang Y. Y. 1981. Assimilatory and dissimilatory nitrate reduction: perspectives and methodology for simultaneous measurement of several nitrogen cycle processes. In *Terrestrial Nitrogen Cycles* (Eds. F. E. Clark, T. Rosswall). *Ecol.*

CONTROL OF DIFFUSE POLLUTION 143

Woś A., Tamulewicz J. 1996. The macroclimate of the Turew region. In *Dynamics of an Agricultural Landscape* (Eds. L. Ryszkowski, N. R. French, A. Kędziora). Państwowe Wydawnictwo Rolnicze i Leśne, Poznań: 37–44.
Zantua M. I., Bremner J. M. 1977. Stability of urease in soils. *Soil Biol. Biochem.* 9: 135–140.

CHAPTER **6**

Implementation of Riparian Buffer Systems for Landscape Management

Richard Lowrance and Susan R. Crow

CONTENTS

BUFFERS AS CRITICAL COMPONENTS OF MANAGED LANDSCAPES

In recent years efforts have been made to incorporate a landscape perspective into U.S. national policy initiatives for land management. Supporting these efforts, landscape ecology has been regarded as an effective paradigm for organizing and evaluating various approaches to land management. In part, this reflects landscape ecology's focus on applying principles derived from studying landscape elements, their interactions and changes over time, to solving practical problems in the real world. A fundamental premise of landscape ecology is that the pattern of component ecosystems or landscape elements affects ecological processes. Study of landscape

or flows of energy, materials, and species among ecosystems; and changes in landscape structure and function over time.

Within this context, landscape structure may be described as a mosaic of three elements: patch, corridor, and matrix (Forman and Godron 1986). A patch is a unit of a landscape represented by discrete areas or periods of relative homogeneity in environmental conditions, and is perceived by organisms or relevant ecological phenomenon of interest as bounded by discontinuity in environmental character. Patches are dynamic, occurring at a variety of spatial and temporal scales. Thus, a landscape is composed of a hierarchy of patch mosaics across a range of scales. Corridors are narrow landscape elements differing from their surroundings, distinguished by their linear spatial configuration in which corridor width and sinuosity are important characteristics. The landscape matrix is the most extensive and connected of the elements composing a landscape and, therefore, is functionally the most dominant landscape element. The overall pattern of landscape elements may include networks in which patch number and configuration, corridor connectivity, and boundary shapes affect flows and interactions of energy, matter, and species which is particularly relevant to riparian landscapes.

It is important to note that landscape extent, as well as the area associated with each landscape element, is dependent on the organism or ecological phenomenon of interest. Therefore, there is no single definition of landscape. Rather, it is necessary to define a landscape by delineating the land area of interacting landscape elements relevant to a particular organism or phenomenon. There are at least two important implications of defining landscape in this way. The first is that scale and context are critical concerns for landscape description and management. For example, in characterizing a particular landscape, is forest the landscape matrix or is forest an element within an agricultural matrix? A second important implication relates to the degree to which the landscape may be characterized as an open or closed system. If a watershed is defined as the landscape of interest, from a hydrologic perspective at least, the landscape may be considered a relatively closed system in which few outputs leave the system. On the other hand, the system may be characterized as relatively open in terms of species movement through the system. These implications are particularly relevant to the management of landscapes containing typical agricultural production systems because of the significant inputs and outputs associated with these systems.

Landscapes and elements within landscapes may be more or less intensely managed. In an agricultural landscape, for example, upland areas with fertile soils are intensely manipulated, whereas lowlands with hydric soils and streams may not be directly manipulated (though certainly they may receive significant inputs as a result of management of adjacent areas). In these intensely managed landscapes, some landscape elements may be regarded as critical to mitigate exports of other components. In this context, riparian environments are critical components in landscapes managed for agricultural production, as well as other intensively managed landscapes.

Riparian environments include those ecosystems adjacent to rivers and streams. In addition to extensive floodplains, riparian zones include narrow strips along downcutting rivers, islands, and channel landforms (Malanson 1993). These riparian systems represent ecotones between terrestrial and aquatic ecosystems and are characterized as landscape corridors in which lateral water flow is the main force organizing and regulating the function of riparian wetlands and their role in the landscape. These unique environments are of particular interest to land managers and policy makers because of the indirect, but critically important, economic benefit they serve in maintaining water quality. In particular, studies in agricultural landscapes have established the effectiveness of riparian vegetation in filtering agricultural nutrients and trapping agricultural sediments. U.S. agricultural policy has designated riparian buffer zones as eligible for inclusion in the U.S. Conservation Reserve Program, allowing farmers to be paid not to farm these environmentally sensitive lands if forest vegetation is regenerated either through planting or succession.

Agricultural production systems are inherently "leaky" for nutrients, sediment, and agricultural chemicals. The degree of "leakiness" is generally measured by comparing agricultural systems to nonagricultural systems, typically forested landscapes in humid regions. The leakiness is caused by numerous interacting factors, including (1) tillage to produce seedbeds and control weeds; (2) application of nutrients from fertilizer and manure; (3) application of xenobiotics to control pests; (4) hydrologic modifications of drainage systems to remove excess water; and (5) application of water to the root zone to relieve soil moisture deficits. Buffer zones, established as landscape corridors in agricultural systems and existing as integral landscape elements in agricultural watersheds and other landscapes, primarily have been implemented to control the unplanned leaks from agricultural management systems before they impact adjacent terrestrial, wetland, and aquatic ecosystems (Lowrance et al. 1997). In addition, as discussed above, buffers are needed in agricultural landscapes to serve other landscape functions such as wildlife habitat, visual screens, and as setbacks from vulnerable ecosystems.

In addition to the more conventional view of buffers as primarily linear ecosystems or ecotones juxtaposed between vulnerable ecosystems and an agricultural production system, it is also useful to consider agriculture and buffer systems in a more general way. Applying the concept of a buffer as any less intensively managed portion of a landscape producing fewer external outputs than other landscape components, agriculture may serve as a buffer. For instance, in urbanizing landscapes, agriculture may serve as a transition between high density developments with a significant percentage of impervious surfaces and nearby naturally functioning eco-

This situation is especially true in ground water-dominated landscapes where most lands are equally likely to provide recharge to ground water. In these sorts of systems, water quality problems from agricultural production systems often may be avoided by having a substantial part of the landscape in nonagricultural uses.

RIPARIAN ECOSYSTEM FUNCTIONS
IN AGRICULTURAL LANDSCAPES

Riparian ecosystems originally referred to streamside or riverside ecosystems that developed due to the influence of the stream itself. The definition has now been expanded to include almost any land adjacent to a freshwater system, including ponds, lakes, ditches, and canals. As has been documented in numerous reports, riparian ecosystems control many of the interactions among terrestrial and aquatic ecosystems. Most receive subsidies in the form of water and nutrients from adjacent uplands. Many are flooded periodically, but whether the flood acts as a subsidy or not is a function of how often it is flooded and the severity of the flood. In areas of infrequent scouring floods, the flood acts less as a subsidy and more as a perturbation to produce new flow channels and to reset both soil and vegetation succession of the systems. Although riparian systems are usually considered primarily as horizontally connected to the stream and upslope ecosystems, there are also important vertical connections with the hyporheic zone beneath the stream. This point is of particular importance in regions where the hyporheic zone consists of gravel and coarser materials and where there are large interstitial spaces for water and organisms.

Although typically referred to as ecosystems, riparian zones actually form ecotones between terrestrial and aquatic ecosystems. They may or may not be wetlands, depending on both frequency and duration of flooding and hillslope hydrology. Floodplain wetlands tend to develop in the riparian zone along larger streams and rivers, while seepage wetlands tend to develop at the base of hillslopes along smaller streams. As a result of their position in the landscape, riparian ecosystems often serve as natural buffer zones. Most riparian ecosystems serving as buffers in agricultural landscapes are functioning essentially as natural buffers with very little management.

Most riparian buffers that have been studied, and that form the basis for our knowledge of buffer system functions, are naturally occurring ecosystems that developed over time as forests were converted to agricultural landscapes (Jacobs and Gilliam 1986, Lowrance et al. 1984, Peterjohn and Correll 1984). The water quality functions of these naturally occurring buffers have been summarized in numerous reviews and are represented from most to least general in Figure 6.1 (Lowrance et al. 1997). We have just recently begun to consider both the management and re-

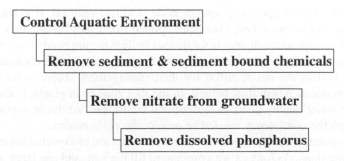

Figure 6.1   Most general to least general water quality functions.

PURPOSES OF RIPARIAN ECOSYSTEM BUFFERS

Whether we realize it or not, we expect and receive numerous goods and services from natural riparian ecosystems. We do not yet know which of these goods and services may be sustained from restored or established riparian ecosystems. Natural riparian ecosystems provide, at least, the following goods and services: retain sediment; retain dust and windblown particles, including plant propagules; retain and transform nutrients; retain and transform toxics; shade streams and margins of rivers and lakes; provide leaf litter and woody debris as energy sources; provide woody debris as substrate and channel structures; provide sources of biodiversity; increase micro-topographic variability; provide carbon sinks; provide energy sinks, especially for moving wind or water; provide sources of plant propagules; provide terrestrial wildlife habitat; provide products such as wood and forage; and provide recreation, hunting, and landscape aesthetics.

As fundamental components of agricultural landscapes, riparian buffers mitigate off-site effects of production systems, including increased nutrient, sediment, and pesticide loading. Agricultural landscapes, especially annual crops, decrease permanent cover, wildlife habitat, and biodiversity, and they create undesirable habitats for noncrop species. Buffers protect critical areas, including wetlands, recharge areas, aquatic ecosystems, and endangered species habitat, and link landscape components, providing corridors between critical areas and less disturbed areas. Buffers also may decrease the intensity of agricultural production at the edges of critical areas. Riparian buffers provide specialized features for human use in managed landscapes, including recreation, woodlots, and aesthetic experiences.

In a sense, riparian buffers serve as insurance to protect critical areas from both unexpected and expected external effects of production systems. The expected effects of production systems depend largely on the intensity of production and the extent to which production systems dominate more natural systems. As noted above, if the

would be needed for specific purposes such as nitrate removal. It is reasoned that we now have the ability to keep chemicals applied where they belong in the landscape to enhance crop production, and if we do that buffers are not needed to control these chemicals. In short, we are capable of controlling crop production chemicals on site; therefore, buffers are not required for their management. There is no doubt that keeping chemicals where they belong, in the root zone or in plants, is the best way to achieve water quality goals in agricultural landscapes. At doubt is our ability to do this with the consistency needed to protect the environment.

At this point, what are expected effects and what are unexpected effects become relevant questions. As long as we continue to till the soil, and use large amounts of nutrients and xenobiotic chemicals, and cannot control the weather, conservative management practice requires us to assume that the expected effects of agriculture will be soil, nutrient, and chemical movement significantly beyond both the background from natural ecosystems and the ability of aquatic ecosystems to assimilate and process. Given these conditions, implementing a combination of buffer systems and proper field management will provide both the best regional environmental quality and ecologically sustainable agriculture.

MAKING BUFFERS SUSTAINABLE AT MULTIPLE SCALES IN AGRICULTURAL LANDSCAPES

The goal of establishing buffer systems integrated with conservation practices in agricultural production areas is to ensure sustainable, productive landscapes. Buffers mainly contribute to ecological sustainability (Lowrance et al. 1986) but may also contribute to economic sustainability. Three biophysical factors influence the ability of buffers to contribute to ecological sustainability: size, time, and management. The ecological sustainability of the buffer systems, and in turn the ecological sustainability of the agricultural landscape, ultimately will be governed by properties of the buffer systems interacting with properties of the production system.

Given the linear nature of riparian ecosystems, size involves both the linear extent along a stream and the width of the buffer relative to specific pollutant sources. Size matters in riparian ecosystems. Although it is clear that a continuous buffer is best, it is not clear that there is an optimum buffer width, given the opportunity costs associated with buffers relative to other more economically important land uses. Simulation studies using the Riparian Ecosystem Management Model have shown that most sediment and nutrient removal occurs in about 15m riparian buffer systems in the U.S. Coastal Plain (Williams, et al. 2000). The generality of these results needs to be tested in other regions. In the absence of government subsidy, optimum buffer width, although lacking more specific definition, is the minimum width that

streamside fields, establishing a buffer will be a very large change in chemical loading, tillage, etc., and establishment of buffer functions should be rapid relative to the agricultural uses. Even in areas of surface drainage with ditches, providing coarse woody debris to the stream in the first few years of buffer establishment should have effects on flow regimes. The three agricultural scenarios most likely to need longer time to re-establish buffer functions are areas with extensive subsurface drainage, areas with intensive streamside use by domestic animals, and steep areas with eroded soils.

Managing riparian buffers for multiple benefits is a new concept in modern production agriculture. Although one approach to buffer design for pollution control allows intensive management to be substituted for size, this may not be a realistic approach in practice. Although many farmers may spend noncritical times (especially winter) managing buffer areas for multiple benefits (especially as wood lots and wildlife habitat), managing buffers for pollution control during the growing season will not be a high priority for most farmers. Active growing-season buffer management is probably not a realistic expectation for most modern farmers. Therefore, buffers should be largely self-regulating, be highly resistant to failure during the growing season, and be suitable for largely nongrowing season management.

To date, riparian buffers have been implemented largely on a field-by-field basis. Future efforts should establish and evaluate riparian buffers at multiple spatial scales. At the field scale, it is possible to adjust inputs and management to gain desired outputs. Where there are likely to be greater outputs from a given field, for instance an area in intensive vegetable or animal production, the farmer may implement a larger riparian buffer to interact with in-field buffers such as contour filter strips or grass waterways. At the farm-scale, the goal should be to integrate production components where possible — rotations, animal manures, agroforestry — to buffer critical areas on and adjacent to the farm. At the landscape scale, practices should be coordinated to buffer and protect critical areas and link habitat resources with buffer systems serving as landscape corridors. In most cases, it will be possible to restore critical areas only through such coordinated efforts.

USDA PROGRAMS FOR RIPARIAN ECOSYSTEM BUFFERS

Riparian buffers are an important part of the USDA Conservation Buffer Initiative (CBI), intended to have 2 million miles of total buffers installed by the year 2002. Although the CBI includes many types of buffers, the focus of CBI financial incentive programs has been to provide financial support for long-term riparian forest buffers through the USDA Conservation Reserve Program (CRP) and the Conservation Reserve Enhancement Programs carried out as state/USDA partnerships. Riparian

riparian forest buffers provides a one-time sign-up bonus and increased cost-share for installation of the practice. With the extra incentives, there is very little initial cost to the landowner.

A comparison of the use of riparian forest buffers under CS-CRP in two states provides insights into both the need for riparian forest buffers and the effectiveness of USDA financial incentives for them. Iowa has extensive acreage of cropland along streams and rental rates for CS-CRP land is from $150 to $175 per acre per year. Georgia has much less row cropland along streams, more pastureland along streams, and lower rental rates for riparian lands, typically $40 to $50 per acre per year. Rental rates for both states are based on soil productivity for typical riparian soils. The difference in rental rates is reflected in dramatic differences in both the overall use of riparian forest buffers under CS-CRP and the distribution of new riparian forest buffers under CS-CRP within the two states. From October, 1999 through April, 2001, over 22,500 acres of riparian forest buffer were installed under CS-CRP in Iowa (USDA-NRCS-PRMS 2001). In contrast, only 1586 acres were installed under CS-CRP in the same time period in Georgia (USDA-NRCS-PRMS 2001).

The distribution of CS-CRP riparian forest buffers in the two states is very different also. In Iowa, CS-CRP riparian forest buffers are being installed in the majority of counties. Although counties in the more humid eastern part of the state have higher acreages, the practice is widespread under CS-CRP (Figure 6.2). In Georgia, the practice is much less widespread geographically with the limited acreage concentrated in a few counties (Figure 6.2). Of the total CS-CRP riparian forest buffer acreage in Georgia, 70% has been installed in five counties (out of 163 counties total) with 38% in one county.

Although the CS-CRP recognizes the value of riparian buffers by providing for continuous sign-up, the rental rates are tied to the agricultural productivity of the land based on the soil type. This is a relatively successful approach in cases where riparian soils are supporting productive croplands because the rental payment is relatively high. However, where riparian soils are supporting pastures, and the soil productivity is generally low, the program has not been as successful in re-establishing riparian buffers. Yet many streams in Georgia are impaired due to low dissolved oxygen and high fecal coliform levels (SRWMD, 2000). Although cause and effect are not well understood, both low dissolved oxygen and high fecal coliform may be related to streamside pastures. These conditions indicate that some important benefits to society are not being realized in CS-CRP riparian forest buffers on pasturelands, and they suggest that adjusted rental rates might better capture the overall societal values of riparian forest buffers. At this point, one of the most pressing concerns for CS-CRP is to acquire more riparian pastureland for permanent riparian forest buffers.

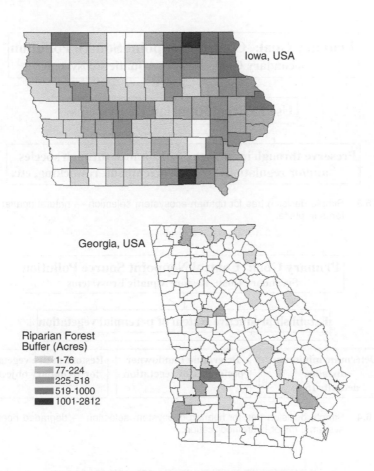

Figure 6.2   Comparison of Georgia and Iowa riparian forest buffer acres enrolled in the USDA
Conservation Reserve Program from October 1999 to April 2001.

here to illustrate possible interactions between riparian ecosystem conditions and
human goals to guide buffer design at a site or field scale.

If riparian systems are largely intact, the goal will be to use and enhance existing
riparian ecosystems (Figure 6.3). Unfortunately, incentives generally are not available to preserve existing riparian ecosystems. A few regulations exist, especially for
wetlands, but programs are needed to reward landowners for maintaining riparian
systems. In many cases, riparian systems could be enhanced for multiple purposes
via species enrichment, restocking, etc. However, very limited financial incentives
are available for these practices.

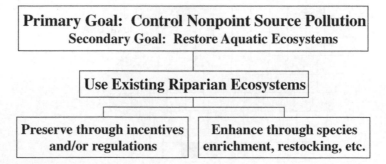

Figure 6.3   Simple decision tree for riparian ecosystem selection — natural riparian ecosystems in place.

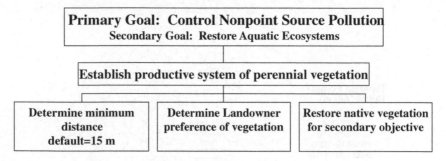

Figure 6.4   Simple decision tree for riparian ecosystem selection — degraded riparian ecosystems, minor hydromodification.

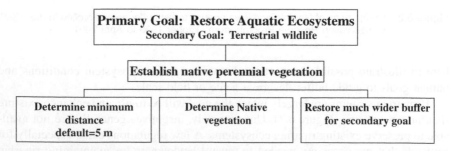

Figure 6.5   Simple decision tree for riparian ecosystem selection — degraded riparian ecosystems, minor hydromodification.

tested. Practices implemented beyond the minimum distance for cost-share programs

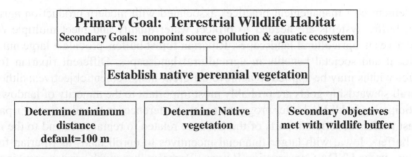

Figure 6.6   Simple decision tree for riparian ecosystem selection — degraded riparian ecosystems, minor hydromodification.

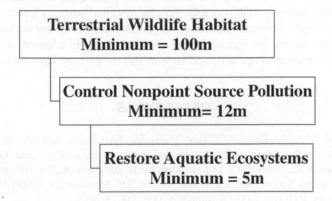

Figure 6.7   Hierarchy of widths, based on minimal science.

needed. Alternatively, if restoring terrestrial wildlife habitat is the primary goal, it is likely that all other objectives will be met as secondary goals (Figure 6.6).

The decision tree in Figure 6.7 reflects a hierarchy of buffer widths for achieving different riparian ecosystems functions. Although not well tested, the hierarchy is being implemented in USDA programs. Overall experience has shown that most landowners establishing and maintaining riparian buffers in agricultural landscapes are interested in buffers much wider than the minimum requirement. In general, the first landowners participating in these programs are ones with broader stewardship goals and desire to provide for terrestrial wildlife needs, as well as nonpoint source pollution control and aquatic ecosystem restoration. When wildlife management goals are met, it is almost certain that pollution and aquatic restoration goals also will be met, if they can be met. In reality, the nonpoint source pollution goal, especially relative to nitrate and dissolved phosphorus, probably is less certain to be met and depends on hydrologic flow paths that are unknown for most systems.

implement best management practices associated with modern production agriculture, buffer systems are needed to protect water quality and meet multiple other objectives in agricultural landscapes. Riparian forest buffers provide a large number of local and societal benefits in agricultural landscapes. Different riparian forest buffer widths may be applied to achieve different management objectives, although overall stewardship goals are probably most important to the majority of landowners participating in USDA buffer programs today. Differences in the adoption of riparian forest buffers in different parts of the U.S. are related to rental rates and to the need for buffers. Iowa, with larger financial incentives to enroll lands as riparian forest buffers in the USDA Conservation Reserve Program, has much greater participation than Georgia, a state with lower financial incentives in the same program. It is likely that rental rates based on soil productivity do not capture the entire suite of societal values derived from riparian forest buffers on marginal pastures and lower productivity soils. Larger financial incentives are needed to enroll riparian lands that are not highly productive for agriculture into the programs. In addition, new incentive programs are needed to reward good stewardship of existing riparian buffers.

REFERENCES

Barrett, G.W. 1992. Landscape Ecology: Designing Sustainable Agricultural Landscapes, pp. 83–103, in *Integrating Sustainable Agricultural, Ecology, and Environmental Policy*, R.K. Olson (Ed.). The Haworth Press, Binghamton, NY. 161 pp.

FSA. 1997. Farm Service Agency Handbook 2-CRP (Revision 3). USDA-Farm Service Agency, P.O. Box 2415, Washington, D.C. 20013-2415.

Forman, R.T.T. and M. Godron. 1986. *Landscape Ecology*. John Wiley & Sons, New York. 619 pp.

Jacobs, T.J. and J.W. Gilliam. 1985. Riparian Losses of Nitrate from Agricultural Drainage Waters. *J. Environ. Quality* 14:472–478.

Lowrance, R., R. Todd, J. Fail, Jr., O. Hendrickson, Jr., R. Leonard, and L. Asmussen. 1984. Riparian forests as nutrient filters in agricultural watersheds. *BioScience* 34: 374–377.

Lowrance, R., P.F. Hendrix, and E.P. Odum. 1986. A hierarchical approach to sustainable agriculture. *Am. J. Alt. Agric.* 1:169–173.

Lowrance, R., L.S. Altier, J.D. Newbold, R.R. Schnabel, P.M. Groffman, J.M. Denver, D.L. Correll, J.W. Gilliam, J.L. Robinson, R.B. Brinsfield, K.W. Staver, W. Lucas, and A.H. Todd. 1997. Water quality functions of riparian forest buffers in the Chesapeake Bay Watershed. *Environ. Manage.* 21:687–712.

Malanson, G.P. 1993. *Riparian Landscapes*. Cambridge University Press, Cambridge, U.K. 296 pp.

Peterjohn, W.T. and D.L. Correll. 1984. Nutrient dynamics in an agricultural watershed: Observation on the role of a riparian forest. *Ecology* 65:1466–1475.

Schultz, R.C., J.P. Colletti, T.M. Isenhart, W.W. Simpkins, C.W. Mize, and M.L. Thompson.

Welsch, D.J. 1991. *Riparian Forest Buffers*. USDA-Forest Service Publ. No. NA-PR-07-91. Radnor, PA. 20 pp.

Williams, R.G., R. Lowrance, and S.P. Inamdar. 2000. Simulation of nonpoint source pollution control using the riparian ecosystem management model. In *Proceedings Riparian Ecology and Management in Multi-Land Use Watersheds*. American Water Resource Association, Middleburg, VA, pp. 433–438.

CHAPTER 7

Dissolved Organic Substances in Water Bodies of Agricultural Landscapes

Barbara Szpakowska

CONTENTS

INTRODUCTION

of nitrates and pesticides are most frequently subjected to analysis since the presence of these components in water supplies constitutes a direct threat to human health. Long-term studies on landscape ecology carried out by the Research Centre for Agricultural and Forest Environment in Poznań have indicated that the spatial pattern of landscape elements can modify migration of mineral components. By restructuring uniform field landscapes into mosaic ones through installing biogeochemical barriers such as shelterbelts, meadows, small ponds, and other nonagricultural components, it is possible to decrease the output of phosphates and nitrates from a watershed (Ryszkowski 1994, Ryszkowski et al. 1997).

On the other hand, there is little data on the leaching and migration of dissolved organic matter (DOM) in waters situated in agricultural areas. Recent studies indicate a quite disturbing phenomenon of growing concentrations of DOM in ground and surface waters. DOM ought to be analyzed not only because of elevated concentrations in water but also because the organic compounds, mainly humic substances (HS), are able to bind and distribute various toxic organic and inorganic compounds. In the presence of HS, precipitation of some nutritionally important anions, such as phosphates, is delayed or modified (Jansson 1998). Dissolved HS are also carriers of toxic organic compounds such as aromatic polycyclic hydrocarbons with carcinogenic activity. These compounds are adsorbed on humus surfaces or react with humus (Meyer 1990). Binding of metal ions by HS through the process of complexing or chelating seems to be one of the important properties of HS. As a result of this process, the geochemical mobility of metals, particularly of heavy metals, is changed. HS binding of heavy metals into metal-organic complexes increases their solubility, which results in higher amounts of heavy metals dissolved in water. In the presence of HS, toxicity of trace heavy metals decreases and availability of trace elements necessary for proper functioning of water organisms increases.

TERMINOLOGY AND METHODOLOGY

Organic substances in water can be divided according to fraction of dissolved organic carbon (DOC) and particulate organic carbon (POC). DOC and POC may be used to monitor seasonal inputs of plant and soil organic matter to streams and rivers. For example, the melting snows of spring leach carbon into water, increasing DOC. During the summer, algal growth combined with decreased discharge increase the concentration of DOC (Thurman 1985). DOC is chemically more reactive because it is a measure of individual organic compounds in a dissolved state, while POC is both discrete plant and animal organic matter and organic coatings on silt and clay.

The main obstacle encountered in studies on organic substances dissolved in

is determined by oxidation to carbon dioxide and by measurement of carbon dioxide by infrared spectrometry.

Decomposition of DOC, principally by microorganisms, is a dynamic and important process in aquatic ecosystems. In the case of POC, both animal decomposition and microbial decay occur simultaneously (Thurman 1985). Organic matter dissolved in water constitutes 60 to 90% of organic substances in natural waters.

SOURCE AND ORIGIN OF DISSOLVED ORGANIC SUBSTANCES

Organic substances dissolved in water can be categorized according to origin: autochtonous (e.g., originating in the aquatic environment) and allochthonous (e.g., from land, or outside the aquatic system). There are two sources of allochthonous organic carbon: soil and plants.

In the aquatic environment algae are the major source of autochthonous organic matter. Algae do excrete DOM, probably as a result of cellular destruction or lysis, and this DOM is actively metabolized by heterotrophic bacteria. Another important source of autochthonous organic matter is excretion of organic compounds by animals. Ruptured cells of algae or other aquatic organisms release compounds that undergo further biogeochemical degradation and mineralization. Some of them, as a result of condensation, form macromolecules. High molecular compounds of autochthonous origin are mainly polysaccharides and polyamides, a product of phytoplankton decomposition (Meyer 1990).

Streams and rivers have allochthonous organic matter from both plants and soil as a major source of their organic matter. Lakes contain considerable amounts of autochthonous organic matter which may range from 30% for lakes with inputs from rivers flowing through and high flushing rates, to nearly 100% for eutrophic lakes fed by ground water. The larger the lake the more important autochthonous input of organic matter is thought to be (Thurman 1985).

CONCENTRATIONS OF DISSOLVED ORGANIC SUBSTANCES IN WATERS

Concentrations of dissolved organic substances in ground waters change significantly. In some ground waters, concentrations of DOM are below 2 $mg \cdot dm^{-3}$ (Gaffney et al. 1996). Concentrations of DOM in ground waters are small because those compounds are quickly decomposed. Organic carbon is a good energy supply for heterotrophic microbes in the ground water. Moreover, organic carbon adsorbs

Table 7.1   Estimation of DOM Concentrations in Waters of Agricultural Area
in the Turew Landscape [mg·dm⁻³]

Period	Ecosystem	Mean Concentration	References
Ground Water			
1987–1989	Under field	22.4	Szpakowska and
	Under meadow	31.1	Życzyńska-Bałoniak (1994)
	Under meadow on slope	41.4	
	Under shelterbelt	31.2	
1990–1991	Under field	43.6	Szpakowska (1999)
	Under meadow	28.9	
	Under meadow	24.3	
1993	Under field	34.2	Szpakowska and
	Under meadow	44.7	Życzyńska-Bałoniak (1996)
1995	Under field	10.5	Szpakowska and
	Under meadow	19.4	Karlik (1996)
1996	Under field	18.3	Karlik et al. (1996)
1998	Under field	10.5	Karlik and Szpakowska (2001)
Surface Water			
1987–1989	Pond	47.6	Szpakowska and Życzyńska-Bałoniak (1994)
1990–1991	Canal water (1.5-m width)	48.7	Szpakowska (1999)
1993–1994	Drainage canal	48.7	Szpakowska (1999)
	Pond	46.1	
1995	Lake water	22.7	Szpakowska and
	Mid–field pond	42.6	Karlik (1996)
	Model reservoir	19.8	
	Drainage canal	24.8	
	Melioration ditch	51.2	
1996	Drainage water	17.7	Karlik et al.(1996)
	Pond	19.7	
1997	Pond	26.0	Karlik and Szpakowska (1997)
1998	Small pond	38.2	Karlik and Szpakowska (2001)

soils, concentrations of DOM are also very high, confirming phenomena observed
by other authors (Meili 1992). The study area of the Turew landscape was an
agricultural drainage basin with an area of 11 ha, divided into two parts: area A
included a small, field pond, surrounded by a narrow belt of meadow where wells
were installed, and area B included a drainage canal, starting from the pond in area
A, and a wide belt of meadow separating the canal from cultivated fields. The soils,
consisting of hapludalfs originating from boulder clay (rich in calcium carbonate)

Surface waters of the Turew landscape also contained high concentrations of DOM (Table 7.1). Mean concentration of DOM in water of the mid-field pond in dry years amounted to 47.6 mg·dm^{-3}. Strong influence of precipitation on DOM concentration in the pond water was observed ($r = 0.44$; $p = 0.04$), which indicated the input of DOM from surrounding areas. Flushing of soil and plant organic matter by precipitation caused the observed increase in DOM. Moreover, high concentrations of DOM in analyzed water were due to the location of the pond in the natural depression into which surface runoff was collected. Additionally, increased DOM concentrations may be caused by decomposition of dead plants in the pond, as well as excretion of DOM by living macrophytes. According to Meyer (1990), excretion of DOM by living macrophytes (often neglected in quantitative estimations of DOM) is substantial. Also, Rutherford and Hynes (1987) noted increase of DOM concentrations in waters situated in agricultural areas as an effect of intensive primary production of algae. Influx of DOM to the pond through ground water can also be an important element in the total carbon budget. Inflow of ground waters can modify DOM concentrations in surface waters by dissolution of DOM. Removal of organic substances from surface waters as suspended sediment also occurs (Rutherford and Hynes 1987).

In agricultural areas drainage channels also contain high concentrations of DOM (Table 7.1). The average concentration of DOM in a small channel situated in the examined area was 48.7 mg·dm^{-3}. Seasonal concentrations of DOM varied greatly. In summer this concentration rose to 104.3 mg·dm^{-3}, whereas during spring it declined, which can be related to a dissolution effect from intensive spring rainfall and snowmelt. Maximum concentrations of DOM that occurred in the summer may result from decomposition of biomass, which is most intensive in summer. Other authors also indicated a strong relationship between precipitation and DOM concentration. Lush and Hynes (1978) noted that not only rainfall but also snow, thaw, and leaf input may be of some importance to organic carbon budgets. Moreover, aquatic macrophytes, which consist mainly of cellulose, are degraded and in autumn contributed to the increase of DOM concentrations. Release of organic carbon by macrophytes and algae caused a high concentration of DOM, which is well documented in the literature (Meyer 1990).

Huge fluctuations in concentration of DOM observed in these investigations in winter can be related to freezing and thawing. Frozen soil hinders quick DOM inflow, and ice cover limits microbial activity. Irregular melting processes intensify influx of DOM and occasionally increase DOM concentrations. Apart from microbial activity having strong influence on DOM concentrations, Muenster (1992) also found adsorption of organic compounds on clay minerals when there is slow inflow and weak water mixing. Because of accumulation of organic compounds in the drainage channel, their free migration in the agricultural area is hindered, and this process plays an important role in matter cycling and transformation.

of dissolved organic carbon (mean 22.5 ± 9.6 mg·dm^{-3}), probably the highest among the retention reservoirs in Poland that are situated in less modified forest-meadow catchment. During the year the highest DOM concentrations were observed in May, and concentrations then gradually decreased to a low in winter. The amounts of DOM in this water were directly proportional to the flushing effect and were inversely proportional to the rate of water exchange in the reservoir.

Comparisons of surface, subsurface, and ground waters in south-central Minnesota, done by Maier et al. (1976), show that organic C concentrations are highest in surface waters. During May and June, organic C levels in river water were 5 to 17 mg·dm^{-3}, whereas subsurface water had 1 to 6 mg·dm^{-3} organic C level and ground water had less than 2 mg·dm^{-3}. These results indicated that organic materials were decomposed or retained in the upper layers of the soil, and that residence time in the aquifer facilitates removal of trace organics by geochemical and biochemical processes. The most likely explanations for the presence of high organic C concentrations in surface waters are carbon's solubilization from humified soil organics and biodegradation of vegetation and animal tissues in stream channels or at the soil surface.

HUMIC SUBSTANCES IN AQUATIC ENVIRONMENTS

The main fraction of dissolved carbon in waters (50 to 75%) constitutes aquatic humic substances (HS) (Thurman 1985, Szpakowska 1999). One of the difficulties encountered in the study of HS is terminology. HS do not conform to a unique chemical entity and cannot be described in unambiguous structural terms. As a result, HS are described operationally as a mixture of such organic acids as fulvic and humic acids (Thurman et al. 1988). These terms originate in the field of soil science and determine the part of soil organic matter extracted by severe chemical treatment. It seems that it would be more reasonable to classify these aquatic materials based on chromatographic methods used to extract them from water.

In recent years, it has been found that high recoveries of HS from water are possible with nonionic macroporous sorbents such as the Amberlite XAD resin series. By these resins, HS can be easily extracted from any aquatic sample: from waters with very low HS values, such as sea water or ground water, as well as from more concentrated systems (Thurman et al. 1988).

AMOUNTS OF HUMIC SUBSTANCES IN GROUND WATERS

There are relatively few studies on HS in ground water (Kramer et al. 1990, Hessen and Tranvik 1998)). Low concentration of HS in ground waters (usually less

contain high concentrations of HS (Szpakowska and Życzyńska-Bałoniak 1989, Szpakowska 1999).

On the basis of rank Spearman correlation, a strong dependence of HS concentration on rainfall was indicated. It was particularly clear in ground water under fields ($r = 0.45$; $p = 0.04$). The highest HS concentrations were found in ground water under fields in summer months both in 1990 (29.8 mg·dm^{-3}) and 1991 (44.5 mg·dm^{-3}) (Figure 7.1). Also during these periods very high concentrations of HS in ground water under meadow were observed; HS concentration was 26.0 mg·dm^{-3} in 1990, and 36.9 mg·dm^{-3} in 1991. Leaching of organic compounds from surrounding terrain and litter during the spring flush was also observed by other authors (Thurman 1985, Easthouse et al. 1992). High concentrations of HS were observed in ground water after thaws and heavy rainfall. The highest HS values appeared in ground water not immediately after rainfall but after 2 months delay. This flush carries HS into the saturated zone of ground water and is a major contributor of organic carbon to ground water (Szpakowska 1999).

Górniak (1996) studying ground waters found fluctuations in HS concentrations during seasonal sampling. The autumn-winter period was characterized by huge variability in concentrations, which was connected with a change in thermal regime and in amounts of rainfall. According to Górniak (1996), the highest HS concentrations in ground waters were observed in spring. Investigations on seasonal variations of organic substances concentration are very important for aquatic ecosystems, particularly when arable fields are close to surface waters that are recharged by ground waters. However, ground water–surface water interactions are rarely quantified, probably because of perceived difficulties in measuring seepage rates. In addition, many factors affect ground water-surface water interaction, e.g., fluctuations in hydraulic gradients near the surface, watershed morphometry, and heterogeneity of porous media near the surface.

AMOUNTS OF HUMIC SUBSTANCES IN SURFACE WATERS

Small ponds and drainage canals situated in agricultural areas are poorly recognized surface reservoirs — there is a lack of water supplies assessment. Their estimation is also difficult because of the astatic character of small ponds. Examining surface waters in the Wielkopolska Lowland high concentrations of HS in the midfield pond were observed. These concentrations were 91% higher than in ground water under meadow in close vicinity of the pond and 78% higher than in ground water under field. Mean concentration of HS in the mid-field pond during the time of investigations was 34.0 mg·dm^{-3}, which accounted for 71.4% of DOM. Using rank Spearman correlation indicated that not only the concentration of DOM but also HS were dependent on precipitation ($r = 0.53$; $p = 0.04$) (Figure 7.2). The highest HS

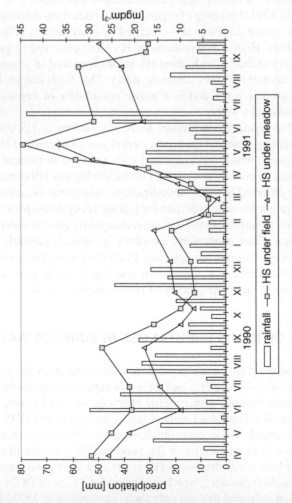

Figure 7.1 Seasonal variation in concentrations of humic substances in ground water under field and under narrow belt of meadow.

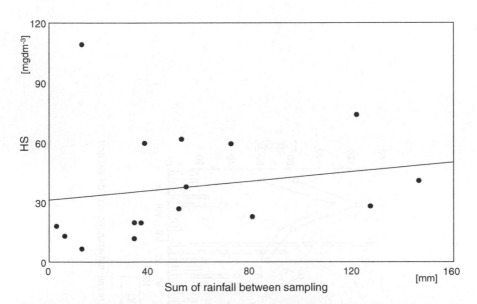

Figure 7.2   Relationship between humic substance concentrations in surface water of pond
and intensity of rainfall in dry years. (After Szpakowska 1999.)

dynamic relationship between both variables. It was found that using a linear model,
one can predict the effect of rainfall on HS concentration. The measured HS concentrations in water of the pond during 2 years, expressed as a variable in this mode,
were similar to those estimated theoretically (Figure 7.4), which can be essential for
prognosis of system behavior, particularly changes taking place in ecosystems
(Kundzewicz et al. 1994).

In comparing the concentration of HS from the analyzed pond with water examined earlier from a eutrophic lake situated in the same area (Życzyńska-Bałoniak
and Szpakowska 1989), much higher concentrations of HS were found in the small
mid-field pond. In the lake water, maximum HS concentration was 25 mg·dm^{-3}.
Although this lake has a broad littoral zone, the input from the littoral zone decreases
when lake size increases.

In water of the mid-field pond, relatively high concentrations of organic substances, particularly HS, were observed. HS do not undergo rapid changes, and their
increased concentrations in the pond water can be interpreted as a result of accumulation. Because HS are resistant to microbial and chemical decomposition, they
accumulate even in aerobic conditions (Gaffney et al. 1996). High increases in HS
concentrations in water from the pond were observed, particularly in years with low
intensity of rainfall. After drought, precipitation flushes HS into ground water, and,
therefore, there is a lag in response between the rainfall and appearance of HS in

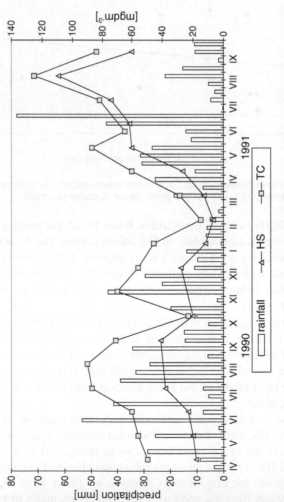

Figure 7.3   Seasonal variation in concentrations of total carbon and humic substances in mid-field and pond water in dry years.

DISSOLVED ORGANIC SUBSTANCES IN WATER BODIES   169

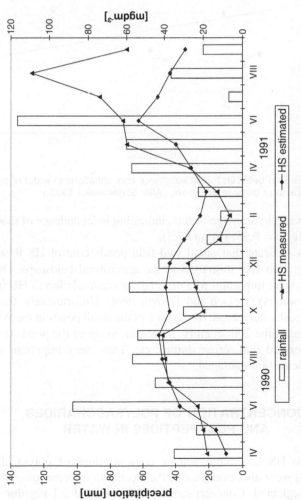

Figure 7.4 Simulation of concentrations of humic substances in mid-field pond water. (After Szpakowska 1998.)

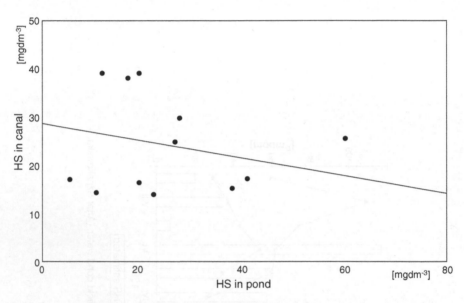

Figure 7.5   Relationship between humic substance concentrations in water of mid-field pond
and drainage canal in dry years. (After Szpakowska 1999.)

between HS in pond and drainage waters, indicating independence of these analyzed
waters ($r = -0.016$; $p = 0.96$) (Figure 7.5).

These studies indicate that small mid-field ponds control HS flow and are a
barrier preventing their free migration in the agricultural landscape. Other studies
on HS also indicate an important role of ponds in accumulation of HS (particularly
well-oxygenated waters) (Hessen and Tranvik 1998). Unfortunately, the number of
ponds has dramatically decreased. About 80% of the small ponds in the Wielkopolska
region disappeared in the 20th century. Moreover, many of the ponds that still exist
have been transformed into refuse dumps, etc. Thus, very important elements of
agricultural landscape are damaged.

CONCENTRATION OF POLYSACCHARIDES
AND POLYPEPTIDES IN WATER

In addition to HS in ground waters, concentrations of polysaccharides and
polypeptides (PP) were also examined. PP concentrations were lower than HS, but
they cannot be neglected. Concentrations of PP averaged 2.1 mg·dm^{-3} in ground
water under field and 5.2 mg·dm^{-3} in ground water under meadow where roots
reached the water table. The difference between these fractions in both analyzed

time of organic compounds in ground water. Concentrations of PP do not vary seasonally. There are relatively few studies on PP in ground water.

Blooming of phytoplankton in surface waters has a strong effect on PP concentrations (Sundh and Bell 1992). In previous studies (Życzyńska-Bałoniak and Szpakowska 1989) on dissolved organic compounds in a eutrophic lake situated in an agricultural landscape also observed variability of PP. After phytoplankton bloom the highest PP concentrations, up to 95 mg·dm⁻³ (average 27.0 mg·dm⁻³), were noted. In the studies presented, no pronounced changes in PP concentrations were found. The greatest concentrations of PP amounted to 27.1 mg·dm⁻³, with an average of 8.2 mg·dm⁻³, and this constituted 17.2% of DOM. The pond bottom was covered by vascular plants — up to 90% *Ceratophyllum submersum* L. It is known that if reservoirs are dominated by such macrophytes, phytoplankton appears in very low amounts. A similar phenomenon was observed in this mid-field pond.

Mean concentration of PP in drainage water during wet years was much higher (11.9 mg·dm⁻³) in comparison with this fraction concentration in the dry season (3.7 mg·dm⁻³). Usually concentrations of PP do not exceed 10% of DOM (Thurman 1985). These substances easily undergo enzymatic degradation. The enzymatic hydrolysis of polysaccharides releases simple monosaccharides and oligosaccharides that are quickly utilized by microorganisms, and therefore concentrations of simple carbohydrates are very low.

EFFECT OF MEADOW WIDTH ON HUMIC SUBSTANCES CONCENTRATION IN DRY YEARS

Studies of ground waters flowing through the 5- to 9-m belt of meadow carried out on area A indicated that the composition of organic compounds is altered during filtering of ground water from field through narrow meadow. There was a twofold increase in PP concentration with the flow of ground water through meadow. At the same time, a 7% fall in HS concentration in ground water under meadow was observed (Table 7.2) but the difference was not statistically significant, which suggests

Table 7.2   Changes in Humic Substances (HS) Concentrations in Ground Waters According to Width of Meadow Belt in Dry Years (1990–1991)

Area of Study	Ground Waters	Concentrations of HS [mg·dm⁻³]	Changes of HS in %	Significant Differences (on the basis of Wilcoxon test)
Area A (narrow belt of meadow 5–9 m)	Under field	19.1		N = 20, Z = 1.49, p = 0.14
	Under meadow	17.8	7% decrease	Difference not statistically significant

that the narrow belt of meadow was not effective in reducing HS migration from fields. The proportion of the concentration of HS to PP in the ground water under the field was different (9.1:1) from that in the ground water under meadow (3.4:1).

Studies carried out simultaneously on area B indicated that the wide belt of meadow had a strong effect on ground water chemistry. With the ground water flow through the 20-m belt of meadow, concentrations of DOM were reduced by 45%. A similar phenomenon was noted for concentrations of HS migrating with ground water from under field through the wide meadow. This decrease averaged 52% (Table 7.2). This decrease is probably related to the longer time of microbial activity and greater effect of biodegradation of organic compounds. Moreover, less leaching of HS from the upper part of meadow soil is possible and can be related to slower velocity of water flow through rusty soils.

From the studies we can conclude that the narrow belt of meadow separating arable fields has an effect on composition of DOM but was not effective in limiting free migration of HS in the years with low intensity of rainfall. The wide, 20-m belt of meadow is much more effective in reducing transport of chemical compounds, particularly HS and DOM. Thus, especially in dry years, a wide belt of meadow can be an important biogeochemical barrier for controlling the spreading of chemical components from cultivated fields in ground water. Taking into consideration the disturbing phenomenon of increasingly hard-to-control nonpoint source pollution, appropriate planning of agricultural environments should include the formation of such biogeochemical barriers as meadow enclaves or small field ponds.

EFFECT OF MEADOW WIDTH ON ORGANIC SUBSTANCES CONCENTRATION IN WET YEARS

The present studies indicated that there were no pronounced changes in organic compounds after ground water passed across a narrow or wide belt of meadow in wet years. The rainfall intensity in 1993 was different. The first half of 1993 had very low precipitation (total for the period was 218 mm) and a 23% reduction of HS in the ground water under the narrow meadow, as compared to the ground water under field, was observed. A more pronounced decrease in HS concentration was observed in the first half of 1993 after ground water flow across a wide 20-m meadow belt. The fall in HS concentrations was on average 35% (Table 7.3). The second half of 1993 had significantly more precipitation. Total precipitation in the second half of 1993 was 190% higher than in the first half, and there was a 29% rise in HS concentration in ground water under the narrow belt of meadow, and a 60% rise in HS concentrations in ground water under the wide belt of meadow, as compared to those in ground water flowing from under the field. Rainfall intensity was similar

DISSOLVED ORGANIC SUBSTANCES IN WATER BODIES

173

Table 7.3 Changes in Humic Substances Concentration in Ground Waters According to Width of Meadow Belt in 1993

Area of Study	Intensity of Rainfall	Ground Waters	Concentrations of HS [mg·dm⁻³]	Changes in HS in %	Significant Differences (on the basis of Wilcoxon test)
Area A (narrow belt of meadow 5–9 m)	218 mm (I–VI)	Under field Under meadow	18.7 14.4	23% decrease	N = 5, p = 0.001 Difference statistically significant
	635 mm (VII–XII)	Under field Under meadow	11.5 14.9	29% increase	N = 5, p = 0.47 Difference statistically insignificant
Area B (wide belt of meadow 20 m)	218 mm (I–VI)	Under field Under meadow	31.9 20.7	35% decrease	N = 5, p = 0.001 Difference statistically significant
	635 mm (VII–XII)	Under field Under meadow	25.2 40.3	60% increase	N = 5, p = 0.01 Difference statistically significant

Source: Szpakowska 1999.

found that the decrease in these concentrations depends not only on the width of meadow but also on precipitation (Table 7.3). A more significant decrease in organic compounds was observed in a dry period in ground water situated on area B with a 20-m belt of meadow. It was shown that in dry months the washing of organic compounds from arable fields and post-bog soils decreases, whereas in wet months the outflow of both DOM and HS from soils is much higher.

According to Gron et al. (1992), about 22% of dissolved organic carbon undergoes biodegradation in ground water, and degradation of the hydrophilic fraction is even higher (27%). Inflow of ground waters can modify the concentration of organic compounds in surface waters, particularly with HS, which come, for example, from humification processes taking place in soil. Soil temperature, humidity, and aeration can also be important elements having an effect on the dissolution rate and degradation of other soil components.

EFFECT OF HUMIC SUBSTANCES ON HEAVY METALS SPECIATION

Many chemical and biological properties of metals in natural waters can hardly

species (such as carbonate), complexes of metal with macromolecular ligands (such as humic acids), and metal ions adsorbed on a variety of colloidal particles (for example, clay minerals). Differentiation of these forms is very important because free ionic metal is the form responsible for the metal's biological effect, whereas total metal concentration provides no information on its toxicity and prospect for uptake because a mix of various speciation forms often has contrasting properties. For example, it is generally accepted that ionic copper is much more toxic than copper in a form bound to organic compounds (Tubbing et al. 1994).

Organic compounds, mostly HS, are particularly important in speciation of metals. Because of the presence of functional groups such as carboxylic and hydroxylic, these compounds can bind large amounts of cations, modifying their availability to water organisms. Generally, there are two directions in analysis concerning speciation of metals. The first — analytical — is directed to selection of an appropriate method or technique of determination, while the second — functional — deals with the problem of explaining binding and transport of metals in the environment. Such an approach requires parallel application of separation techniques: chromatographic separation membrane techniques and nonseparation ones — ion-selective electrodes or anodic stripping voltammetry (DPASV). The DPASV method enables discrimination between electroactive forms of metals (hydrated ions, labile forms) and those that are electro-nonactive (organic complexes and colloidal mixtures) (Nürnberg 1981, Szpakowska and Karlik 1996). Determination of heavy metal forms by DPASV is possible even when they appear in very low concentration (0.05–1.0 ppb) in samples containing high concentrations of other salts, e.g., calcium or magnesium (Mart et al. 1978). Modification of the DPASV method done by Szpakowska and Karlik (1996) was the basis for presenting the analytical scheme for chemical speciation of heavy metals in natural waters.

Because humic substances are a mixture of compounds it was necessary to separate them into humic and fulvic acids for further analysis. This separation was done by acid precipitation at pH 2. The results indicated that fulvic acids dominated (62.7%) in surface water, whereas in ground waters these acids constituted only 28.8% (Figure 7.6).

Many more humic acids, which have strong chelating properties, were found in ground waters (71.2%) than in surface water (37.3%). In waters of the Turew landscape (particularly in the mid-field pond) situated in agricultural areas, an increase in humic acid concentrations was observed in early spring. This increase can result from freezing in winter. During freezing, when ice cover is formed, a rise in HS concentrations was observed. Increased humic acid concentration in spring in dystrophic lakes was also observed by Górniak (1996).

TOTAL CONCENTRATION OF HEAVY METALS

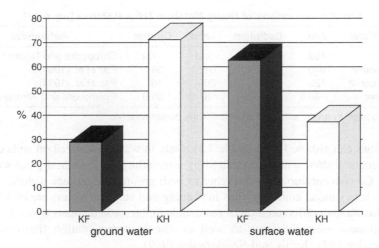

Figure 7.6   Percent of humic acids and fulvic acids in ground and surface waters.

Table 7.4   Total Concentrations of Heavy Metals in Waters of Drainage Catchment of the Turew Landscape [μg-dm⁻³]

Waters Examined	Zinc	Cadmium	Lead	Copper
Lake	123.1	0.72	10.3	28.9
Mid-field pond	92.6	0.22	13.6	8.2
Artificial reservoir	61.1	0.72	5.4	22.4
Drainage canal	87.6	0.40	13.1	14.6
Melioration ditch	145.6	0.19	7.4	16.5
Ground water under field	57.7	0.39	4.2	25.7
Ground water under meadow	132.0	0.34	7.8	21.5

Source: Szpakowska and Karlik 1996.

forms. Analyzing the total concentrations of heavy metals in the ground and surface waters of the Turew landscape, the highest mean concentrations were found for zinc (99.9 μg·dm⁻³), and the lowest for cadmium (0.42 μg·dm⁻³) (Table 7.4). The range of zinc concentrations in ground water was from 57.7 to 132 μg·dm⁻³ and was similar to the concentrations of this metal in the surface waters (61.1 to 145.6 μg·dm⁻³). The concentrations of copper and lead had intermediate values and ranged from 4.2 to 7.8 μg·dm⁻³ for lead in ground waters and 5.4 to 13.6 μg·dm⁻³ for surface waters. Comparing copper concentrations in ground and surface waters, it was found that Cu concentrations were 23% higher in ground water. Copper concentrations in surface waters differed considerably, depending on the kind of reservoir or water course (steady, running, differentiation of plant communities). The

Table 7.5   Total Concentrations of Heavy Metals in Natural Waters [$\mu g \cdot dm^{-3}$]

Type of Water	Zinc	Cadmium	Lead	Copper	References
Rainfall	159	0.48	4.9	1.9	Chłopecka and Dudka (1991)
Ground water[a]	1940	13	64	39	Pilc et al. (1994)
Mid-field pond[a]	120	8	26	16	Pilc et al. (1994)
River water	88.6	0.628	8.18	35.8	Golimowski and Sikorska (1983)

[a] Water located in areas with high chemical soils contamination.

similarities can also be found (Table 7.5). Only in waters localized on soils considerably contaminated with dust emitted by a metallurgical industrial plant were the Pb and Cd concentrations much higher in both ground and surface waters.

The heavy metal concentrations in running and stagnant waters are highly variable. Quantitative differentiation of the metals in the studied waters depends mainly on catchment area utilization as well as rate of sedimentation (Pasternak and Antoniewicz 1971, Karlik and Szpakowska 1999).

As emphasized earlier, heavy metals in waters occur not only in ionic forms but also are able to create bonds with, for example, ligands such as aliphatic acids, amino acids, and other compounds.

CONCENTRATION OF DIFFERENT FORMS OF METALS

Ionic forms, labile forms, and the sum of stable complexes and chelates for Zn, Cd, and Pb were determined in the surface and ground water of the Turew landscape. In both types of water, great differences in percentages of each form of the metals were noted, particularly concerning ionic forms (Tables 7.6 and 7.7). Clearly, a lower

Table 7.6   Chemical Forms of Zinc, Cadmium, and Lead in Ground Water under Field

Metal Examined	Concentration [$\mu g \cdot dm^{-3}$]			% of Form		
	Ionic Forms	Labile Complexes	Stable Complexes	Ionic Forms	Labile Complexes	Remain (stable)
Zinc	17.9	17.2	36.3	25.2	24.1	50.8
Cadmium	0.003	0.088	0.248	0.9	25.9	73.2
Lead	0.28	1.78	1.02	9.1	57.8	33.1

Source: Szpakowska and Karlik 1996.

Table 7.7   Chemical Forms of Zinc, Cadmium, and Lead in Surface Water

Metal	Concentration [$\mu g \cdot dm^{-3}$]			% of Form		
	Ionic Forms	Labile Complexes	Stable Complexes	Ionic Forms	Labile Complexes	Remain (stable)

proportion of ionic forms was found in ground water (from 0.9 to 25.2%), while in surface water the amount of metal in this form ranged from 28.4 to 42.3%. Labile forms of metal occurred in greater quantities in ground water (24.1 to 57.8%), significantly due to a very high percentage of Pb (57.8%) and relatively high proportions of Zn (24.1%) in these forms.

The forms of heavy metal bonds with organic compounds and their concentrations in waters can be affected not only by chemical composition of dissolved organic matter but also by quantity of dissolved organic matter (Karlik et al. 1996, Karlik and Szpakowska 1999). The surface and ground waters studied had a high concentration of organic matter. Therefore, a high proportion of metals in the form of stable complexes was found there. The stable forms (permanent complexes and strong chelates) were higher in ground water than in surface water (Tables 7.6 and 7.7) (Szpakowska and Karlik 1996). Such high concentrations of metals in the form of stable complexes in ground waters resulted from the higher concentration of humic acids in these waters (Figure 7.6). Humic acids are known to possess strong chelating properties and are able to form stable bonds that were destroyed after UV radiation.

Estimation of metal form is very important for forecasting toxicity and transport of metals in the aquatic environment. For the study of the detoxification of waters it is not enough to determine the metal form bound to organic compounds; it is also necessary to estimate the stability of such binding. According to Tubbing et al. (1994), labile complexes, for example copper, with natural ligands such as humic acids can also be toxic to algae.

ASSOCIATION OF METAL IONS WITH HUMIC SUBSTANCES IN NATURAL WATER

The solubility of several metals in water is much greater than expected from calculations based on the inorganic ions of the medium. For example, about 90% of soluble iron is bound to HS, and its solubility is greater than one can conclude from solubility constants (Weber 1988). Because this increase in solubility was related to complexation, adsorption, or reduction by HS, the study tends to estimate the proportion of heavy metals in a form bound to HS. In ground waters the amounts of bound lead and copper were higher than in surface water (in the case of Pb by 84.5% and by 49.9% for Cu) (Table 7.8). In the case of Cd, strong complexing ability was found for humus isolated from ground water under field, whereas five

Table 7.8   Binding Ability toward Heavy Metals of 1 g Humic Substances Isolated from Waters of Area A [µg·g⁻¹]

times lower binding of Cd was observed for HS obtained from ground water under meadow and surface water of mid-field pond.

In estimating the amount of heavy metals bound to 1 g of HS, similar complexing abilities of HS isolated from ground water under meadow and field toward copper and lead were found. A smaller amount of copper in a form bound to HS in surface water may result from uptake of this metal by plants. Copper is easily taken by plants, in both ionic form and in complexes (Ozimek 1988). Research indicated that 1 g of HS isolated from ground water had stronger chelating properties toward heavy metals than HS obtained from surface waters. This finding may be connected with the different structure of HS in these waters, for example, higher proportion of humic acids in HS from ground water.

HS have a reduction potential range of about 0.5 to 0.7 volts (vs. the standard hydrogen electrode) and are unlikely to affect oxidation. Undoubtedly there is a synergic process in which extremely low concentrations of metal ions are strongly complexed by humic acid, thus shifting the equilibrium toward the complexed metal ion. The total binding of HS for metal ions is 200 to 600 $\mu mol \cdot g^{-1}$. Approximately one third of this total is cation exchange sites and the remainder is complexing sites (Weber 1988). Complexations of metals by HS carboxylic and phenolic groups is very important for water migration of heavy metals, influencing their concentrations in water as well as in sediment.

Bonds of high molecular organic substances with metals were determined on a sample of copper, which is often chosen for analyses. Using the method of sorption on Amberlite XAD-2 enabling separation of organic substances into polysaccharides, polypeptides, and humic substances, the degree of their binding with copper was determined (Table 7.9).

Copper in ionic form constitutes 39.1% of its total concentration and in colloidal form 24.8%. The bound copper (with HS and PP) contribution was 32% of total concentration. A study by Linnik and Nabivanets (1984) showed that protein-polypeptides fraction released by macrophytes can bind large amounts of copper forming metal-organic complexes. As was indicated in other studies (Thurman 1985, Szpakowska 1999), organic substances contain acidic functional groups, of which about one half are carboxylic acids, which are primarily responsible for binding metals and forming complexes (Shuman et al. 1990). Weber (1988) suggests that complex stability is highest for copper, intermediate for lead, and weakest for cadmium. Higher complex stability explains higher affinity of copper than cadmium and the displacement of the latter from complexes (Tubbing et al. 1994).

Table 7.9   Chemical Forms of Copper Occurrence in Surface Water

Metal Forms	Concentration $[\mu g \cdot dm^{-3}]$	% of Total Concentration

DISSOLVED ORGANIC SUBSTANCES IN WATER BODIES   179

Table 7.10 Mean Concentration of Different Forms of Heavy Metals

| | | Ground Water | | | | Surface Water | |
| | | Under Field | | Under Meadow | | of Mid-Field Pond | |
Water Examined		$\mu g \cdot dm^{-3}$	%	$\mu g \cdot dm^{-3}$	%	$\mu g \cdot dm^{-3}$	%
Cd	Inorganic	0.300	98	0.300	94	0.23	48
	Organic	0.006	2	0.019	6	0.25	52
	Total	0.306	100	0.319	100	0.48	100
	Range	(0.2–0.3)		(0.1–0.4)		(0.2–1.3)	
Pb	Inorganic	1.76	68	2.15	73	1.42	50
	Organic	0.83	32	0.78	27	1.42	50
	Total	2.59	100	2.93	100	2.84	100
	Range	(1.6–5.0)		(1.3–0.4)		(1.6–5.1)	
Cu	Inorganic	3.02	64	2.50	55	2.26	33
	Organic	1.68	36	2.05	45	4.58	67
	Total	4.72	100	4.56	100	6.84	100
	Range	(2.4–14.2)		(1.8–13.6)		(2.8–27.2)	
Zn	Inorganic	7.16	93	7.12	88	9.68	60
	Organic	0.56	7	0.94	12	6.43	40
	Total	7.52	100	8.06	100	16.11	100
	Range	(5.9–11.5)		(3.8–24.6)		(5.4–35.8)	

Source: Życzyńska-Bałoniak et al. 1993.

THE EFFECT OF MEADOW BELT ON HEAVY METAL SPECIATION

The study carried out in the Wielkopolska region indicated that heavy metals migrating with ground water from the field under the meadow to the pond undergo speciation. The cation forms are bound by DOM to metal-organic complexes. For example, 50% of cadmium, a very toxic element, is transformed from the inorganic ionic form to organic form during its movement to the pond (Table 7.10). The concentration of inorganic copper passing from the field and meadow into the pond drops, while the organic form rises by 30%. Similar changes occur in the other cations. In the mid-field pond, the concentrations of lead and zinc in the organically bound form increase significantly. These observations confirm that the ecosystem character not only considerably affects the composition of DOM in waters, but also changes the form of heavy metals (speciation) carried away from the fields.

Analysis of metal complexation competition for binding sites ought to be considered. For example, divalent ions, such as calcium, would compete for the copper binding site. The binding strength for calcium would be 100 to 1000 times less, but

CONCLUSION

The literature on soil organic substances is immense, but the organic matter of waters has received far less attention. The studies carried out in the Turew landscape indicated high concentrations of dissolved organic substances in waters situated in the agricultural areas. Land-use practices are a significant factor affecting the increase in the transport of organic matter. The water migration of DOM from agroecosystems and the methods of limiting it constitute the basic problem in the protection of aquatic environment. From the study presented, one can conclude that biogeochemical barriers, such as meadow belts, are very effective in limiting free DOM migration.

Organic matter in water has great ecological and geochemical significance. On the one hand, DOM is an important energy source for microbially based aquatic food webs but is also very essential to the interaction of DOM with such elements as the heavy metals. Particularly important in this phenomenon are humic substances, which can bind large amounts of zinc, cadmium, lead, and copper. In the case of cadmium, strong complexing ability was found for humus isolated from ground water under fields, whereas five times less binding of cadmium was observed for HS isolated from surface water. There are two main reasons for studying the interactions of organic material with metal ions: to understand its chemical nature and to understand its effects on the environment. Such studies are very important for prediction of the reactivity, toxicity, and migration of metal ions in the aquatic environment where type of ecosystem plays a vital role. In particular, humic acids are very active in life cycles of flora and microorganisms. There is no doubt that these substances compete with definite organic compounds of known structure (e.g., polysaccharides), which are much more effective in their specific reactions. On the other hand, the latter compounds take part in only specific reactions while humic acids have an immense spectrum of effects on biota. Identification of all organic compounds is not possible at present, but the description of properties (e.g., binding ability of metals) may help in differentiating organic matter from different sources.

REFERENCES

Chłopecka A. and Dudka S. 1991. Wpływ osadów ściekowych na zawartość metali śladowych w glebie i roślinach oraz na wymywanie tych pierwiastków z gleby. Zeszyty Naukowe AGH, Sozologia i Sozotechnika 31: 79–85.

Easthouse K. B., Mulder J., Christophersen N., Seip H. M. 1992. Dissolved organic carbon fractions in soil and stream water during variable hydrological conditions at Birkenes, Southern Norway. *Water Resources Res.*, 28 (6): 1585–1596.

Gaffney J. S., Marley N. A., Clark S. B. 1996. Humic and fulvic acids. Isolation, structure

Gron C., Torslov J., Albrechtsen H. J. 1992. Biodegradability of dissolved organic carbon in ground water from an unconfined aquifer. *The Science of the Total Environment* 117/118: 241–251.

Hessen D. O., Tranvik L. J. 1998. Aquatic humic substances. *Ecological Studies* 133, Springer-Verlag, Berlin: 178–195.

Jansson M. 1998. Nutrient limitation and bacteria-phytoplankton interactions in humic lakes. In *Aquatic Humic Substances.* D. O. Hessen, L. J. Tranvik (Eds.), Ecological Studies 133, Springer-Verlag, Berlin: 178–195.

Karlik B., Szpakowska B., Otabbong E., Siman G. 1996. Migration of mineral elements bound to dissolved organic compounds in agricultural watershed. *Pol. Ecol. Stud.* 22, 3–4: 95–103.

Karlik B., Szpakowska B. 1997. Dissolved organic matter and chemical forms of heavy metal in waters of agricultural basin. *Roczniki Akademii Rolniczej w Poznaniu*, 194: 39–46.

Karlik B., Szpakowska B. 1999. Effect of dissolved organic matter on metal speciation in waters from agricultural area. In *Paysages agraires et Environnement.* S. Wicherek (Eds.). CNRS, Paris: 253–259.

Karlik B., Szpakowska B. 2001. Labile organic matter and heavy metals in waters of agricultural landscape. *Pol. J. Environ. Stud.* 10 (2): 85–88.

Kramer J. R., Brassard P., Collins P., Clair T. A., Takats P. 1990. Variability of organic acids in watersheds. In *Organic Acids Aquatic Ecosystems.* E. M. Perdue, E. T. Gjessing (Eds.). John Wiley & Sons, New York: 127–139.

Kundzewicz Z., Szpakowska B., Sibrecht R. 1994. Modelling of chemicals dissolved in waters in agricultural landscape. In *Hydrochemistry: Hydrological, Chemical and Biological Processes of Transformation and Transport of Contaminants in Aquatic Environments.* Proceedings of the Rostov-on Don Symposium 219: 231–239.

Linnik T. N., Nabivanets B. J. 1984. The state of metal ions in natural waters. *Acta Hydrochim. Hydrobiol.* 12: 335–361.

Lush D. L., Hynes H. B. N. 1978. Particulate and dissolved organic matter in a small partly forested Ontario stream. *Hydrobiologia* 60: 177–185.

Maier W. J., Gast R. G., Anderson C. T., Nelson W. W. 1976. Carbon contents of surface and under ground waters in south-central Minnesota. *J. Environ. Qual.* 5 (2): 124–128.

Malcolm R. L. 1985. Geochemistry of stream fulvic and humic substances. In *Humic Substances in Soil, Sediment and Water.* G. R. Aiken, D. M. McKnight, R. L. Wershaw, P. MacCarthy (Eds.). John Wiley & Sons, New York: 181–209.

Mantoura R. F. C. 1981. Organo-metallic interactions in natural waters. In *Marine Organic Chemistry.* E. Duursma, R. Dafson (Eds.). Elsevier, Amsterdam: 179–224.

Mart L., Nürnberg H. W., Valenta P., Stoeppler M. 1978. Determination of levels of toxic metals dissolved in sea water and inland waters by different pulse anodic stripping voltammetry. *Thalassia Jugoslavica*, 14 (1/2): 171–188.

Meili M. 1992. Sources, concentrations and characteristics of organic matter in softwater lakes and streams of the Swedish forest region. In *Dissolved Organic Matter in Lacustrine Ecosystems: Energy Source and System Regulator.* K. Salonen, T. Kaire-salo, R. I. Jones (Eds.). Kluwer Academic Publishers, Dordrecht: 23–41.

Meyer J. L. 1990. Production and utilization of dissolved organic carbon in riverine ecosys-

Muenster U. 1992. Microbial extracellular enzyme activities in Humex Lake Skjervatjern. *Environment International,* 18: 637–647.

Nürnberg H. W. 1981. Voltammetric studies on toxic metal speciation in natural waters. In *Heavy Metals in the Environment.* International conference, Amsterdam. Edinburgh, U.K.: 635–641.

Ozimek T. 1988. Rola makrofitów w krążeniu metali ciężkich w ekosystemach wodnych. *Wiadomości Ekologiczne,* 34 (1): 31–44.

Pasternak K., Antoniewicz A. 1971. Variability of copper, zinc and manganese in water of several rivers, streams and carp ponds. *Acta Hydrobiol.* 13 (3): 251–268.

Pilc L., Rosada J., Siepak J. 1994. Stężnia metali w glebach, wodach i roślinach w sąsiedztwie huty Głogów. In *Materiały,* 34 Sesji Naukowej IOR, Poznań: 228–232.

Rutherford J. E., Hynes H. B. N. 1987. Dissolved organic carbon in steams and ground water. *Hydrobiologia* 154: 33–48.

Ryszkowski L. 1994. Strategy for increasing countryside resistance to environment threats. In *Functional Appraisal of Agricultural Landscape in Europe.* L. Ryszkowski, S. Bałazy (Eds.). Research Centre for Agricultural and Forest Environment, PAS: 9–18.

Ryszkowski L., Bartoszewicz A., Kędziora A. 1997. The potential role of mid-field forests as buffer zones. In *Buffer Zones: Their Process and Potential in Water Protection. Proc. Int. Conf. Buffer Zones.* N. Haycock, T. Burt, K. Goulding, G. Pinay (Eds.). Harpenden, U.K.: 171–191.

Shuman M. S., Calmano W., De Haan H., Fredrickson H. L., Henriksen A., Kramer J. R., Mannio J. Y., Morel F. M. M., Niemeyer J., Ohman L. O., Perdue E. M., Weis M. 1990. How are acid-base properties of "DOC" measured and modeled and how do they affect aquatic ecosystems? Group report. In *Organic Acids in Aquatic Ecosystems.* E. M. Perdue and E. T. Gjessing (Eds.). John Wiley & Sons, New York: 141–149.

Sundh I., Bell R. T. 1992. Extracellular dissolved organic carbon released from phytoplankton as a source of carbon for heterotrophic bacteria in lakes of different humic content. In *Dissolved Organic Matter in Lacustrine Ecosystems: Energy Source and System Regulator.* K. Salonen, T. Kairesalo, R. I. Jones (Eds.). Kluwer Academic Publishers, Dordrecht: 93–106.

Szpakowska B., Karlik B. 1996. Chemical forms of heavy metals in agricultural landscape water. *Pol. J. Environ. Stud.* 5 (6): 67–73.

Szpakowska B., Życzyńska-Bałoniak I. 1989. The effect of environmental pollution on the migration of chemical compounds in water in an agricultural landscape. *Ecol. Int. Bull.* 17: 41–52.

Szpakowska B., Życzyńska-Bałoniak I. 1994. The role of biogeochemical barriers in water migration of humic substances. *Pol. J. Environ. Stud.* 3 (2): 35–41.

Szpakowska B., Życzyńska-Bałoniak I. 1996. Migration of dissolved humic substances in agricultural landscape. *Pol. J. Soil Sci.* 29/2: 139–147.

Szpakowska B. 1999. Występowanie i rola substancji organicznych rozpuszczonych w wodach powierzchniowych i gruntowych krajobrazu rolniczego, Rozprawy, Toruń: 110 pp.

Thurman E. M., Aiken G. R., Evald M., Fisher W. R., Förstner U., Hack A. H., Mantoura R. F. C., Parsons J. W., Pocklington R., Stevenson F. J., Swift R. S., Szpakowska B. 1988. Isolation of soil and aquatic humic substances: Group report. In *Humic Sub-*

Tubbing D. M. J., Admiraal W., Cleven R. F. M. J., Igbal M., Van de Meent D., Verweij W. 1994. The contribution of complexed copper to the metabolic inhibition of algae and bacteria in synthetic media and river water. *Water Res.* 28 (1): 37–44.

Weber J. H. 1988. Binding and transport of metals by humic materials. In *Humic Substances and Their Role in the Environment.* F. H. Frimmel, R.F. Christman (Eds.). John Wiley & Sons, New York: 165–178.

Życzyńska-Bałoniak I., Szpakowska B. 1989. Organic compounds dissolved in water bodies situated in an agricultural landscape and their role for matter migration. *Arch. Hydrobiol. Ergebn. Limnol.* 33: 315–322.

Życzyńska-Bałoniak I., Szpakowska B., Ryszkowski L., Pempkowiak J. 1993. Role of meadow strips for migration of dissolved organic compounds and heavy metals with ground water. *Hydrobiologia* 251: 249–256.

0919 ch07 frame Page 184 Tuesday, November 20, 2001 6:22 PM

CHAPTER **8**

Influence of Landscape Mosaic Structure on Diversity of Wild Plant and Animal Communities in Agricultural Landscapes of Poland

Lech Ryszkowski, Jerzy Karg, Krzysztof Kujawa, Hanna Gołdyn, and Ewa Arczyńska-Chudy

CONTENTS

INTRODUCTION

According to prevailing opinion, agricultural activity eliminates many wild plant and animal species and is among those human actions that impoverish living resources. Attempts to eradicate any plant competitors and the pests and pathogens to cultivars result in enormous simplification of biotic communities in cultivated

moisture and its pH, irrigate fields, introduce fertilizers, level earth surface, and shape fields. All those efforts, together with frequent tillage and other kinds of farmer interference in agroecosystem structures and processes, change the living conditions of many wild organisms and often lead to their disappearance.

Intensification of agriculture also leads to alteration of countryside structure. Simplification of crop rotation patterns due to increasing specialization of plant production and formation of large fields to facilitate mechanization of work are frequently observed. Eradication of patches of mid-field forests, shelterbelts (rows of mid-field trees), hedges, field margins, stretches of meadows, and riparian vegetation strips is performed on a large scale during field consolidation. Drainage of mid-field small wetlands or small ponds also leads to the simplification of the agricultural landscape structure. All these activities eliminate refuge sites for many organisms in the agricultural landscape. One can conclude, therefore, that the interests of agriculture and nature conservancy are contradictory, and this conclusion was frequently used and broadly disseminated in general as well as in specific discussions on nature protection problems.

The problem of protecting living resources became the central theme not only among biologists but also in political and administration circles when evidence was presented that the world's flora and fauna are disappearing at an alarming rate (e.g., Wilson and Peter 1988, Reaka-Kudla et al. 1997, Vitousek et al. 1997). These concerns culminated in the Biodiversity Convention during the World Summit in 1992.

Following the Biodiversity Convention, several policies were recommended by the Council of Europe as well as by the European Commission, such as the Pan-European biological and landscape diversity strategy, the European Ecological Network, and Nature 2000. All these policies stress integrating nature protection with sectoral activities, indicating a substantial change from the previous point of view that nature should be shielded against human activity in order to ensure its successful protection. That change, still opposed by many biologists, was stimulated by a slowly growing consensus that the way in which resources have been used, rather than the fact that they are used at all, has caused the threats to nature. The possibility that agriculture could be integrated with biodiversity protection is related to changing cultivation technologies (Srivastava et al. 1996) and to managing agricultural landscape structures to provide survival sites for biota (Baldock et al. 1993, Ryszkowski 1994, 2000). There is no doubt that high-input modern farming practices frequently pollute water and soil, compact soil, and stimulate erosion. But inappropriate agricultural practices can be modified to mitigate their adverse effects on the biota and environment. Diversification of the agricultural landscape pattern through introduction of refuge sites can mitigate biota impoverishment due to intensive farming, at least with respect to some plant and animal communities. To evaluate that prospect

IMPOVERISHMENT OF PLANT COMMUNITIES
RESULTING FROM AGRICULTURE

According to the Main Statistical Office more than 60% of Poland crop production is based on cereals (wheat, rye, barley, oats, maize, and triticale, which is the product of wheat and rye hybridization used for fodder). The production of potatoes, sugar beets, rape seed and common agrimony, leguminous crops for human consumption and fodder (peas, beans, clover, alfalfa), and other crops constitute the rest of plant production. In 1998 cereals comprised 70.7% of arable land, cultivated on 31.9% of the total territory of Poland. The total land area used to cultivate cereals is greater than the total forest area, comprising a total of 28.2% of the entire country. The arable land in Poland is dominated by light soils, and, when cereals are cultivated too frequently on the same field, depleted nutrients and decreased soil organic matter can result if pulse crops are not included in the crop rotation pattern or if organic fertilizers are not applied. Thus, overall simplification of the plant cover structure in Poland because of agricultural activity resulted in the dominance of cereals. The simplification of plant cover structure is even more advanced because wheat and rye cultivations cover 55.5% of the total area under cereals, and during 1988–1998 the contribution of area under wheat increased by 20.7% while the area under cultivation of rye, barley, and oats decreased. Wheat fields constitute, therefore, not only the dominant element of the countryside, but they also influence distribution of many organisms in the agricultural landscape and influence their prospects for migration and survival.

Growing alongside cultivated plants are weeds, which are inevitable components of agroecosystems but controlled to a large extent by farmers. During their long coexistence with cultivars under a regular sequence of tillage activities, weeds adapted to survive and thrive in agroecosystems. Their potential to adapt to cultivation measures is great, and despite the use of highly effective herbicides the diversity and even density of weeds has increased recently (Ghersa and Roush 1993, Cousens and Mortimer 1995). Chemical control limitations led to the development of integrated weed management (IWM), which combines chemical control tactics with mechanical and biological measures. The goal of IWM is to use such control measures to reduce and prevent weed community adaptation to field management (Fick and Power 1992, Swanton and Murphy 1996, Johnson et al. 1998).

Herbicides, crop rotation, and tillage practices are considered the most important elements of weed control programs in the literature. The ranking of those factors varies according to different studies, especially when development of resistance to herbicides was discovered; nevertheless, it is believed that the IWM practices relying on several control methods can control weed populations below the economical threshold of harmful effects on yields (Huffaker et al. 1978, Fick and Power 1992,

In Poland, about 400 species of weeds were detected in various communities characteristic for main cultivars (Adamiak and Zawiślak 1990). During the last decades, the impoverishment of weed communities was observed in Poland by many scientists (Borowiec et al. 1992, Korniak 1992, Fijałkowski et al. 1992, Sendek 1992, Trzcińska-Tacik 1992, Warcholińska 1992, Stupnicka-Rodzynkiewicz et al. 1992a, 1992b). Greater impoverishment of weed communities is observed in regions with more intensive agriculture.

The following factors are considered to limit abundance and diversity of weed communities:

- Use of herbicides — Eliminated are stenotic species well adapted to specific cultivars which are substituted by ubiquitous species communities mainly composed by monocotyledonous species or dicotyledonous ones resistant to herbicides (Borowiec et al. 1992, Gould 1991, Warwick 1991, Korniak 1992a, Trzcińska-Tacik 1992, Warcholińska 1992, Fijałkowski et al. 1992, Stupnicka-Rodzynkiewicz et al. 1992a, 1992b).
- Type and amounts of fertilizers — Increasing inputs of fertilizers eliminates oligothrophic species. Applications of higher amounts of mineral nitrogen or farm manure bring about expansion of nitrogenous species (Adamiak and Zawiślak 1990a, Korniak 1992a, Warcholińska 1992).
- Simplification of the crop rotation pattern — When crop rotation is simplified, weed communities likewise become less diverse; a simplified and stable community develops, composed of a few abundant species. Observed dominance of cereals in crop rotation pattern leads to simplified weed communities composed mainly of such abundant species such as *Apera spica-venti*, *Centaurea cyanus*, *Galium aparine*, *Matricaria perforata*, *Avena fatua*, and *Echinochloa crus-galli*.
- Clearing cultivar seeds from weed's diaspores in winnowing machines — This clearing limits the dispersion of adopted species to dissemination together with the cultivated plant. Thus, for example, due to the winnowing process such species as *Bromus arvensis*, *B. secalinus*, and *Agrosemma githago* are disappearing from weed communities (Warcholińska 1992).
- Abandonment of cultivars — Abandoning cultivars results in disappearance of weed communities associated with that cultivation. Thus, for example, evanescence of the weed community associated with common flax (*Linum usitatissimum*) cultivation observed recently is the direct effect of abandoning that cultivar.

Diversification of plant cover structure in landscape influences the richness of weed communities. Agricultural landscapes with less intensive tillage practices and field-mosaics intersected by many uncultivated refuges, such as field margins, stretches of meadows, small afforestations, and wetlands, have a higher diversity of weed communities (more than 300 weed species) than do areas with more intensive cultivation (less than 200 species of weeds) (Figure 8.1). In the Turew agricultural landscape (an area studied for 30 years by scientists from the Research Centre for

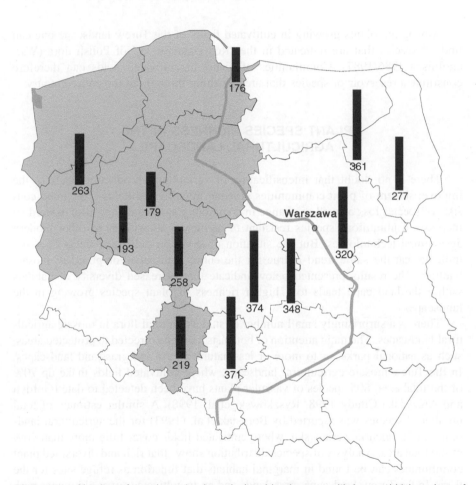

Figure 8.1   Number of weed species in agricultural landscapes with low and high regime of mineral fertilization (Gołdyn Arczyńska-Chudy unpublished data, Hołdyński 1991, Kutyna 1988, Latowski et al. 1979, Labza 1994, Skrzyczyńska 1998, Skrzyczyńska and Skrajna 199, Szotkowski 1973, Warcholinska 1976, 1983, 1997, Wika 1986).

of the species are archaeophytes, plants that invaded Poland before the 15th century, and have been associated with cultivations for a very long time. The newcomers (kenophytes) that became naturalized during the last few centuries make up 7% of the total weed community. All species, in addition to cultivars, found in cultivated fields also exist in surrounding nonproductive habitats, from where they invade fields. Thus, the influence of the plant community living in the total landscape on species composition in cultivated fields is substantial. Similar results were found by other scientists (Skrzyczyńska 1998, Warcholińska and Potebska 1998, Skrzyczyńska and

Among the plants growing in cultivated fields of the Turew landscape one can find 18 species that are indicated in the "Red Data Books" of Polish flora (Warcholińska 1986/1987). Communities of plants in cultivated fields can therefore constitute a reservoir of species that are vanishing throughout the entire country.

PLANT SPECIES RICHNESS IN THE AGRICULTURAL LANDSCAPE

There is no doubt that intensification of agricultural production leads to the impoverishment of plant communities growing in cultivated fields. This trend consists of species loss as well as change in community species composition toward an increase of ubiquitous species resistant to herbicides as well as to other modern agricultural technologies. But the situation observed in cultivated fields does not indicate that the same trend is true for the entire landscape composed of mosaic habitats. The results presented below indicate that increased diversity of habitats within the landscape leads to a higher richness of plant species growing in the landscape.

There is a surprisingly small number of studies on total flora in mosaic agricultural landscapes. The main attention of botanists has been directed to protected areas, such as national parks, or to more or less natural forest and grassland landscapes. In the Turew mosaic agricultural landscape, where cultivated fields make up 70% of the total area, 805 species of vascular plants have been detected to date (Gołdyn and Arczyńska-Chudy 1998, Ryszkowski et al. 1998). A similar estimate of total number of species was reported by Borysiak et al. (1993) for the agricultural landscape of Szwajcaria Zerkowska, where cultivated fields cover little more than 70% of the total area. Analysis of species distribution shows that rich and diversified plant communities can be found in marginal habitats that function as refuge sites for the flora. In the Turew landscape, grasslands and an uncultivated, very old manor park harbor more than 300 species each (Table 8.1). The highest diversity of flora was

Table 8.1   Number of Vascular Plant Species in Various Habitats of the Turew Agricultural Landscape

Habitat	Total	Archaeophytes	Kenophytes	Diaphytes[a]
Grasslands	321	22	14	0
Shelterbelts and afforestations	266	13	16	2
Manor park	308	32	20	7
Roadsides	220	49	27	26
Water reservoirs with rushes	211	2	5	2
Cultivated fields	193	54	13	16
Total landscape	805	85	55	39

Table 8.2   The Number of Threatened and Protected Species of Vascular Plants in Various Habitats of the Turew Landscape

Habitat	Threatened Species		Protected Species	
	Poland	Wielkopolska	Totally	Partially
Water reservoirs with rush	3	18	2	3
Grasslands	6	24	7	6
Afforestations and shelterbelts	—	5	1	4
Manor place	—	4	5	3
Roadsides	—	2	1	1
Cultivated fields	—	2	—	—
Total landscape	7	45	14	9

found in grasslands located mainly in the lower parts of the landscape close to small water reservoirs or along the drainage system of the landscape. In mowed grasslands, the common species prevail while in rush associations and in water reservoirs the threatened or protected species can also be found.

The Turew landscape is located in the Wielkopolska region, where the intensity of agricultural production is greater than in most other areas of Poland. The higher anthropogenic pressure on nature leads to lower survival rates of the plants in the region compared to the entire territory of Poland. The International Union for Conservation of Nature and Natural Resources (IUCN) developed criteria for evaluation of species survival status. The term *threatened* used in Table 8.2 corresponds to the categories of *endangered* and *vulnerable* in IUCN standards. The list of threatened species for the Wielkopolska region was published by Żukowski and Jackowiak (1995) and for the whole of Poland by Zarzycki et al. (1992). The status of vascular plant species found in the Turew landscape was determined with this information. A much higher number of threatened species was found according to the Wielkopolska Red Data Book than according to the Red Data Book for all of Poland, which shows that despite higher anthropogenic pressure, 45 threatened species survive well in the mosaic agricultural landscape of Turew. Water reservoirs and grasslands located in the agricultural landscape are the habitats that provide refuge sites for the greatest number of threatened species (Table 8.2). For example, in sedge communities, endangered and almost vanished species such as *Carex davalliana*, *Gentiana pneumonanthe*, *Viola persicifolia*, and *Primula elatior* were found. Those landscape elements are also overgrown by the plant communities most resistant to invasion by newcomer species (kenophytes and diaphytes). Among the 321 species constituting the grassland plant communities, only 14 newcomers to native flora succeeded in establishing themselves in those communities and none of the diaphytes that disseminate their propagules in the whole landscape succeeded (Table 8.1). A similar situation is observed in water plant communities. Only four species of newcomers are associated with the rich communities of native plants. Four types of water bodies occur in the Turew landscape: lakes, mid field ponds, peat holes, and ditches discharging water to drainage channels

Table 8.3   Number of Vascular Plant Species in Water Reservoirs of the Turew Landscape

Type of Water Body	Number of Species	Species Restricted only to Examined Habitat
Lake	111	29
Small ponds	108	18
Peat-holes	95	24
Drainage channels and ditches	115	31
All water bodies	211	211

Undergrowth plants are important contributions to the shelterbelt list of plant species. Among trees, the two widely distributed species of newcomers are *Robinia pseudoaccacia* and *Prunus serotina*, which were introduced in the 19th century. In shelterbelts, especially old ones with Robinia, the high incidence of therophytes (21%) indicates increased human pressure (Ratyńska and Szwed 1997). Under the conditions of intensive human management, the total number of kenophytes in shelterbelts is even higher than that found in cultivated fields (Table 8.1), which shows that new-comers easily settle in intensively managed afforestations. Nevertheless, the contribution of the native species is high, making up 88.0% of the total recorded in those habitats. Kenophytes make up 6% and the contribution of archaeophytes is 5%.

In cultivated fields, 193 species were found. Weed communities were composed of common species, the majority of which are associated with cereal cultivations. Kenophytes make up 7% of the total weed species list, archaeophytes make up 28%, and diaphytes comprise 8%. Native species constitute 57% of the total number of species living in cultivated fields. Floristic analysis shows that out of 110 native species found in weed communities, 33 (30%) come from grasslands, 28 (25%) from water reservoirs, 24 (22%) from afforestations, and 10 (9%) from xerothermic swards that appear infrequently in the Turew landscape. Plants from all seminatural habitats found in the landscape have their input into weed communities, which indicates the importance of plant dissemination processes for building and maintaining weed diversity in cultivated fields.

Very high plant diversity was found in stressed habitats, such as roadsides. Again, there was a high number of kenophytes species, amounting to 27 (11%), as well as a very high number of archaeophyte species, equal to 49 (19%) (Table 8.1). The highest number of diaphytes of all habitats was found in roadsides, amounting to 26 species (10% of total). The manor's 20-ha park is characterized by a very high diversity of plants, surpassing the total diversity of plants found in the 2200 ha of shelterbelts and afforestations of the landscape studied.

After evaluating the taxonomic status of various ecosystems in the studied agricultural landscape, one can state that seminatural grasslands and water plant communities harbor rich associations of native plants that show high resistance to

Table 8.4   The Species of the Most Abundant Families
in the Turew Agricultural Landscape (TAL)
and Wielkopolski National Park (WNP)

Family	TAL %	WNP %
Asteraceae	10.4	10.9
Poaceae	9.5	8.9
Rosaceae	5.9	7.3
Cyperaceae	5.8	4.8
Fabaceae	5.6	6.1
Caryophyllaceae	3.9	4.6
Brassicaceae	3.9	4.1
Scrophulariaceae	3.5	3.7
Lamiaceae	3.5	3.6
Polygonaceae	3.2	1.8
Ranunculaceae	3.0	2.8
Apiaceae	2.6	3.5
Contribution to total floristic list of species	60.8	62.1

In the mosaic agricultural landscape studied, 805 species of vascular plants were detected, while Żukowski et al. (1995) estimated the number of species present in the nearby Wielkopolski National Park at 1120. Keeping in the mind that the National Park has more types of habitat than does the landscape studied, it is nonetheless clear that the mosaic Turew landscape is characterized by a rich plant community. Studying the higher taxonomic categories, one can find 494 genera and 116 families of plants in the Wielkopolski National Park, while in the mosaic Turew landscape 350 genera and 101 families were detected. The similarity of the list of Turew landscape flora to the list of plants in the National Park is underscored by the fact that the 12 families richest in species (with 60% of the total recorded species in each area) are the same in both locations (Table 8.4). The order of family importance, estimated by the percentage of the total for each species, is almost the same in both locations. Thus, the mosaic agricultural landscape reflects well the potential for species diversity in the region.

In comparison to other studies on landscape species diversity, it can be stated that the lack of large forest complexes plus intensive human interference very drastically decreased the number of species in Turew afforestations and shelterbelts. Thus, in the forest complexes of Wielkopolska Park, Żukowski et al. (1995) estimated 508 species, and 581 species were found in forest complexes of Kraków-Wieluń Upland (Wika, 1986) while in Turew's afforestations only 266 species can be found.

Native species comprise 75% of the species list of the Wielkopolski National Park and almost the same amount (77%) in the Turew mosaic agricultural landscape.

these estimates are not exact, they indicate, nevertheless, the high diversity of plant communities in the agricultural landscape studied in Turew.

These comparisons of the number of plant species do not indicate the qualitative differences between mosaic agricultural landscapes and natural parks. In the Wielko-polski National Park, botanists found 51 protected species (38 totally protected and 13 partially). According to the Wielkopolska Red Data Book, the number of threat-ened species was 184 and according to the threatened species list of Poland, there were 26 species threatened. These estimates are higher than those obtained in the Turew agricultural landscape but not as high as one could presume, relying on the prevailing opinion that agriculture exerts widespread negative impacts on nature protection. Diversification of the agricultural landscape structure amends, to some extent, the negative effects of agricultural activities on the plant diversity.

INFLUENCE OF AGRICULTURE ON ANIMALS

To increase production, farmers subsidize energy to simplify the plant cover structure both within cultivated fields (selection of cultivars well adopted to a narrow set of environmental conditions which increase production under controlled growth conditions and also eliminate weeds) and within the agricultural landscape (eradi-cation of small afforestations, shelterbelts, mid-field ponds, wetlands, and others). Use of fertilizers and pesticides, ploughing of soils, and drainage of fields as well as application of other modern agriculture technologies affect the survival rates of animals living in cultivated fields, often leading to impoverishment of animal com-munities. There are many publications on the impact of various agricultural tech-nologies on particular groups of animals, but there is a scarcity of studies concerning reactions of the total set of animals living in agroecosystems. A few compilations of estimates soil fauna data from different periods or places were performed by Russel (1977), Hendrix et al. (1986), Hansson et al. (1990), Prasad and Gaur (1994). The long-term studies on ecology of agricultural landscapes, carried out by the Research Centre for Agricultural and Forest Environment in Turew (Ryszkowski et al. 1996), included complex investigations on total above- and belowground ani-mal communities and aimed at evaluation of their functional reactions to agricultural activity (Karg and Ryszkowski 1996).

Animals, as compared to vascular plants and microorganisms, comprise a small part of total organic matter, amounting to less than 1% of the total. Total organic mass during the plant growing season was almost 3 times higher in the meadow than in the wheat field studied in the Turew landscape but the differences in animal biomass amounted to 4.4 times in favor of the meadow (Table 8.5). Thus, in culti-vated fields not only is a decline of animal biomass observed, but the rate of animal community suppression is greater than in the entire organic system.

Table 8.5 Distribution of Biomass (mg d.w.·m⁻²) in Components of a Wheat Field and a Meadow Ecosystem (to 30 cm soil depth)

	Wheat	Meadow
Above ground (total)	265014	158647
Plants	264660	158400
Animals	354	247
Below ground (total)	830979	2915167
Plants (roots)	104580	937500
Animals	2109	10637
Bacteria	172000	460000
Fungi	516000	1380000
Litter	36290	127030
Total	1095993	3073814

Source: Modified from Ryszkowski et al. 1989.

Table 8.6 Season-Long Mean Biomass (mg d.w.·m⁻²) of Soil Invertebrates in Soil under Continuous Rye and Rye in 4-Year Rotation Pattern

Site of Study	Wielichowo (1983–1984)		Jelcz-Laskowice (1986–1989)	
Kind of cultivation	Rye at 13th year of continuous cropping	Rye in 4-year rotationᵃ	Rye at 12th till 15th year of continuous cropping	Rye in 4-year rotationᵃ
Mean values of biomass (Protozoa, Nematoda, Annelida, Acarina Insecta)	1108.6	1327.2	2265.6	6502.0

ᵃ Mean values of rye after potatoes, and rye after pea in Norfolk rotation.
Source: Calculated from data of Karg et al. 1990.

animal community is impoverished by agricultural activity. This conclusion can be supported by studies on the effects on soil biota of continuous cropping of rye, which can be considered a more intensive system than crop rotation (Ryszkowski et al. 1998). Simplification of plant cover to continuous cultivations of rye resulted in further intensified impoverishment of total soil animals (Table 8.6).

More striking than changes in biomass due to cultivation are changes in functions of the soil biota. Smaller body sizes were detected in many groups of edaphon (soil animals) when the body weight distribution of animals living in the cultivated fields was compared with those populating the meadow (Table 8.7). With the exception of

Table 8.7   Mean Individual Body Weight (mg live weight) Distribution of Edaphon in Wheat Field and Meadow

Taxon	Winged Insect Larvae	Collembola	Lumbricidae	Enchytraeidae	Acarina	Nematoda
Wheat field	5.75	0.0052	506.0	0.161	0.0104	0.000637
Meadow	42.75	0.0055	317.0	0.192	0.0107	0.000788

Source: Ryszkowski 1985.

especially distinct in winged insect larvae; in the meadow soil they were 7.4 times heavier in mean body weight than larvae found in the wheat field soil. Thus, it can be inferred that agricultural activity of farmers creates conditions that can be tolerated by smaller animals, and by that token agriculture favors r-strategy species, in the terminology of MacArthur and Wilson (1967) — species having high reproduction rates and characterized by small body sizes survive in field soil although they are exposed to high mortality rates.

Because of the well-known inverse relationship between body weight and metabolic rate, smaller animals use higher amounts of energy per mass unit to maintain their activity than do larger ones. According to estimates of energetic costs to maintain the total set of animals in the studied wheat and meadow ecosystems, the animals in meadow expended $0.23 \text{ kJ} \cdot \text{mg}^{-1} \cdot \text{m}^{-2}$ during 225 days of the plant growing season, while in the wheat field they expended $0.75 \text{ kJ} \cdot \text{mg}^{-1} \cdot \text{m}^{-2}$ (Ryszkowski and Karg 1996). Thus, due to the changes in body sizes of individual species, maintenance of the standing biomass of the entire wheat field community requires more intensive dissipation of energy than in the meadow community.

Further simplification of plant cover to continuous cropping of rye can increase the energetic expenses for maintenance of the biomass unit in the plant growth season from $0.82 \text{ kJ} \cdot \text{mg}^{-1} \cdot \text{dw} \cdot \text{m}^{-2}$ in animal communities living in soil under rye grown in rotation to $0.93 \text{ kJ} \cdot \text{mg}^{-1}\text{dw} \cdot \text{m}^{-2}$ in rye cultivated in continuous cropping in Wielichowo studies.* The studies carried out in Jelcz-Laskowice showed much greater differences. In this last situation, maintenance costs of unit biomass of edaphon during the vegetation season in rye grown in rotation was equal to $0.38 \text{ kJ} \cdot \text{mg}^{-1}\text{dw} \cdot \text{m}^{-2}$, and in continuous cropping this cost increased to $0.79 \text{ kJ} \cdot \text{mg}^{-1}\text{dw} \cdot \text{m}^{-2}$. In all of the above energy estimates of maintenance costs of total communities, the metabolism rates were calculated by measuring specific live body weights for the studied taxonomic groups and using specific equations relating body weights and respiration rates for each taxonomic group, including corrections for changes in temperature (Ryszkowski and Karg 1996). In order to make all biomass estimates comparable, the dry weight was used for calculations of efficiency index of energy use for biomass maintenance. Estimates of total community dissipation of energy per unit of biomass

Table 8.8 Mean Studied Biomass (mg dw·m⁻²) and Energy Costs Maintenance
(kJm⁻²) of Soil Invertebrates during the Growing Season

	Wheat Field		Meadow	
Taxon	Biomass	Energy Maintenance	Biomass	Energy Maintenance
Protozoa	440.0	1190.0	525.0	1360.0
Nematoda	557.4	314.6	816.6	499.5
Lumbricidae	940.0	56.0	3840.0	254.2
Enchytraeidae	50.9	26.5	198.6	88.7
Acarina	11.2	1.3	475.7	22.6
Collembola	47.4	20.5	65.6	21.1
Winged insects mainly larvae	109.6	14.6	4883.0	223.7
Total	2156.5	1623.5	10,704.5	2469.8

Source: Modified from Ryszkowski and Karg 1996.

dividing that value by the sum of dry weights of populations composing community (Ryszkowski and Karg 1996).

In all the ecosystems studied, Protozoa made the greatest contribution to energy flow through animals. Second in the rank, but much lower than Protozoa, is the contribution of Nematoda to dissipation of energy. The soil larvae of insects with very low contributions to energy flow through total soil community consisted of the group of large invertebrates that react very strongly to the intensity of farmer interference with ecosystems. Greater intensity of agricultural interference has resulted in elimination of large insects. The mean body size of winged insect larvae in meadow was about 7.4 times heavier than the mean body weight of larvae found in the wheat field (Table 8.7). Insect larvae in meadow made up 45.6% of the total community biomass, but contributed a total of only 9.0%. In wheat cultivation their contribution to biomass structure was 5.0%, but in terms of dissipated energy their share was 0.8% (Table 8.8). The same was true for earthworms. In terms of biomass, their contribution was substantial, while they dissipated only a small amount of energy.

All these comparisons show that agricultural activity impoverishes animal communities. More simplified plant cover structure accompanied by frequent farmer interference with soil conditions relates to more intensive energy flow through edaphon communities. This observation means that the use of organic matter by animal communities is more intensive in cultivated fields than in other ecosystems of the agricultural landscape and that energy storing capacities are diminished by elimination of large and long-lived species. Turnover of the organic matter in agro-ecosystems is therefore increased.

Among other functional changes appearing under agricultural pressure, Karg and Ryszkowski (1996) and Ryszkowski and Karg (1996) found that belowground fauna

factor is discussed later in this chapter when influence of landscape structure on insect and bird communities is shown.

Another functional characteristic of agroecosystems is that herbivores constitute a higher proportion of trophic structure in the wheat field than in the meadow (Ryszkowski and Karg 1996). The same situation is observed in many other crops. For example, in potato cultivation, Colorado beetles (*Leptinotarsa decemlineata*), a pest to this cultivar, constitute 84% of the total biomass of the aboveground insect community (Ryszkowski and Karg 1977). In the rapeseed cultivations, *Meligethes aeneus*, which is also a pest, makes up more than 60% of the total aboveground biomass of the insect community (Ryszkowski and Karg 1986). Cereal cultivations have increasing populations of aphids and *Lema melanopa* because of the increasing trend of domination of cereals in crop rotation in Poland.

Agricultural measures influence survival of particular groups of animals in different ways. There are numerous publications discussing reactions of particular animal populations to pesticides, fertilizers, ploughing, drainage, and other measures (Lowrance et al. 1984, Mills and Alley 1973, Tischler 1965, Braman and Pendley 1993, Tucker and Heath 1994, Goh Hyun Gwan et al. 1995, Lopez Fando and Bello 1995, Edwards and Bohlen 1996, Hu Feng et al. 1997, Ellsbury et al. 1998, Krooss and Schaefer 1998, Heath and Evans 2000).

According to a thorough review of literature by Karg and Ryszkowski (1996), agricultural measures do not significantly impoverish Protozoa, Enchytraeidae, or Collembola. However, that finding does not mean that these groups are immune to agricultural influences. Some species show decreased numbers while others can increase their abundance. For example, Collembola in cultivated fields are represented by a fewer number of species and display large variation over time and low stability of species composition, but their long-term density is similar to that observed in forests or grasslands, with the exception of that found in some forests rich in litter. Similar results were recently published by Alvarez et al. (2001) indicating changes in particular species densities but no significant differences in total Collembola assemblages in organic, integrated, and conventional farming systems. Protozoa also show significant density fluctuations. Paprocki (1992) has shown a statistically significant correlation between soil humidity, the content of organic matter in soil, and the density of Protozoa. Nevertheless, comparisons of different estimates of the Protozoa population density shows that their abundance in cultivated fields is similar to that in forest and meadow ecosystems. There are even indications that tillage stimulates their abundance.

Groups suppressed by agricultural measures include nematodes, earthworms, mites, spiders, and winged insects. Especially dramatic reduction in density and species composition is observed in earthworms, mites, and winged insects.

Attempts to explain the causes responsible for invertebrate elimination from cultivated fields indicate an interplay of various factors rather than a single agent

1995, Castro et al. 1996, Weber et al. 1997). Majority of earthworm populations are eradicated from cultivated fields because of lack of litter accumulation and frequent ploughing, that is by change of abiotic and food conditions than by application of agrochemicals with probable exception of fungicides (Paoletti 1988, Tucker 1992, Edwards and Bohlen 1996, Makulec 1997). Landscape structure diversity also has an important influence on the persistence of animals under agricultural stress, as discussed below.

ABUNDANCE AND DIVERSITY OF ANIMAL COMMUNITIES IN AN AGRICULTURAL LANDSCAPE

Two aspects of landscape structure influence on animal communities can be distinguished. Distribution of species in the landscape reflects their adaptation to habitat conditions, e.g., some species can live in forest but not in grassland. Dispersion of animals from refuge sites can compensate for the losses caused by agricultural measures in a particular field. Thus, for example, in the Turew landscape the highest abundance as well as species diversity of earthworms was found in meadows (Table 8.9).

Meadows or shelterbelts in the landscape result, therefore, in high diversity of those animals. Dominance of older specimens in cultivated fields, in contrast to meadows, could be explained by limited reproduction in the soil of agroecosystems and migration of observed mature individuals in fields from refuge sites such as meadows or shelterbelts. A similar situation is observed in other groups of animals. Thus, the number of Thysanoptera species in the whole Turew landscape amounts to 41, while in cultivated fields 13 to 18 species can be found (D. Szeflińska, personal communication). Very high densities of Thysanoptera populations are recorded in cereal cultivations, although their species diversity is low (Table 8.10). This is the

Table 8.9   Distribution of Earthworms in the Turew Landscape

Habitat	Cultivated Fields			Meadow	Shelterbelt
	Rapeseed	Wheat	Alfalfa	Meadow	Shelterbelt
Density of individuals·m⁻²	8.7	10.0	23.8	74.0	15.0
Biomass g l.w.·m⁻²	2.2	3.8	5.5	15.2	7.7
Number of species	4	3	2	7	4

Source: Modified after Ryl 1984, Karg and Ryszkowski 1996.

Table 8.10   Density and the Number of Thysanoptera Species in Turew Agricultural Landscape

	Density	Number

Table 8.11 Mean Annual Density and Biomass
of Enchytraeidae in Soil of Various
Ecosystems in the Turew Landscape

Cultivation	Number of Estimates	Density Individuals·m^{-2}	Biomass mg d.w.·m^{-2}
Maize	4	2000	46.8
Spring cereals	3	2400	84.6
Winter crops	4	6450	180.0
Alfalfa	7	3700	127.8
Sugar beets	4	3900	93.6
Potatoes	5	8000	225.0
Road side	2	8400	—
Meadows	5	8800	241.2
Shelterbelts	4	9700	—

Source: Ryl 1977, 1980.

Table 8.12 Biomass of Aboveground Insects
in the Agricultural Landscape of Turew

Agroecosystem	Number of Estimates	Mean Year-long Biomass mg·m^{-2} Dry Weight
All spring cereals	13	11
All rows crops	13	20
All winter cereals	10	21
Rapeseed	5	28
Alfalfa	11	36
Meadows	8	45

Source: Karg and Ryszkowski 1996.

result of increasing contribution of cereals in rotation pattern. A quite different
pattern of distribution of Enchytraeidae is observed. In cultivations of crops for
which manure was used, such as potatoes, the density and biomass of those animals
are comparable to those in meadow or shelterbelt habitats (Table 8.11).

Total biomass of aboveground insects reaches highest values in agroecosystems
subjected to the smallest agricultural stress. Comparison of densities of total insect
communities between different ecosystems is meaningless because of great differences in body weights in various species (from 0.002 mg d.w. per ind. to 750 mg
d.w. per ind.). In meadows, the mean year-long standing biomass was estimated at
45 mg m^{-2}, which is four times larger than biomass observed in spring cereals
(Table 8.12).

Using the same method of quick-trap sampling as used in the Turew landscape,
aboveground insect fauna was studied in Romania, Italy, and Russia (Ryszkowski

INFLUENCE OF LANDSCAPE MOSAIC STRUCTURE 201

Table 8.13 Mean Biomass of Above-ground Insects in Annual Crops and Grasslands in Different Regions of Europe (mg·m⁻² dry weight)

Country	Annual Crop	Grassland
Poland	39.4	45.2
Italy	63.7	92.3
Russia	47.2	149.1
Romania	240.1	279.3

Source: Karg and Ryszkowski 1996.

The differences in the levels of mean standing biomass among countries to some extent reflect the adaptation of insects to prevailing environmental (climatic) conditions. The taxonomic structure of the insect community in the southern regions of Europe is characterized by the domination of large-sized Orthoptera species, which make up 60% of the total mean biomass in Romania and 48% in Italy. In Poland and Russia, Diptera, which are much smaller than Orthoptera dominate. Nevertheless, in all situations, lower biomass was observed in annual crops in comparison to grasslands.

Flying insects in the Turew landscape were sampled by catching them in nets fixed at various heights on motorcycle (Karg 1980). The distribution of flying insects over the whole landscape could be evaluated in a very short time span (e.g., during one day) by this method.

Analyzing records of flying insect distribution in all habitats of the Turew landscape, it is possible to distinguish three groups of habitats (Karg and Ryszkowski 1996). To the first group belong villages and spring crops in which the abundance of flying insects is very low. The average season-long biomass per one cubic meter of air in these habitats varied from 0.10 to 0.13 mg of dry weight. The second category of habitats consists of winter crops and perennial crops with biomass ranging from 0.21 to 0.25 mg of dry weight. The highest biomass, ranging from 0.26 to 0.29 mg dry weight, was found in meadows, shelterbelts, and afforestations. Taking into account the whole landscape, one can detect high taxonomic diversity of the insect community in mosaic landscapes. Of a total 3483 samples, Karg (1989) distinguished 172 taxonomic families from 18 orders of insects.

Animal species of the Turew mosaic landscape make a considerable contribution to the faunistic lists of the region and even the country.* For instance, despite a relatively poor water network in the studied area, preservation of small mid-field water bodies allows for the appearance of 36 dragon fly species (Odonata), 50% of the country list, 40 species of water bugs (Heteroptera), 80% of the total province list, and 90 water beetle species (Coleoptera), 62% of the province list. Among the

Table 8.14 A Comparison of Aboveground Insect Communities
in Mosaic and Uniform Agricultural Landscapes

Characteristics	Landscape		Statistical Significance
	Mosaic	Uniform	
Mean number of families	59.6	49.4	P < 0.001
Mean density (ind/m²)	61.9	40.7	P < 0.01
Mean biomass (mg d.w./m²)	55.0	40.3	P < 0.05

Source: Karg and Ryszkowski 1996.

terrestrial insects in the superfamily of bees (Apoidea), 258 species were recognized
making 80% of the province list. More than 40 species of bark-beetles (Scolytidae)
occur, about 70% of species list recorded in the province. Among Thysanoptera,
41 species were recorded. More than 600 species of Lepidoptera were identified.
Earthworms (Lumbricidae) are represented by at least 7 species, Enchytraeidae by
16 species, nematodes by 40 species, spiders by 49, and Acarina by 216 species.
Vertebrates are represented by 12 species of amphibia, 4 reptiles, 119 birds, and
47 mammals. Recorded terrestrial vertebrate species in the Turew mosaic landscape
make up almost 100% of the province list with the exception of birds, which make
up 62% of the regional list for breeding species.

The above data clearly indicate that changes in the land-use pattern of a landscape
have an enormous effect on density, biomass, and taxonomic diversity of animals.
An increase of area under cereal cultivation will lead to overall impoverishment of
animal resources in the landscape. This process is under way presently in Poland
because of ongoing simplification of crop rotation patterns in favor of cereals. The
area under cereal cultivation is currently 31.9% of the total territory of the country.

Existence of numerous refuges for animals from which they can disperse into
cultivated fields plays an important role in the survival of many species of animals.
Analyzing the same cultivations in two types of landscapes showed that in the mosaic
landscape not only density and biomass but also taxonomic diversity are greater
than in a uniform landscape composed mainly of cultivated fields (Table 8.14). A
similar relationship between the structure of landscape and the richness of aboveground insect communities was found by Karg (1997) in the northwest region of
France (vinicity of Soissons). Higher biomass and taxonomic diversity was observed
in mosaic agricultural landscapes than in uniform ones.

Both Turew landscapes under investigation were situated close to each other
(about 10 km apart), and the intensity of field cultivation (fertilizers, tillage) and
application of pesticides were similar. The results show that 20% higher diversity
of insect community could be maintained in a mosaic agricultural landscape than
in a uniform one, despite intensive agriculture. Dispersion of insects from refuge

Table 8.15 Taxonomic Diversity of Aboveground Insect Communities in Uniform and Mosaic Landscapes — Number of Families in Particular Crops

Crop	Number of Estimates	Landscape Uniform	Landscape Mosaic
Spring crops			
Oats	1	48	56
Barley	6	61	65
Maize	6	65	77
Winter crops			
Wheat	5	90	121
Perennial crops			
Alfalfa	11	100	121

Table 8.16 Mean Biomass (mg d.w.·m^{-2}) and Its Share Indicated as Percentage (in parentheses) of Main Trophic Groups of Aboveground Insect Communities Located in Mosaic and Uniform Agricultural Landscapes

Trophic Group	Cereal Crops in Landscape Mosaic	Cereal Crops in Landscape Uniform	Sugar Beets in Landscape Mosaic	Sugar Beets in Landscape Uniform	Alfalfa in Landscape Mosaic	Alfalfa in Landscape Uniform
Herbivores	16.3 (44.5)	9.5 (45.7)	23.1 (45.7)	19.4 (53.1)	48.7 (55.5)	33.8 (55.9)
Saprovores	8.4 (22.9)	5.2 (25.0)	11.1 (22.0)	7.0 (19.1)	20.0 (22.8)	15.7 (25.9)
Predators	10.5 (28.8)	5.4 (26.0)	13.8 (27.4)	8.5 (23.4)	13.9 (15.9)	7.7 (12.7)
Parasites	1.4 (3.8)	0.7 (3.3)	2.5 (4.9)	1.6 (4.4)	5.1 (5.8)	3.3 (5.5)
Total	36.6 (100.0)	20.8 (100.0)	50.5 (100.0)	36.5 (100.0)	87.7 (100.0)	60.5 (100.0)

groups. Almost two times more standing biomass of predators can be observed in fields located in mosaic landscape than in uniform agricultural landscape devoid of refuge sites (Table 8.16).

Predatory and parasitic species of insects disperse from refuge sites and populate cultivated fields, enhancing the biological control of herbivorous populations. Taking the ratio of total standing biomass of herbivores and saprovores to total standing biomass of predators and parasites as an index of predator pressure on prey populations one can infer that biological control is more effective in a mosaic landscape than in a uniform one. Although the intensity of predator pressure on prey varies in different types of cultivations, the trend is nevertheless consistent in all cases, showing that the lower biomass of herbivores and saprovores falls on biomass unit of predators and parasites in fields located in mosaic landscapes. The lower this ratio is, the higher is the pressure of predators on prey (Table 8.17).

In studies carried out in Romania by the same method as in Turew (Ryszkowski

Table 8.17   Index of Predator Pressure on Prey Populations Measured as the Ratio
             of Herbivores and Saprovores (prey) Standing Biomass (mg d.w.·m^{-2})
             to Predators and Parasites (predator) Biomass

Kind of Landscape	Cereal Crops in Landscape		Sugar Beets in Landscape		Alfalfa in Landscape	
	Mosaic	Uniform	Mosaic	Uniform	Mosaic	Uniform
Prey	24.7	14.7	34.2	26.4	68.7	49.5
Predator	11.9	6.1	16.3	10.1	19.0	11.0
Predator pressure	2.1	2.4	2.1	2.6	3.6	4.5

Table 8.18   Mean Biomass (mg d.w.·m^{-2}) of Aboveground Insects in Uniform
             and Mosaic Landscapes in Romania

Cultivation	Wheat	Barley	Maize	Sugar Beet	Alfalfa	Mean
Uniform landscape	98.7	375.9	83.3	111.5	73.0	148.5
Mosaic landscape	101.2	584.0	174.5	100.8	199.2	231.9

Source: Modified from Ryszkowski et al. 1993.

The fields in the mosaic landscape were small and refuge sites covered about
12% of the studied area. In a uniform landscape there were no shelter habitats for
a few kilometers from the studied fields. Crop rotation was also very simplified,
showing in many cases of cereal cultivation almost continuous cropping of the same
cultivar. Under such conditions, no correlation was found between predator and
parasitic biomass and their potential prey (herbivores and saprovores). The pests to
crops, such as *Eurygaster integriceps* (Heteroptera), had very high biomass and
species from the Iassidae and Chysomelidae families had very high herbivore biomass in relation to existing predators. In such cases of very high uniformity of the
agricultural landscape, the effectiveness of biological control is very small.

The range of field penetration by various insects is differentiated, and therefore
much higher densities of predatory forms are observed close to the refuge site than
in the middle of a field (Thomas et al. 1991, Karg and Kundzewicz 1992, Karg and
Szeflińska 1996). Studies on aphids and their predators clearly illustrate that point
(Figure 8.2). Aphids that overwinter in shelterbelts show an almost exponential
decline in densities as the distance increases from shelterbelts. Their distribution
in the field is mimicked by predators (Syrphidae). In the ecotone zone (up to 20
m from the shelterbelt) where densities of aphids were high, eight species of
Syrphidae as well as 3.5-fold higher densities of their pupae were observed in
contrast to three species and much lower pupal densities recorded in the middle of
the field located 50 to 100 m from the shelterbelt. In addition, Chalcidoidea and
Ichneumonidae (10 species) parasites to Syrphidae showed threefold higher inva-

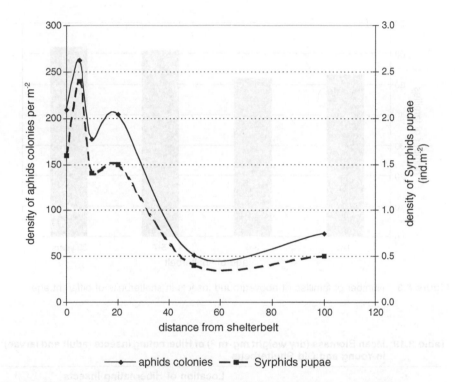

Figure 8.2   Aphids and predatory Syrphids in relation to distance from shelterbelt.

rich animal communities depends on the presence of refuge sites. The habitats less disturbed by agricultural measures have better conditions for survival. The soils of the spring crops with the most frequent impacts of tillage activities usually show less abundance of animals than do winter crops and perennial crops, while the highest abundance is detected in meadows, shelterbelts, and mid-field forest patches.

When a new shelterbelt is planted in a cultivated field, mobile animals, such as insects, very quickly colonize that newly created refuge site. In the first or second year a very high biomass, reaching 95.2 mg d.w.m^{-2}, was observed. With the growing shelterbelt the diversity of the community increases, reaching after 5 to 7 years similar number of taxonomic families similar to that observed in 80-year-old shelterbelts (Figure 8.3). As early as in the first or second winter after planting of the shelterbelt, numerous insect communities can be found in its soil.

The newly created refuge site in the landscape is populated very quickly by mobile animals such as insects. The biomass of hibernating insects in the newly planted shelterbelts is almost 15 times higher than that in a cultivated field (Table 8.19). After 3 to 5 years, the biomass of hibernating insects is only 3% lower

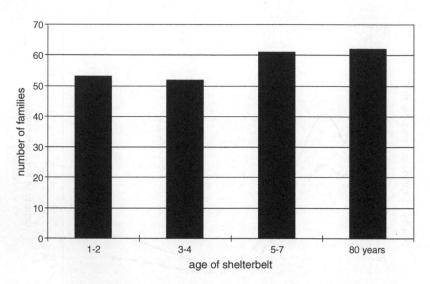

Figure 8.3   Number of families of aboveground insects in shelterbelts of different age.

Table 8.19   Mean Biomass (dry weight mg·m⁻²) of Hibernating Insects (adult and larvae) in Young and Old Shelterbelts

Habitat	Age of Shelterbelt (years)	Soil (10 cm deep)		Litter and Aboveground Dry Plants		Total	
		Mg·m⁻²	Percent	mg·m⁻²	Percent	mg·m⁻²	Percent
Shelterbelt	1–2	681.3	88.9	85.1	11.1	766.4	100.0
	3–5	842.9	86.5	131.9	13.5	974.8	100.0
	80	845.8	83.7	164.8	16.3	1010.6	100.0
Cultivated fields 50–100 m from shelterbelts	—	51.9	100	—		51.9	100

A well-developed mosaic pattern of shelterbelts provides refuge sites where animals can overwinter and find shelter from harmful agricultural measures such as the application of pesticides. The fields where animals were eliminated by harmful agents can be recolonized from unaffected refuges situated in the mosaic landscape. Thus one can suppose that the main factors counteracting biodiversity decline caused by agriculture are the mosaic structure of landscape and dispersal properties of species. Successful biodiversity protection programs in agricultural landscapes

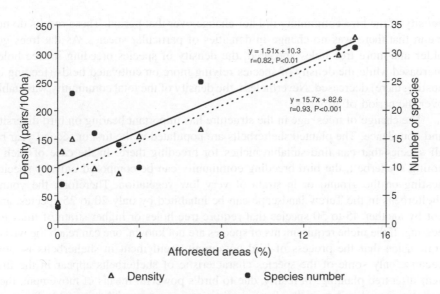

Figure 8.4   Relation between afforested areas (%) and number of species and total bird
density in the Turew agricultural landscape.

Table 8.20   Breeding Bird Communities in Various Mid-Field Afforestations

Characteristics	Tree Patches (N = 21)	Shelterbelts (N = 33)	Alleys (N = 20)
Number of species	60	51	32
Mean number of species/ha	14.1	16.8	7.6
Density (pairs·ha⁻¹)	14.9	18.3	9.8

N – number of plots.
Source: Ryszkowski, et al., 1999.

almost 80 species of breeding birds. The diversity of bird communities depends on
the contribution of area under shelterbelts or afforestations in a total landscape
territory (Kujawa 1996, Kujawa and Tryjanowski 2000). The density of birds in rural
areas is also related to the proportion of area covered by afforestations (Figure 8.4).
The share of area under afforestations in total landscape area is the most important
factor influencing the richness of bird communities. The type of afforestation also
influences diversity and density of bird communities (Table 8.20).

In the afforestations analyzed, avian fauna is richest in tree patches and poorest
in alleys (single row of trees along roads, ditches, etc.). Kujawa (1997) showed that
the number of species and density of birds in various afforestations depends on the

density of the bird community did not change over that period. These results do not mean that there was no change in densities of particular species. As the trees got older and more tree holes appeared, the density of species breeding in tree holes increased while the density of species relying more on cultivated fields (feeding or nesting there) decreased. Nevertheless, the density of the total community was stable over the period of 30 years.

The change of trees' age in the structure has important bearing on bird diversity and abundance. The planted shelterbelts are populated in the first or second year by all species that can find suitable niches for breeding there. In the case of such a young shelterbelt, the bird breeding community can be composed only by species nesting on the ground or in strata of very low vegetation. Therefore, the young shelterbelts in the Turew landscape can be inhabited by only 20 to 25 species, and not by another 45 to 50 species that require tree holes or higher strata of trees for nesting. If the niche requirements of species are not known, one can reach the wrong conclusion that the process of bird community settlement in shelterbelts is slow because only some of the species characteristic of shelterbelts appear in the first year after tree planting. In reality, due to bird's powerful means of movement, they populate new habitats in the landscape as soon as new niches suitable for them appear. Probably the same interpretation applies in the case of insects populating newly planted shelterbelts, but their specific niche requirements are poorly known.

The importance of shelterbelts as well as other nonarable habitats of agricultural landscape to bird species diversity and abundance has been demonstrated in many studies (see, for example, Green et al. 1994, Best et al. 1995, Fuller et al. 2001, Jobin et al. 2001 among others). All these studies provide information complementary to studies reported on the Turew mosaic agricultural landscape indicating that shelterbelts provide important habitats for birds and, by that token, can counteract impoverishment of their communities in rural areas. Although some woodland species of birds do not inhabit shelterbelts, many of the other species survive well in agricultural landscapes due to introduction of shelterbelts. To this last category belong such endangered species as the red-backed shrike (*Lanius collurio*), great gray shrike (*Lanius excubitor*), ortolan bunting (*Emberiza hortulana*), and corn bunting (*Miliaria calandra*). These two last species nest on sheltered ground and need trees as singing posts. With a more complex structure of shelterbelt, more species can find suitable niches for breeding. Nevertheless, some woodland species do not nest in shelterbelts, for example, goshawk (*Accipiter gentilis*), hobby (*Falco subbuteo*), middle spotted woodpecker (*Dendrocops medius*), wood warbler (*Phylloscopus sibilatrix*), pied flycatcher (*Ficedula hypoleuca*), among others. That situation may change in the future with respect to some of them, as recently happened with prior shy species such as the raven (*Corvus corax*), crane (*Grus grus*), gray lag goose (*Anser anser*) which now use tree patches (raven) or small mid-field ponds

changed environmental conditions although some are unable to adapt and therefore vanish. The mosaic agricultural landscape probably better facilitates adaptations to agriculture than does the open landscape composed only of large cultivated fields.

PROSPECTS FOR THE BIOLOGICAL DIVERSITY MANAGEMENT IN AGRICULTURAL LANDSCAPES

The Convention on Biological Diversity issued in 1992 at the World Summit in Rio de Janeiro strongly emphasized the need to stop and reverse the impoverishment of biological diversity. To overcome some operational shortcomings of this policy, which did clearly indicate the reciprocal links between protection of biodiversity and economic activities and did not recognize the importance of landscape ecological functions for species protection, the Pan-European Strategy for Biological and Landscape Diversity Strategy was adopted in 1995 by the Environment Ministers of the European Union. Two points are stressed clearly in this last document. The first one emphasizes initiatives promoting the integration of a biodiversity strategy with social and economic sectors. Successful living resource protection can be achieved when nature is not separated from or shielded against socioeconomical activities, but when human activities are reconciled with protection of biodiversity. The second one stresses the need not only for conservation but also for management of protected areas. Failure to manage structures and processes supporting biota results in unsuccessful execution of protection goals. The landscape approach, which embodies all provisions to sustain biological diversity, is thus gaining the utmost significance. This approach is becoming more important as humanity's effects on nature increase and the area of the earth free from human interferences rapidly dwindles. The success of such a policy relies on the use of adequate tools, including the important role played by landscape ecology recognizing countryside structures, processes, and services enabling maintenance of the biological diversity in rural areas.

This brief of the studies carried out in the Turew landscape demonstrated that diversification of the landscape structure is an important tool for the maintenance of rich plant and animal communities, including some protected or endangered species. The importance of nonarable habitats for biodiversity protection in the agricultural landscape was also recognized by other scientists. For example, Bunce and Hallam (1993) showed that various nonarable components of landscape are important reservoirs for protection of plant species because they can function as sources of propagules for recolonization of the agricultural landscapes.

The enrichment effect of nonarable patches in landscape on populations or assemblages of taxonomically close species are well documentated in ornithological

and Łukaszewicz 1994, Lagerlöf et al. 1992, Paoletti et al. 1992, Thomas and Marshall 1999, Varchola and Dunn 2001 and others). The important role of shelterbelts and other linear features of permanent vegetation was shown for dispersion of mammals in the landscape (Merriam 1991, Dobrowolski et al. 1993). Linear features of vegetation in the case of mobile species facilitate movement among habitat patches and ensure their recolonization. The interrelations between persistence of population and pattern of corridors connecting separate habitats were described by the metapopulation theory proposed by Opdam (1988).

Thus it seems that the importance of the mosaic landscape structure for survival of plant or animal populations is evident, at least in the case of some populations, and implementation of nonarable patches into the open fields should be proposed for programs of nature protection.

The studies carried out in the Turew landscape clearly show the impoverishment of the total set of animals in cultivated fields although some taxonomic groups, such as Protozoa, Enchytraeidae, and Collembola, do not show distinct changes under pressure of agriculture. Herbivores are more numerous and some species, such as crop pests (e.g., Colorado beetles [*Leptinotarsa decemlineata*] in potato cultivations), may even achieve the level of 84% of total aboveground insect biomass (Ryszkowski and Karg 1977). With that background, it is clearly seen that when agricultural landscape is properly managed in a mosaic, losses of biodiversity can to some extent be counteracted. The introduced refuge sites, such as newly planted shelterbelts, are very quickly populated by many species — an important finding for successful programs of biodiversity restoration. In the case of landscapes composed only of large cultivated fields, herbivores are almost free from predator pressure (Ryszkowski et al. 1993). With increasing eradication of refuge sites, an important functional change is observed among soil invertebrates; species having smaller body sizes, higher reproduction success, and shorter developmental cycles are dominating. This finding means that energetic costs of maintenance of soil animal community is higher than in the mosaic landscape, and outbreaks of population densities can appear more frequently in the uniform agricultural landscape.

Plant diversity in agricultural landscape contributes to overall diversity not only by its modifications of habitat conditions, enabling persistence of other plant species, but also by its strong influence on animal diversity. A complicated structure of shelterbelt not only consists of a diversified plant community but also harbors more animal species than does a cultivated field. Long-term studies on the biota in uniform and in diversified agricultural landscapes show that the mosaic character of landscape plays an important role in maintaining the richness of plant and animal communities. This function is a result of the existence of numerous refuges for plants and animals and for their potential dispersal. The mosaic agricultural landscapes can therefore support biodiversity protection in reserves, national parks, and other territorial forms of nature conservancy that are usually surrounded by cultivated fields. The strategy of biodi-

REFERENCES

Adamiak E., Zawiślak K. 1990. Changes in weed communities under continuous cropping of cereals and maize [in Polish] L. Ryszkowski, J. Karg and J. Pudełko. (Eds.) In *Ekologiczne procesy w monokulturowych uprawach zbóż.* Zakład Badań Środowiska Rolniczego i Leśnego, Poznań: 47–75.

Adamiak E., Zawiślak K. 1990a. Weed communities in cereal grown in rotations and long-term monoculture. Part II. Winter wheat. In *Badania Monokultur Zbożowych.* Szkoła Główna Gospodarstwa Wiwjskiego, 16, Warszawa: 27–42.

Alvarez T., Frampton G. K., Goulson D. 2001. Epigeic Collembola in winter wheat under organic, integrated and conventional farm management regimes. *Agric. Ecosyst. Environ.* 83: 95–110.

Baldock D., Beaufoy S., Bennett G., Clark J. 1993. *Nature Conservation and New Directions in the EC Common Agricultural Policy.* Institute for European Environmental Policy. London. 224 pp.

Banaszak J. 1983. Ecology of bees (Apoidea) of agricultural landscape. *Pol. Ecol. Stud.* 9: 421–505.

Barberi P., Silvestri N., Bonari E. 1995. Weed flora development in different cropping systems. *Rivista di Agronomia* 29: 523–532.

Barberi P., Silvestri N., Bonari E. 1997. Weed communities of winter wheat as influenced by input level and rotation. *Weed Res.* (Oxford) 37: 301–313.

Berger L. 1987. Impact of agriculture intensification on Amphibia [in Polish]. In *Proceedings of the 4th Ordinary General Meeting of the Societas Europaea Herpetologica,* J.J. van Gelder H., Strijbosh H. and Bergers P.J.M. (Eds.), Nijmegen: 79–82.

Best L. B., Freemark K. E., Dinsmore J. J., Camp M. 1995. A review and synthesis of habitat use by breeding birds in agricultural landscapes of Iowa. *Am. Midland Nat.* 134: 1–29.

Borowiec S., Kutyna I., Leśnik T. 1992. The weedy state of private farm fields located in south-western part of Szczecin Lowland in the years 1970–1974 and 1986–1990. *Zeszyty Naukowe Akademii Rolniczej w Krakowie* 33: 7–14.

Borysiak J., Brzeg A., Kasprowicz M. 1993. Interesting values of plant cover of "Szwajcaria Zerkowska" landscape protected area [in Polish]. *Badania Fizjograficzne nad Polską Zachodnią,* B 42: 169–200.

Braman S. K., Pendley A. F. 1993. Relative and seasonal abundance of beneficial arthropods in centipedegrass as influenced by management practices. *J. Econ. Entomol.* 86: 494–504.

Buhler D. D., Doll J. D., Proost R. T., Visocky M. R. 1995. Integrating mechanical weeding with reduced herbicide use in conservation tillage in production systems. *Agron. J.* 87: 507–511.

Bunce R. G. H., Hallam C. J. 1993. The ecological significance of linear features in agricultural landscapes in Britain. In *Landscape Ecology and Agroecosystems.* R. G. H. Bunce, L. Ryszkowski, M. G. Paoletti (Eds.). Lewis Publishers, Boca Raton, FL: 11–19.

Burel F. 1996. Hedgerows and their role in agricultural landscapes. *Critical Rev. Plant Sci.* 15: 169–190.

Castro J., Campos P., Pastor M. 1996. Soil arthropods under the influence of soil management

Cyrul D. 2000. *Polską-Parki Narodowe*. Muza SA, Warszawa, 319 pp.

Dennis P., Thomas M. B., Sotherton N. W. 1994. Structural features of field boundaries which influence overwintering densities of beneficial arthopod predators. *J. Appl. Ecol.* 31: 361–370.

Dobrowolski K., Banach A., Kozakiewicz A., Kozakiewicz M. 1993. Effect of habitats barriers on animal populations and communities in heterogenous landscapes. In *Landscape Ecology and Agroecosystems*. R. G. H. Bunce, L. Ryszkowski, M. G. Paoletti (Eds.). Lewis Publishers, Boca Raton, FL: 61–70.

Edwards C. A., Bohlen P. J. 1996. *Biology and Ecology of Earthworms*. Chapman and Hall, London, 426 pp.

Ellsbury M. M., Powell J. E., Forcella F., Woodson W. D., Clay S. A., Riedell W. E. 1998. Diversity and dominant species of ground beetle assemblages (Coleoptera: Carabidae) in crop rotation and chemical input systems for the northern Great Plains. *Ann. Entomol. Soc. Am.* 91: 619–625.

Exner D. N., Thompson R. L., Thompson S. N. 1996. Practical experience and on-farm research with weed management in an Iowa ridge-based system. *J. Prod. Agric.* 9: 496–500.

Fick G. W., Power A. G. 1992. Pest and integrated control. In *Field Crop Ecosystems*. C. J. Pearson (Ed.). Elsevier, Amsterdam: 59–83.

Fijałkowski D., Taranowska B., Sawa K. 1992. Changes in weed infestation of selected fields in the Lublin region in the years 1973 and 1986 [in Polish]. *Zeszyty Naukowe Akademii Rolniczej w Krakowie* 33: 15–26.

Foissner W. 1992. Comparative studies on the soil life in ecofarmed and conventionally farmed fields and grasslands of Austria. In *Biotic Diversity in Agroecosystems*. M. G. Paoletti, D. Pimentel (Eds.). Elsevier, Amsterdam: 207–218.

Fuller R. J., Chamberlain D. E., Burton N. H. K., Gough S. J. 2001. Distribution of birds in lowland agricultural landscapes of England and Wales: How distinctive are bird communities of hedgerows and woodland? *Agric. Ecosystems and Environ.* 84: 79–92.

Ghersa C.M., Roush M.L. 1993. Are weed problems caused by competition or dispersion? *Bioscience* 43: 104–109.

Głowaciński Z. 1998. Magurski Landscape Park [in Polish]. In *Encyklopedia Biologiczna*, VI, OPRES, Kraków: 237–238.

Goh Hyun Gwan, Choi Dong Ro, Kim Hong Sun, Lee Young Hwan, Hwang Kwang Nam. 1995. Survey on the microanimals in crop fields farmed organically. *RDA J. Agric. Sci. Crop Prot.* 37: 371–375.

Gołdyn H., Arczyńska-Chudy E. 1998. Plant diversity in Landscape Park of General D. Chłapowski and its protection [in Polish]. In *Kształtowanie Środowiska rolniczego na przykładzie Parku Krajobrazowego im. Gen. D. Chłapowskiego*. L. Ryszkowski and S. Bałazy (Eds.). Zakład Badań Środowiska Rolniczego i Leśnego PAN, Poznań: 123–132

Good J. A., Giller P. S. 1991. The effect of cereal and grass management on staphylinid (Coleoptera) assemblages in south-west Ireland. *J. Appl. Ecol.* 28: 810–826.

Gould F. 1991. The evolutionary potential of crop pests. *Am. Sci.* 79: 496–501.

Green R. E., Osborne P. E., Sears E. J. 1994. The distribution of passerine birds in relation to characteristic of the hedgerow and adjacent farmland. *J. Appl. Ecol.* 31: 677–692.

Gromadzki M. 1970. Breeding communities of birds in mid-field afforested areas. *Ekol. Pol.*

Hendrix P. F., Parmelee R. W., Crossley D. A., Coleman D. C., Odum E. P., Groffman P. M. 1986. Detritus food webs in conventional and no-tillage agroecosystems. *BioScience* 36: 374–380.

Hołdyński Cz. 1991. Segetal flora, floristic and ecologic differentiation, and changes in plant cover of cultivated fields in current agroecologic conditions of Żuławy Wiślane. *"Acta Academiae Agriculturae ac Technicae Olstenensis" Agricultura* 51, B, 51 pp.

Hu Feng, Li Hui Xin, Wu Shan Mei. 1997. Differentiation of soil fauna populations in conventional tillage and no-tillage red soil ecosystems. *Pedosphere* 7: 339–348.

Huffaker C. B., Shoemaker C. A., Gutierrez A. P. 1978. Current status, urgent needs and future prospects of integrated pest management. In *Pest Control Strategies*. E. H. Smith, D. Pimentel (Eds.). Academic Press. New York: 237–259.

Jobin B., Choiniere L., Belanger L. 2001. Bird use of tree types of field margins in relation to intensive agriculture in Quebec, Canada. *Agric. Ecosystems Environ.* 84: 131–143.

Johnson G. A., Hoverstad T. R., Greewald R. E. 1998. Integrated weed management using narrow corn spacing, herbicides and cultivation. *Agron. J.* 90: 40–46.

Kajak A., Łukaszewicz J. 1994. Do semi-natural patches enrich crop fields with predatory epigean arthropods. *Agric. Ecosystems Environ.* 49: 149–161.

Karg J. 1980. A. Method of motor-net for estimation of aeroentomofauna. *Pol. Ecol. Stud.* 2: 345–354.

Karg J. 1989. Differentiation in the density and biomass of flying insects in the agricultural landscape of the Western Wielkopolska [in Polish]. *Roczniki Akademii Rolniczej w Poznaniu 188*: 1–78

Karg J. 1997. Preliminary studies on the above-ground insect communities in agricultural landscapes in France and Poland. In *Ecological Management of Countryside in Poland and France*. L. Ryszkowski and S. Wicherek (eds.). Research Centre for Agricultural and Forest Environment. Poznaniu: 93–100.

Karg J., Czarnecki A., Witkowski T., Paprocki R. 1990. Density and biomass of edaphon in continuous cropping of rye and crop rotation. In *Ekologiczne Procesy w Monokulturowych Uprawach Zbóż*. L. Ryszkowski, J. Karg, J. Pudełko (Eds.). Zakład Badań Środowiska Rolniczego i Leśnego PAN. Poznań: 187–195.

Karg J., Kundzewicz Z. 1992. Preliminary evaluation of the ecotone effect on insect biomass in agroecosystems. In *Produkcja Pierwotna, Zasoby Zeierząt i Wymywanie Materii Organicznej w Krajobrazie Rolniczym*. Eds. S. Bałazy, L. Ryszkowski. Zakład Badań Środowiska Rolniczego i Leśnego PAN. Poznań: 113–126.

Karg J., Ryszkowski L. 1996. Animals in arable land. In *Dynamics of an Agricultural Landscape*. L. Ryszkowski, N. French, A. Kędziora (Eds.). Państwowe Wydawnictwo Rolnicze i Leśne. Poznań: 138–172.

Karg J., Szeflińska D. 1996. Predator-prey relationships in forest-cultivated field ecotone [in Polish]. In *Ekologiczne Procesy na Obszarach Intensywnego Rolnictwa*. L. Ryszkowski, and S. Bałazy (Eds.). Zakład Badań Środowiska Rolniczego i Leśnego PAN. Poznań: 45–52.

Kasprzak K., Ryl B. 1978. The influence of agriculture on the occurrence of the Oligochaeta in arable soils. *Wiad. Ekol.* 24: 333–366.

Koehler H. H. 1992. The use of soil mesofauna for the judgment of chemical impact on ecosystems. In *Biotic Diversity in Agroecosystems*. M. G. Paoletti, D. Pimentel (Eds.).

Kreuter T. 1995. Investigations on the effects of changing farming system and landscape structure on the ground beetles (Coleoptera, Carabidae) of a loess site in the Central German dry region. *Mitteilungen der Deutschen Gesellschaft für Allgemeine und Angewandte Entomologie* 10: 501–504.

Krooss S., Schaefer M. 1998. The effect of different farming systems on epigeic arthropods: a five-year study on the rove beetle fauna (Coleoptera: Staphylinidae) of winter wheat. *Agric. Ecosystems Environ.* 69: 121–133.

Kujawa K. 1996. Influence of agricultural landscape structure on nesting birds communities [in Polish]. Ph.D. thesis, Wrocław University.

Kujawa K. 1997. Relationships between the structure of midfield woods and their breeding bird communities. *Acta Ornithologica* 32: 175–184.

Kujawa K. 2000. Bird community of D. Chłapowski Landscape Park. In Birds of Wielkopolska Landscape Parks. *Wielkopolskie Prace Ornitologiczne* 9: 89–121.

Kujawa K. Population density and species composition changes for breeding species in farmland woodlots in west Poland between 1964 and 1994. *Agric. Ecosyst. Environ.*, in press.

Kujawa K., Tryjanowski P. 2000. Relationships between the abundance of breeding birds in Western Poland and the structure of agricultural landscape. *Acta Zoologica Acadamiae Scientarum Hungaricae* 46: 103–114.

Kutyna I. 1988. Weeds in cultivated fields and their associations in the western part of Kotlina Gorzowska and adjoining territories [in Polish]. *Rozprawy,* Szczecin 116. 107 pp.

Labza T., 1994. Ecological and agriculture aspects of cereals and crops weediness in the province of Kraków [in Polish]. Zeszyty Naukowe Akademii Rolniczej im. H. Kołłątaja w Krakowie 194. 122 pp.

Lagerlöf J., Stark J., Svensson B. 1992. Margins of agricultural fields as habitats for pollinating insects. In *Biotic Diversity in Agroecosystems.* M. G. Paoletti and D. Pimentel (Eds.). Elsevier. Amsterdam: 117–124.

Latowski K., Szmajda P., Żukowski W. 1979. Characteristic of ploughland flora in Great Poland exemplified by chosen research stations [in Polish]. *Badania Fizjograficzne nad Polską Zachodnią* 31, B: 65–88.

Lopez Fando C., Bello A. 1995. Variability in soil nematode populations due to tillage and crop rotation in semi-arid Mediterranean agroecosystems. *Soil Tillage Res.* 36: 59–72.

Lowrance R., Stinner B. R., House G. J. (Eds.). 1984. *Agricultural Ecosystems.* Wiley & Sons. New York. 233 pp.

Łuczak J. 1979. Spiders in agrocenoses. *Pol. Ecol. Stud.* 5: 151–200.

MacArthur R. H., Wilson E. O. 1967. *The Theory of Island Biogeography.* Princeton University Press. Princeton. 203 pp.

Makulec G. 1997. Density and biomass of earthworms (Lumbricidae) on leys and permanent meadows. *Ekologia Polska* 45: 815–823.

Merriam G. 1991. Corridors and connectivity: animal populations in heterogenous environments. In *Nature Conservation 2: the Role of Corridors.* D. A. Saunders, R. J. Hobbs (Eds.). Surrey Beatty and Sons. Chipping Norton, N.S.W. Australia. 133–142.

Mills J. T., Alley B. P. 1973. Interactions between biotic components in soil and their modification by management practices in Canada. *Can. J. Plant. Sci.* 53: 425–441.

Paoletti M. G., Pimentel D., Stinner B. R., Stinner D. 1992. Agroecosystem biodiversity: matching production and conservation biology. In *Biotic Diversity in Agroecosystems*. M. G. Paoletti and D. Pimentel (Eds.). Elsevier. Amsterdam: 3–23.

Paprocki R. 1992. Influence of abiotic factors on Protozoa in cultivated soils. *Acta Univ. Nicolai Copernici* XII, 80: 75–84.

Prasad D., Gaur H. S. 1994. *Soil Environment and Pesticides*. Venus Publishing House. New Delhi, India. 400 pp.

Ratyńska H., Szwed W. 1997. Anthropogenic changes of forest communities in the Agricultural Landscape Park near Turew in mid-west Poland. *Fragmenta Floristica et Geobotanica* 42: 131–146.

Reaka-Kudla M. L., Wilson D. E., Wilson E. O. 1997. *Biodiversity*. Joseph Henry Press. Washington. 551 pp.

Russel E. J. 1977. *Soil Conditions and Plant Growth*. Longman. London. 849 pp.

Ryl B. 1977. Enchytracidae on rye and potato fields in Turew. *Ekol. Pol.* 25: 519–529.

Ryl B. 1980. Enchytraeid (Enchytracidae, Oligochaeta) populations of soil of chosen cropfields in the vinicity of Turew (Poznań Region). *Pol. Ecol Stud.* 6: 277–291.

Ryl B. 1984. Comparison of communities of earth worms (Lumbricidae) occurring in different ecosystems of agricultural landscape. *Ekol Pol.* 32: 155–165.

Ryszkowski L. 1982. Structure and function of the mammal community in an agricultural landscape. *Acta Zool. Fennica* 169: 45–59.

Ryszkowski L. 1985. Impoverishment of soil fauna due to agriculture. *Intecol Bull.* 12: 7–17.

Ryszkowski L. 1994. Strategy for increasing countryside resistance to environment threats. In *Functional Appraisal of Agricultural Landscape in Europe*. L. Ryszkowski and S. Bałazy (Eds.). Research Centre for Agricultural and Forest Environment. Poznań: 9–18.

Ryszkowski L. 2000. The coming change in the environmental protection paradigm. In *Implementing Ecological Integrity*. P. Crabbe, A. Holland, L. Ryszkowski, L. Westra (Eds.). Kluwer Academic Publishers. Dordrecht: 37–56.

Ryszkowski L., Karg J. 1977. Variability in biomass of epigeic insects in the agricultural landscape. *Ekol. Pol.* 25: 501–517.

Ryszkowski L., Karg J. 1986. Impact on agricultural landscape structure on distribution of herbivore and predator biomass. In *Impact of Structure of Agricultural Landscape on Crop Protection*. J. Missionier and L. Ryszkowski (Eds.). Les Colloques de'INRA 36. INRA Publ. Rte de St-Cyr 78000 Versailles: 39–48.

Ryszkowski L., Karg J. 1996. Changes in animal community functions due to intensity of farming impact. In *Dynamics of an Agricultural Landscape*. L. Ryszkowski, R. French, A. Kędziora (Eds.). Państwowe Wydawnictwo Rolnicze i Leśne. Poznań: 173–184.

Ryszkowski L., French N. R., Kędziora A. (Eds.). 1996. *Dynamics of an Agricultural Landscape*. Państwowe Wydawnictwo Rolnicze i Leśne. Poznań. 223 pp.

Ryszkowski L., Gołdyn H., Arczyńska-Chudy E. 1998. Plant diversity in mosaic agricultural landscapes: a case study from Poland. In *Planta Europa*. H. Synge and J. Akeroyd (Eds.). Plantlife. London: 281–286.

Ryszkowski L., Szajdak L., Karg J. 1998. Effects of continuous cropping of rye on soil biota and biochemistry. *Critical Rev. Pol. Sci.* 17: 225–244.

Ryszkowski L., Karg J., Szpakowska B., Życzyńska-Bałoniak I. 1989. Distribution of phosphorus in meadow and cultivated fields ecosystems. In *Phosphorus Cycles in Terrestrial and Aquatic Ecosystems*. H. Tissen (Ed.) Turner-Warwick Communications,
Saskatoon, Canada: 178–192.

Ryszkowski L., Karg J., Margarit G., Paoletti M. G., Zlotin R. 1993. Above-ground insect
biomass in agricultural landscapes of Europe. In *Landscape Ecology and Agroecosystems*. R. G. H. Bunce, L. Ryszkowski, M. G. Paoletti (Eds.). Lewis Publishers,
Boca Raton, FL: 71–82.

Sendek A. 1992. The weed communities in the cereal cultures of the Middle Triassic Ridge
of the Silesian Upland [in Polish]. *Zeszyty Naukowe Akademii Rolniczej w Krakowie*
33: 61–72.

Skrzyczyńska J. 1998. Flora of the cultivated fields of the Sidlecka Upland [in Polish].
Fragmenta Agronomica 4: 47–66.

Skrzyczyńska J., Skrajna T., 1999. The segetal flora of the Kałuszyn Upland [in Polish]. *Acta
Agrobotanica*, 52: 183–202.

Srivastava J. P., Smith N. J., Forno D. A. 1996. *Biodiversity and Agricultural Intensification*.
The World Bank, Washington, D.C. 128 pp.

Stevenson F. C., Legere A., Simard R. R., Angers D. A., Pagean D., Lafond J. 1998. Manure,
tillage and crop rotation: effects on residual weed interference in spring barley
cropping systems. *Agron. J.* (Canada) 90: 496–504.

Stupnicka-Rodzynkiewicz E., Łabza T., Hochół T. 1992a. The present weediness of cereal
crops in selected mezoregions of Kraków, Tarnów and Nowy Sącz provinces compared to the situation from 10 years ago [in Polish]. Zeszyty Naukowe Akademii
Rolniczej w Krakowie 33: 93–105.

Stupnicka-Rodzynkiewicz E., Łabza T., Hochół T. 1992b. Changes in weediness of crops in
selected fields in Pogórze Wiśnickie during the years 1977–1984. Part I. The dynamics
of weediness of cereal crops and root crops [in Polish]. Zeszyty Naukowe Akademii
Rolniczej w Krakowie 33: 107–114.

Swanton C. J., Murphy S. D. 1996. Weed science beyond weeds: the role of integrated weed
management in agroecosystem health. *Weed Sci.* 44: 437–445.

Szeflińska D. 1996. Thysanoptera association in alfalfa cultivation in comparison to black
locust shelterbelt [in Polish]. In *Ekologiczne Procesy na Obszarach Intensywnego
Rolnictwa*. Zakład Badań Środowiska Rolniczego i Leśnego. L. Ryszkowski and S.
Bałazy (Eds.). Poznań: 53–67.

Szotkowski P. 1973. Weeds of overwintering cereals and row crops in Śląsk Opolski [in
Polish]. Prace Opolskiego Towarzystwa Przyjaciół Nauk, Nauki Przyrodnicze, 32 pp.

Thomas C. F. G., Marshall E. J. P. 1999. Arthopod abundance and diversity in differently
vegetated margins of arable fields. *Agric. Ecosystems Environ.* 72: 131–144.

Thomas M., Wratten S., Sotherton N. 1991. Creation of "island" in farmland to manipulate
population of beneficial arthropods: predator densities and emigration. *J. Appl. Ecol.*
28: 906–917.

Tischler W. 1965. *Agrarokologie*. VEB Gustav Fischer Verlag. Jena. 499 pp.

Trzcińska-Tacik H. 1992. Two types of changes in cereal weed communities in southern part
of Małopolska Upland (Southern Poland) [in Polish]. *Zeszyty Naukowe Akademii*

Varchola J. M., Dunn J. P. 2001. Influence of hedgerow and grassy field borders on ground beetle (Coleoptera: Carabidae) activity in fields of corn. *Agric. Ecosystems Environ.* 83: 153–163.

Vitousek P. M., Mooney H. A., Lubchenco J., Melillo J. M. 1997. Human domination of earth's ecosystems. *Science* 277: 494–499.

Warcholińska A.U. 1976. Segetal flora of Piotrowska Plain (Central Polish Lowland Mesoregion) [in Polish]. *Zeszyty Naukowe Uniwersytetu Łódzkiego* 8: 63–95.

Warcholińska A.U. 1983. Materials to the segetal flora of the south-eastern part of the Nizina Południowowielkopolska (Southern Wielkopolska Lowland) [in Polish]. *Badańia Fizjograficzne nad Polską Zachodnią*, B 34: 103–129.

Warcholińska A. U. 1986/87. A list of endangered segetal plant species in Central Poland [in Polish]. *Fragmenta Floristica et Geobotanica* 31–32: 225–231.

Warcholińska A.U. 1992. Changes of segetal vegetation of Southern Mazovian Elevation in 1971–1990 years [in Polish]. *Zeszyty Naukowe Akademii Rolniczej w Krakowie* 33: 157–170.

Warcholińska A.U. 1997. Flora and segetal vegetation of the Sulejów Nature Park. Part I. Segetal vegetation [in Polish]. *Acta Agrobotanica* 50: 163–180.

Warcholińska A.U., Mazurkiewicz U. 1999. Segetal flora of Inowłódź [in Polish]. *Acta Agrobotanica*, 52: 167–182.

Warcholińska A.U., Potębska A. 1998. Segetal flota of Będkowo [in Polish]. *Acta Agrobotanica*, 51: 63–80.

Warwick S. I. 1991. Herbicide resistance in weedy plants physiology and population biology. *Ann. Rev. Ecol. Syst.* 22: 95–114.

Wasilewska L. 1979. The structure and function of soil nematode communities in natural ecosystems and agrocenoses. *Pol. Ecol. Stud.* 5: 97–145.

Weber G., Franzen J., Buchs W. 1997. Beneficial diptera in field crops with different inputs of pesticides and fertilizers. Entomological research in organic agriculture. Selected papers from the European Workshop, Austrian Federal Ministry of Science and Research. Biol. Agric. and Horticul. 15: 109–122.

Wika S. 1986. Geobotanical problems of the central part of the Cracow-Weiluń Upland [in Polish]. *Prace Naukowe Uniwersytetu Śląskiego w Katowicach* 815, 156 pp.

Wilson E. O., Peter F. M. (Eds.). 1988. *Biodiversity.* National Academy Press. Washington, D.C. 521 pp.

Zarzycki K., Szeląg Z. 1992. Red list of threatened vascular plants in Poland. In *List of Threatened Plants in Poland.* K. Zarzycki, W. Wojewoda, Z. Heinrich (Eds.). Instytut Botaniki im Szafera PAN. Kraków: 87–98.

Żukowski W., Jackowiak B. 1995. List of endangered and threatened vascular plants in Western Pomerania and Wielkopolska (Great Poland). In *Endangered and Threatened Vascular Plants of Western Pomerania and Wielkopolska.* W. Żukowski, B. Jackowiak (Eds.). Prace Zakładu Taksonomii Roślin UAM. Poznań 3: 9–96.

Żukowski W., Latowski K., Jackowiak B., Chmiel J. 1995. *The Vascular Plants of Wielkopolska National Park* [in Polish]. Bogucki Wydawnictwo Naukowe. Poznań. 229 pp.

CHAPTER 9

Field Boundary Habitats for Wildlife, Crop, and Environmental Protection

Jon Marshall, Jacques Baudry, Françoise Burel, Wouter Joenje, Bärbel Gerowitt, Maurizio Paoletti, George Thomas, David Kleijn, Didier Le Coeur, and Camilla Moonen

CONTENTS

INTRODUCTION

Agricultural landscapes in Europe are diverse, reflecting their geology, geographical relief, history, and intensity of management. They vary from small-scale, enclosed landscapes, such as the bocage (INRA, 1976), to open prairie types. Within these landscapes, the majority of the land is farmed. Before expansion of the European Union, the agricultural area of 127.32 million ha comprised at least 56% of the land surface. In some countries, a much higher percentage of land is managed or farmed. Within all farmed landscapes, fields are bounded by seminatural margin habitats. The influences of farming practices are not limited to the cropping areas within agricultural landscapes. Likewise, the areas of unfarmed or noncrop land, which form the framework of agricultural land, can have important influences on adjacent fields. Agriculture does not occur in isolation but interacts with these areas, a fact underlined by the problems now being encountered with ground- and surface-water contamination by nutrients and pesticides from agriculture.

Within agricultural landscapes, crop and noncrop features comprise a diversity of habitats. These include arable land, grassland habitats that range from acid to alkaline communities with varying moisture regimes, aquatic, and riparian zones and a variety of boundary and woodland types. The mosaic structure of the farm landscape and its topography give the regional character to most of Europe. Often, these seminatural areas are important refuges for farmland wildlife, including plants and invertebrate and vertebrate animals, some of which may be of agronomic benefit. The conservation of such species may be best achieved by harmonizing land man-

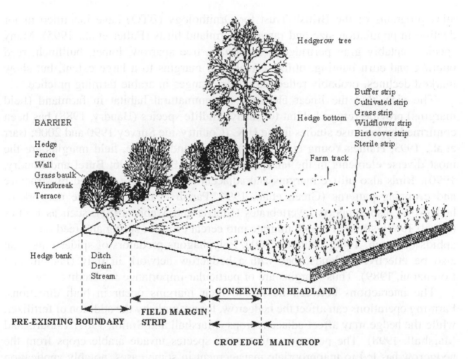

Figure 9.1   The principal components of an arable field margin. (After Greaves and Marshall, 1987.)

ecotone, which may support particular species and may buffer the movement of agrochemicals, water, and soil erosion. Thus the field margin has agricultural, environmental and wildlife attributes, aspects of which may be exploited for more sustainable production and for environmental benefits (Marshall, 1993).

Field margins, as defined by Greaves and Marshall (1987), comprise the field boundary which usually has a structure, such as a hedge, wall, grass bank, or ditch, often a boundary strip, which may be a farm track or sown vegetation strip, and the crop edge (Figure 9.1). The margin is a seminatural habitat, often a hedgerow in the U.K. (Marshall, 1988; Pollard et al., 1974), that contains a range of plant communities. These can include cornfield weeds, grassland, tall herb, scrub, woodland and aquatic communities and often combinations of these. Traditionally, the field margin has agricultural functions, notably impoundment of animals and field delineation. Under intensive arable production, such functions are less important to landowners, and many margins and hedges have been removed since the Second World War (Pollard et al., 1974). The rates of hedge removal declined in the U.K. in the 1970s, and it was not until the Countryside Survey 1990 (Barr et al., 1991; Barr et al., 1993) that more recent data has become available. These indicate that many hedges

atlas programs of the British Trust for Ornithology (BTO) have identified major decline in population size and range of farmland birds (Fuller et al., 1995). Many species, notably gray partridge, song thrush, tree sparrow, linnet, bullfinch, reed bunting and corn bunting, utilize cereal field margins to a large extent, but show marked declines, probably reflecting major changes in arable farming practice.

The view that the linear elements of seminatural habitat in farmland (field margins) provide refuge habitat for many wildlife species (Baudry, 1988) has been confirmed by land use studies in the U.K. (Countryside Survey 1990 and 2000; Barr et al., 1993; Haines-Young et al., 2000). In lowland Europe, field margins are the most diverse elements in the landscape for flora (Burel, 1996; Burel and Baudry, 1990). Birds also utilize margins and adjacent crops and are affected by structure and cropping patterns (Green et al., 1994; Parish et al., 1995). The network of hedgerows also supports invertebrates (Morris and Webb, 1987), such as beetles (Burel, 1989), some of which migrate into cereal crops in spring and feed on cereal aphids (Wratten, 1988; Paoletti, 2001). Polyphagous predators of spider mites can also be effective in landscapes with a hedgerow network in Italy (Paoletti and Lorenzoni, 1989). Thus margins are of particular importance for biodiversity.

The interactions between fields and their margins occur in both directions. Farming operations can affect the hedgerow, for example by the addition of fertilizer, while the hedge may affect adjacent crops (Marshall and Smith, 1987; Tsiouris and Marshall, 1998). The perception that weed species invade arable crops from the hedgerow has led to inappropriate management in some cases, notably application of broad-spectrum herbicides. Detailed studies indicate that most herbaceous plant species associated with field margins do not pose a threat as weeds (Marshall, 1989), though a small number of species can invade adjacent crops. The proximity of the hedgerow to the field in arable cropping has led to many margins containing impoverished flora, reflecting eutrophication from fertilizer additions and disturbance from cultivations and pesticide drift. Techniques of manipulating the field edge to encourage diversity and ameliorate the adverse effects of adjacent farm operations have been studied in the U.K. (Marshall and Nowakowski, 1991; West and Marshall, 1996) and in Europe (Jörg, 1994). Within the crop edge, reduced pesticide application has been used to encourage rare arable weeds (Schumacher, 1987) and, as conservation headlands in the U.K., to increase populations of the gray partridge (Rands, 1985). The impacts of these initiatives on farmland birds as a whole is not clear, and neither are their effect on the range of flora of the field and margin.

Farmers have viewed field margins as the origin of a range of problems within crop land. This was particularly the perception for weed species in the U.K., although pest and disease spread have also been cited. In contrast to farmers' perceptions, the general public views field margins, particularly hedgerows, as important elements in the landscape. The desire to retain traditional farm landscapes and the biodiversity within them is one reason for the designation of environmentally sensitive areas

appropriate management for the benefit of farm wildlife, environmental protection, and sustainable crop production, thus optimizing the use of biological resources on farms. As part of the European Communities Third Framework Programme, a research consortium from five countries investigated the ecology and management of field margins at a range of spatial scales. This chapter summarizes some of the results of this research program (Marshall, 1997).

Specific objectives were:

- To determine the function of field margins in the maintenance of plant and animal communities and the movement of nutrients and pesticides within different community landscape structures and farming systems
- To identify the means of (1) enhancing biological diversity along fields for wildlife conservation and integrated crop protection and (2) exploiting any buffering actions to farm operations
- To develop a generalized model description of the field margin ecotone, including transfer of materials (nutrient, biomass, agrochemicals) and organisms

METHODS

Ecological Impacts of Enlarged Field Boundaries

The study aimed at determining (1) what relationship exists between vegetation development on an extended field boundary and the adjacent (original) boundary, (2) whether these relationships are consistent in contrasting boundary types in different countries, and (3) whether these relationships are consistent between naturally regenerating and grass-sown boundary strips.

In spring 1993, field boundary plots were established next to existing field boundaries near Rennes (France), Wageningen (the Netherlands) and Bristol (U.K.). Similar plots were also established near Göttingen (Germany) and Padova (Italy). The plots were created by taking the outer 4 m of the crop edge out of production and either sowing it to *Lolium perenne* or letting it regenerate naturally. Plots were at least 8 m long. Thus in these plots, the pre-existing field boundary was broadened by 4 m. Stretches of regular field boundary served as control plots. Management in the original boundary remained as it was before the onset of the experiment, while the *L. perenne* sown plots (grass plots) and the plots left regenerating naturally (regeneration plots) were mown once a year in autumn with the cuttings being removed. Alternative seed mixtures, comprising a mixture of grasses and flowers or flowers, alone were included at some sites.

In the original field boundary, 0.5×2 m permanent quadrats (PQ) were established next to each plot type. To relate distance from the original boundary to

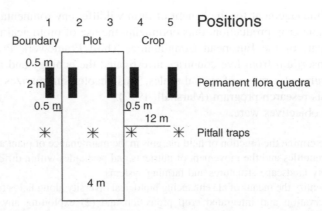

Figure 9.2   Layout of a single boundary plot. Distances in meters. * = pitfall trap for fauna.

(monocots) and dicotyledonous species (dicots). Dry weight was determined after 24 h at 80°C.

The fauna were assessed using pitfall trapping and Dietrich vacuum sampling from the vegetation during the summer. Two pitfall traps were placed approximately 0.5 m apart at each sampling site. The sampling positions were located in parallel rows aligned centrally with each of the field margin plots. Four positions were trapped: the hedgerow (H) or existing boundary (0.5 m from the sown plots); centrally within the sown plot (P); at the crop edge (E) adjacent to the plots (0.5 m from sown plots); and in the cropped area of the field (F)12 m from the plot edge. A total of 96 traps were thus used at each of the field sites in 12 rows of four sampling positions. Pitfall traps were partly filled with trapping fluid (25 ml detergent in 10l of 1:1 water:ethylene glycol antifreeze) to ensure drowning and preservation of captured invertebrates.

D-vac sampling was performed in June. D-vac samples were taken from the hedge position, each of the field margin plot types, the crop edge adjacent to the plots, and 12 m within the field. Each sample comprised three subsamples of a 10- to 15-sec application of the D-vac head (area 0.1 m^2). The D-vac net was 1.5 m long enabling tall vegetation to be sampled without crushing. The samples were placed in labeled plastic bags and stored in a freezer overnight before being transferred to 70% alcohol for preservation prior to identification and analysis. Detailed statistical analyses were made on the Carabidae (ground beetles).

An assessment of invertebrates overwintering in field margin habitats was made at two U.K. sites where soil samples were taken after D-vac sampling of the standing vegetation. Samples were returned to the laboratory, and the invertebrates were extracted by hand from the soil, using a wet sieving technique. Fauna were identified to order or family.

Factors Affecting Flora Diversity in Field Margin Systems in European Landscapes

Landscape Scale Studies

The objective of the studies was to identify the major factors affecting the margin plant communities found in different landscapes and to provide an idea of the different landscape types at a regional scale and their interactions with linear boundary features. Extensive collections of field and cartographic data were made and analyses made using multivariate correspondence packages. The *Bocage* database was specially developed for the data, using the dBase IV program (Denis et al., 1995). The approach to data collecting was as follows:

- Sample areas were selected at random from within, if possible, nationally identified land classes.
- Within the chosen area, all margin units up to a minimum of 50 were surveyed. Each unit comprised a margin length of uniform aspect, usually an entire side of a field.
- A relevée of the plant species present in the undisturbed margin area was made for the entire hedgerow (margin) length, using the five-point Tansley scale (r = rare; o = occasional; f = frequent; a = abundant; d = dominant), to give a semi-quantitative measure of cover/abundance, and repeated on the other side of the margin. In addition, a second relevée of 25-m length of the margin selected at random was made. A third relevée of the weed flora present in the entire field was also made.
- Information on field margin structure, management, size, and on adjacent crop area was also collected.

The flora of field boundaries in ten different farmed landscapes, from France, the Netherlands, and the U.K., were investigated. Data were collected from up to 50 field margins in each area of 50–100 ha, both from whole margins of variable length and from 25-m sections within each. The flora present in the ground, shrub and tree layers were recorded, together with data on the physical structure and orientation of the margin, the adjacent land use, and management of the boundary. These data were subjected to multivariate analyses (PCA, RDA, etc,) on a site basis (Jongman et al., 1995). Selected results from 25-m sections from four areas in the U.K. and France (Table 9.1) are reported below.

Table 9.1   Details of Areas Where Field Margin Flora Have Been Surveyed

Country	Area	Landscape	Margins

Herbaceous Plant Diversity on Two Farms, with and without Sown Grass Strips, in Wiltshire, U.K.

The objectives of this study were (1) to investigate differences in the herbaceous hedge-bottom vegetation of hedgerows with and without a sown strip between the hedge and the field; (2) to assess the effects of different agricultural practices, land use and boundary structure on the hedge-bottom vegetation; and (3) to show any succession of the hedge-bottom vegetation in coppiced hedges. The two farms were chosen for their different approach toward boundary management and for their homogeneity in geology, landscape structure, boundary structure and land use (Moonen and Marshall, 2001). The farms are about 200 ha each, and field size varies between 4 and 40 hectares, mainly occupied by cereals and oil-seed rape. Both are on fine silty clay soil over lithoskeletal chalk. The landscape is largely flat, but with chalk hills close by, and with most fields bordered by hedges. Boundaries on the Manor Farm are characterized by 2-m, 4-m, or 20-m wide sown grass and grass-wildflower strips established between the hedge and the crop. Nine hedges were also coppiced and gapped-up under the former Hedgerow Incentive Scheme (Whelon, 1994). These rejuvenated hedges had been cut to the ground (coppiced) to encourage shrubs and the gaps planted with young hedging plants. Boundaries on Noland's Farm are characterized by a 0.5-m sterile strip (Table 9.1) created with a broad-spectrum herbicide. Differences in agricultural practices between the two farms that are thought to influence the hedge-bottom vegetation are listed in Table 9.2.

The vegetation of the hedge-bottom (excluding the sown strips) was assessed in 25-m long plots in the middle of a field edge, on either side of the hedge. Each side of the hedge was treated as a separate plot, in order to establish the effects of adjacent land use and management. The width of each plot varied with the width of the hedge and associated hedge bottom, excluding any sown strip. Vegetation was assessed using the five-point Tansley scale (1 = rare, 2 = occasional, 3 = frequent, 4 = abundant, 5 = dominant) (Tansley, 1935). Boundary structure, management and adjacent features were recorded according to standards described by Denis et al. (1994). Thus, every vegetation sample was associated with a set of 24 environmental variables, made up of five boundary structure variables, 11 management variables

Table 9.2 Differences in Agricultural Practices between Two Farms

Noland's Farm	Manor Farm
0.5-m sterile strip between hedge and crop	2- to 20-m wide sown strips between hedge and crop
Hedge trimmed annually, cuttings left	Hedge trimmed in alternate years

and eight adjacent features. On Noland's Farm, 23 hedges were examined and 37 on the Manor Farm. As only one side of the hedge was sampled for some sites, there were 43 relevées on Noland's Farm and 74 on Manor Farm.

Following simple tabulation of the data, which comprised 117 relevées for 94 herbaceous and 24 environmental variables, and tests for differences in diversity and abundance between farms, the results were analyzed using multivariate statistics. Principal Component Analysis (PCA) and Redundancy Analysis (RDA) in combination with forward selection and an associated Monte-Carlo permutation test in CANOCO 4.0 (ter Braak, 1987a, 1987b, 1996) were used to assess differences in herbaceous species composition between the two farms. RDA reveals which environmental variables are responsible for those differences and indicates their relative importance. Ordination diagrams were created using CanoDraw 3.1 and CanoPost 1.0 programs. Stepwise linear regression of species richness with environmental variables was also conducted to test which variables influence species richness.

RESULTS

Ecological Impacts of Enlarged Field Boundaries — Flora

Development of the Flora in Margin Strips in Different European Countries

Considering the margins in France, the Netherlands, and the U.K., the vegetation in the different original boundaries was characterized by a large number of the same species despite the fact that there were large differences in boundary types, soil types or even geographical latitude (Kleijn et al., 1998); 49, 59, and 45% of the species in the respective French, Dutch, and English strips were found in one or both of the other countries. None of the species encountered in any of the countries was rare and most species could be classified as common to extremely common.

A comparison between PQ1 next to the control plots and PQ1 next to the grass and regeneration plots (thus buffered from the arable field by a 4-m wide strip of perennial vegetation) did not reveal any significant differences in the similarity index, species numbers, biomass production or abundance of any of the functional groups. Therefore, PQ1 next to the grass and regeneration plots can be considered representative for the field boundary in its original state.

The vegetation in the original field boundary was highly dynamic. Species similarity of the vegetation in PQ1 between 1993 and the following 2 years ranged from 40 to 80%. In the newly established boundary plots, species similarity with the original field boundary in 1993 increased with time and decreased with distance

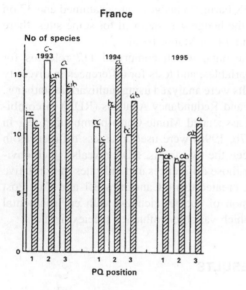

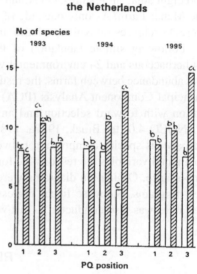

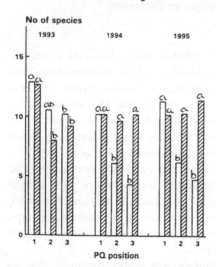

Figure 9.3   Mean number of species (m⁻²) in the original boundary (PQ1) and the adjoining regeneration (shaded bars) and grass (unshaded bars) plots (PQ2 and PQ3). Significances are as in Figure 9.2. (a) France, n = 6; (b) the Netherlands, n = 9; (c) U.K. n = 9.

In the Netherlands, mean total biomass production was consistently higher in regeneration plots compared with grass plots, while in the U.K. the mean grass plot yields were always higher than the yields in the regeneration plots. By 1995, however, differences between PQ position and plot type were not significant. In France, mean total biomass showed a tendency for increased production with increasing distance from the original boundary. Shading by the dense and tall hedgerow in the original boundary may have been the cause of this trend. Mean biomass production of dicotyledonous species showed a similar pattern in all three countries. In the boundary plots large and significant differences were found in 1994 between the grass (low yields) and the regeneration plots (high yields). These differences reduced in size in the following year to become insignificant in France, while in the Netherlands and the U.K. only the differences in the PQ3 position, although considerably decreased in size, remained significant. Biomass production of monocotyledonous species showed just the opposite pattern of the dicotyledonous species.

Finally, in 1995 an extended perennial field boundary had developed which was primarily composed of a limited set of the species found in 1993 in the original boundary. At 0.5 m from the arable field (PQ3) in all three countries, a limited number of species (most notably, *Agrostis stolonifera*, *Poa trivialis*, *Ranunculus repens* and *Trifolium repens*) had become extremely abundant. In the new strip 63, 45, and 63% of the species in, respectively, France, the Netherlands, and the U.K., were found in at least one other country. Total species numbers encountered in the three countries showed a marked decline in France and the U.K. and a sharp increase in the Netherlands. The decline in France and the U.K. was primarily the result of a reduced number of annuals and dicotyledonous species. Most of the dicotyledonous species that were not encountered in the new boundary were woody or woodland species. In the Netherlands the increase in total species numbers was almost entirely caused by the increase in annual species.

Where a diverse seed mixture had been sown, as in the U.K., the Netherlands, and Germany, plant species diversity was consistently higher where the most diverse seed mixture was sown (Gerowitt and Wildenhayn, 1997; Kleijn, 1997). This effect was also reported for a series of sown margin strips in three different areas and on three soil types in the U.K. (Marshall et al., 1998; West and Marshall, 1996, 1997; West et al., 1999). Nevertheless, there were circumstances where undesirable species, notably *Cirsium arvense*, could dominate introduced strips. This effect appears to be markedly reduced where grasses are sown (Smith et al., 1999; West et al., 1997).

Impacts of Fertilizer and Herbicide on the Diversity of Sown Margin Flora

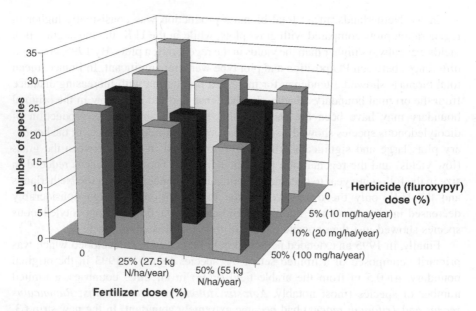

Figure 9.4   Mean number of plant species on plots treated in factorial combinations of fertilizer and herbicide.

Ecological Impacts of Enlarged Field Boundaries — Fauna

Single Year, Single Site Comparisons between Countries — Activity-Density and Diversity: Comparing Carabid Diversity between France, the U.K., and the Netherlands

A species list for a single site and single date (June, 1995) for three countries was developed. Of the 68 carabid species recorded from pitfall traps, 44 occurred in France, 33 in the U.K., and 21 in the Netherlands. Only 7 species were common to all three countries with 24 species appearing only in the French list, 14 only in the U.K. list, and 7 only in the Dutch list. Although sampling effort was similar for each data set, sampling strategies differed slightly among countries making strict comparisons difficult. The absence of a species from the list in one sample did not indicate an absence from the fauna.

No significant differences were found among the countries ($P = 0.57$) and no significant interaction was found between country and plot type ($P = 0.34$). There was, however, a significant effect due to plot type ($P = 0.004$) with highest activity-density in all three countries occurring in the arable plots, where the vegetation was

Table 9.3   Abundance of Invertebrates at the Different Positions in Fields A and B

	Mean Number of Invertebrates per Sample	
	Field A	Field B
Hedge	233.0	223.0
Sown Field Margin	155.8	147.8
Crop Edge	84.5	58.3
Field	96.8	76.9

Table 9.4   Number of Taxa and Plant Species Present in Each Habitat Type

		Crop Edge Ecotone				Field Margin Plots				
	Field	CL	LP	NR	WF	CL	LP	NR	WF	Hedge
Taxa	20	23	22	23	25	26	28	29	29	35
Flora	2.9	6.7	9.7	6.7	9.3	4.0	8.7	12.0	10.2	15.3

CL = arable; LP = *Lolium perenne*; NR = natural regeneration; WF = wild flowers.

Comparisons of Invertebrate Abundance and Composition in the Hedge, Sown Plots, Crop Edge, and Field by Suction Sampling in the U.K.

A total of 21,176 invertebrates were collected. With this number of specimens, identification to species was impractical; identification was therefore restricted to order or family.

The relative distribution and abundance of total invertebrates at different positions (hedge, margin, edge and field) was similar in both fields (Table 9.3). Highest densities were found in the hedge with densities approximately one third less in the sown margin. The lowest densities were at the crop edge ecotone where invertebrate numbers were one fourth to one third that found in the hedge. Invertebrate density in the field was slightly higher than that found at the crop edge.

A total of 45 different taxa were identified in the whole sample. Although each high level taxon (order/family) is composed of several taxa at the level of genus or species, an indication of invertebrate diversity in the different habitat types is given by the number of taxa represented in the total sample from each location (Table 9.4). These results show a pattern similar to that demonstrated for total invertebrate abundance with fewest taxa (20) present in the field, slightly more at the crop edge (22 to 25), higher numbers of taxa in the sown plots (26 to 29), and highest in the hedge samples (35). These results provide evidence for a positive correlation between faunal and floral diversity.

Table 9.5   Mean [log₁₀(n+1)] Numbers of Arthropods per Sample with Standard Errors of Difference[a]

	Carabid Adults		Staphylinid Adults		Coleopteran Larvae		Other Adult Coleoptera		Hemiptera	
	Mean	Sig. Diff.	Mean	Sig. Diff.	Mean	Sig. Diff.	Mean	Sig. Diff.	Mean	Sig. Diff.
					Field 47					
Field	0.00	a	0.03	a	0.43	ab	0.00	a	0.00	a
Cereal margin	0.03	a	0.00	a	0.17	a	0.00	a	0.00	a
Natural regen.	0.12	a	0.05	a	0.40	ab	0.07	a	0.00	a
L. perenne	0.07	a	0.07	a	0.31	ab	0.00	a	0.00	a
Grass and flower	0.17	a	0.00	a	0.44	b	0.00	a	0.00	a
Hedge	0.53	b	1.35	b	1.00	c	0.55	b	0.00	a
SED (48df)	0.10		0.07		0.13		0.06		0.00	

	Dipteran Adults		Dipteran Larvae		Dermaptera		Isopoda and Myriapoda		Araneae	
	Mean	Sig. Diff.	Mean	Sig. Diff.	Mean	Sig. Diff.	Mean	Sig. Diff.	Mean	Sig. Diff.
					Field 47					
Field	0.00	a	0.55	a	0.00	a	0.07	a	0.14	ab
Cereal margin	0.00	a	0.78	ab	0.00	a	0.00	a	0.00	a
Natural regen.	0.00	a	0.89	b	0.00	a	0.13	a	0.12	ab
L. perenne	0.00	a	0.83	ab	0.00	a	0.15	a	0.05	a
Grass and flower	0.00	a	0.80	ab	0.00	a	0.20	a	0.23	bc
Hedge	0.00	a	0.86	b	0.28	b	0.86	b	0.39	c
SED (48df)	0.00		0.14		0.06		0.12		0.08	

[a] Same letter denotes no significant difference at the 5% level, by LSD multiple range test within fields and taxonomic group only.

Significant differences between sites were calculated with an LSD Multiple Range Test at the 5% level. The means, standard errors of difference and significant differences are given in Table 9.5.

In Field 47, Hemiptera and adult Diptera were not found in any sample. There were no significant differences between the field and the field margin cereal plots for any group. Most groups were found in significantly higher numbers in the hedgerow than elsewhere, but no consistent pattern of abundance could be discerned among the grass and natural regeneration plots. However, there was an underlying trend of greatest numbers in the hedgerow, fewer in the mixed grass and wildflower and natural regeneration plots, with the least in the cereal margin plots and field samples. The data show that sown margins were used as overwintering habitat by a number of taxa within 12 months of establishment.

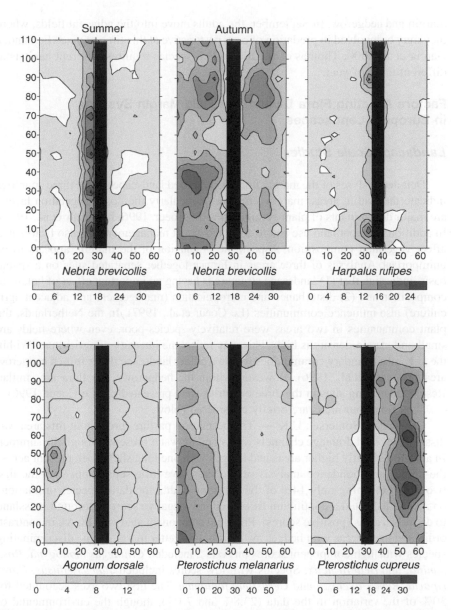

Figure 9.5   Cumulative trapping densities of different Carabidae in a hedge and in two adjacent arable fields over summer and autumn. (After Thomas et al., 2001.)

margin and hedgerow. In September, the adults move into the adjacent fields, where they may be involved in predation of crop pests active at this time of year (Fernandez Garcia et al., 2000; Thomas et al., 2001). This species requires different habitats at different times of year.

Factors Affecting Flora Diversity in Field Margin Systems in European Landscapes

Landscape Scale Studies

Detailed analyses of the margin flora in contrasted landscapes in the three countries indicate that within areas, margin structure, particularly the upper vegetation layers, are major determinants of plant communities (Le Coeur, 1996; Le Coeur et al., 1997). In addition, adjacent land use and to a limited extent management are also factors that affect vegetation composition. Aspect, the orientation of the margin, appears to be unimportant. Analyses of three areas in France together separated sites on a spatial basis, indicating that (1) landscape structure is having an important effect of botanical composition and (2) disturbance and eutrophication (management and adjacent agri-culture) also influence communities (Le Coeur et al., 1997). In the Netherlands, the plant communities in two areas were relatively species-poor, even where fields are small and margin density is high. This may reflect the intensity of agriculture. Within the U.K., the boundary communities within a polder landscape differ from a hedgerow area (Marshall et al., 1996). However, within the hedgerows, the flora are similar. Results comparing sites in the three countries were presented by (Le Coeur, 1996).

Data from four areas are briefly considered below.

Cossington, Somerset, U.K. — The permanent pasture throughout this area was dissected by land drainage channels with standing water present through the summer. In addition, slightly higher areas had hedges with one long shelterbelt present across the area. Correspondence analysis of the sites gave little separation, with the first two axes explaining only 14% of the variation within the data. Species ordinations on Axis 2 indicated a continuum from wetland (negative scores), through grassland to disturbed areas (positive scores). Principal component analysis (PCA), in contrast, ordinated those areas with hedges away from the wetter margins. The discriminating species with high component scores on Axis 2 included *Crataegus monogyna*, *Rosa canina*, and *Hedera helix*. Species with low scores included *Juncus effusus*, *Carex riparia*, *Rumex acetosa*, and *Cerastium fontanum*. The first two axes accounted for 30% of the variation in the data (23.8% and 7.1%), though the environmental or structural variables associated with Axis 1 are not clear.

Corsham, U.K. — Reciprocal averaging correspondence analysis of the field margin ground-layer vegetation indicated that the majority of margins within the study area had similar flora. The ordination showed little grouping of sites, with the first

85 sections, were subjected to TWINSPAN analysis (Marshall et al., 1996). The sites were divided into four groups. Tabulation of the structure of each margin, adjacent land use and management did not indicate major associations, reflecting the similarity of margins in the area. Nevertheless, the separations contain a spatial element in that one group consists of 19 sites, all located to the north of a road through the study area. One group is of 9 sites along the road verges, while the other two groups are mostly to the south of the road. These data indicate that some dissimilarity in plant communities results from location in the landscape. Linear elements such as roads and water courses may thus isolate habitats (Vanruremonde and Kalkhoven, 1991).

Pleine-Fougères bocage — The bocage area of Pleine-Fougères is a mixed agricultural area with hedgerows, woodlots and fields of grassland with some maize and cereals. Correspondence analysis identified eight vegetation groups with a range of characteristic plant species. The results are summarized in Figure 9.6, as a classification of the margins and shows the structure of the margin as an important discriminator. Vegetation groups vary from dense hedges, to open hedgerows and to herbaceous strips that are open to disturbance.

Mont Saint Michel bay polder — The polder is a flat area reclaimed in the 1800s from the Mont Saint Michel bay. The landscape is flat, dissected by dikes, with fields of maize, wheat and vegetables. The margins are usually wet or dry ditches or grassy strips. Analyses of the vegetation of the margins indicated that again the structure of the margin was an important discriminator, though the intensity of farming and management also influenced the flora (Figure 9.7).

Herbaceous Plant Diversity on Two Farms, with and without Sown Grass Strips, in Wiltshire, U.K.

A total of 94 higher plant species were found in the 117 relevées of the herb layer of hedge bottoms (Moonen and Marshall, 2001). Three species were restricted to Noland's Farm and 24 species were found only on the Manor Farm, all of which had low frequencies. These data indicate that the two farms shared the same species pool.

Following RDA analysis, the list of 14 variables, explaining 26% of the total variation in hedge-bottom vegetation, was retained for PCA. Of the 26%, management explains 12%, adjacent features 8%, and boundary structure 6%. Gapped-up hedges (18 relevées) contribute strongly to the diversity in hedge-bottom vegetation, and so do wide boundaries and boundaries with a sown grass strip. Coppiced hedges explain only 1% of the variation in the data set. However, gapping-up and coppicing hedges interact significantly which means that both explain similar variation, and gapping-up and coppicing are both important factors. After fitting the gapped-up hedges to the model, coppiced hedges explain a significant ($P > 0.05$) additional 1% of the variation.

PCA with the 14 explanatory variables as passive variables resulted in three

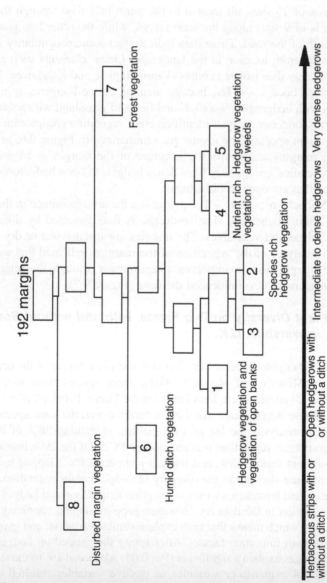

Figure 9.6 Classification of field margins in the bocage. (From Moonen, 1995. With permission.)

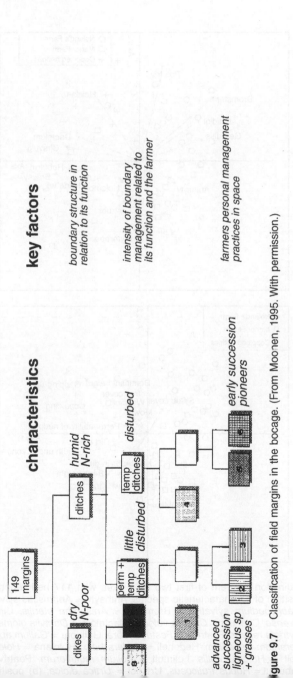

Figure 9.7 Classification of field margins in the bocage. (From Moonen, 1995. With permission.)

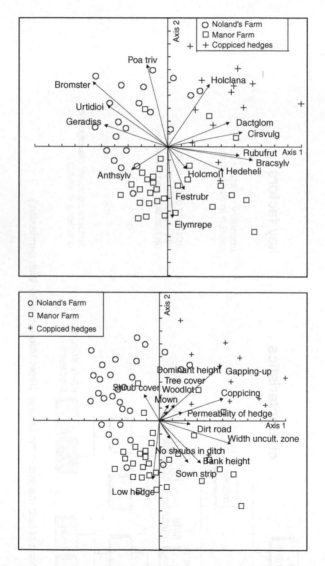

Figure 9.8   Ordination diagram of first two PCA axes with 117 hedge-bottom samples: (a) position of 17 characteristic species: Anthsylv = *Anthriscus sylvestris*, Bracsylv = *Brachypodium sylvaticum*, Bromster = *Anisantha sterilis*, Cirsarve = *Cirsium arvense*, Cirsvulg = *Cirsium vulgare*, Dactglom = *Dactylis glomerata*, Elymrepe = *Elytrigia repens*, Festrubr = *Festuca rubra*, Galiapar = *Galium aparine*, Geradiss = *Geranium dissectum*, Hedeheli = *Hedera helix*, Holclana = *Holcus lanatus*, Holcmoll = *Holcus mollis*, Lolimult = *Lolium multiflorum*, Poatriv = *Poa trivialis*, Rubufrut = *Rubus fruticosus*, Urtidioi = *Urtica dioica*, (b) position on ordination

repens, *Hedera helix*, *Brachypodium sylvaticum*, *Holcus mollis*, and *Cirsium vulgare*. The coppiced and gapped-up hedges on the Manor Farm are characterized by *Holcus lanatus* and to a lesser extent by *Silene dioica*, which occurs in a very high percentage of these samples.

Analysis of variance of the number of herbaceous species per 25-m hedge-bottom length between the hedges on Noland's Farm, the low hedges on the Manor Farm, and the coppiced and gapped-up hedges on the Manor Farm, showed that all three groups differ significantly (P < 0.001) in species richness. The coppiced and gapped-up hedges had the most species (23.2; sed = 1.36), followed by the other hedges on the Manor Farm (17.4) and hedges on Noland's Farm (14.6). Stepwise linear regression of species richness and the 14 explanatory variables show that species richness in the hedge bottom is significantly (P < 0.05) increased by gapping-up of hedges. Also, the vegetation in the hedge bottom side directly next to a dirt road or sown strip had increased species richness. Overall, management of the Manor Farm hedges gives greater botanical diversity, indicating the grass strips reduce disturbance and enhance species richness.

DISCUSSION

Field Margins in European Landscapes

Studies of the flora and fauna of field margins in a range of European countries, farming systems and landscapes demonstrated that a range of factors are important in influencing the abundance and diversity of species (Burel et al., 1998). Consistently, the structure of the margins is important. Margins can contain a range of structural components and therefore a range of species and communities associated with them. Field utilization (cropping) is important, and farm type affects field margins. Farmers are particularly important in affecting margins, via their perceptions and requirements for farming and, thus management. The level of disturbance associated with adjacent farming and management also have profound influences. High disturbance, by both physical and chemical means, can reduce abundance and diversity of species (Kleijn and Snoeijing, 1997; Kleijn and Verbeek, 2000; Tsiouris and Marshall, 1998, Paoletti, 1999). The location of margins within the landscape is also influential (Le Coeur et al., 1997), mediating the opportunities for species dispersal and thus colonization and recolonization of habitats. Hedges connected to woodland are more likely to support woodland species, but other elements, such as roads, may isolate habitats.

Effects of flora and fauna are found at both site and landscape levels. There is a general trend for a reduction in diversity and abundance with increasing landscape simplification and increasing disturbance via farm inputs, cultivation and manage-

of general management prescriptions, e.g., for set-aside, should take account of scale effects as appropriate for the groups under consideration.

A field survey of two farms, one with sown grass strips and one without, showed that introduced margins at arable field edges, together with accurate fertilizer application, could result in less disturbance in the field margin and increase botanical diversity. The technique of introducing field margin vegetation strips thus can increase biodiversity on farms.

Processes Affecting Field Margins

Small-scale experimentation shows that plants are affected by addition of fertilizer and herbicide and by management, including cutting. Fertilizer additions favor the establishment and growth of tall plant species, typically those with a ruderal habit (Marrs, 1993). Such species, including perennials, can colonize margin habitats early on in secondary succession and may dominate the community. In the crop edge, fertilizer effects on the arable flora are modified via the canopy and the light climate. Fertilizer promotes poorer light under the canopy, by favoring the taller crop. Detailed studies of the flora of the field boundary show that plants at the edge, within 20 cm of the tilled soil, are able to exploit the fertility of the crop, producing significantly more biomass than plants further into the margin (Kleijn, 1996). Thus, nutrients are "harvested" from the field into the boundary. This situation favors taller species while low-growing species are reduced (Kleijn, 1997). Further into the margin, reduced fertility allows greater species diversity.

The effects of herbicides on plants are less predictable, partly as there are many chemicals with varying spectra of activity. Some herbicides have direct impacts on species, while others cause changes in plant communities indirectly, by allowing unaffected species to increase. In general, herbicide application or drift from field operations has negative effects on botanical diversity. Management, in the form of vegetation cutting, can also have impacts on the flora of field margins. Cutting and removal, as in hay harvesting, can remove nutrients and ameliorate the adverse effects of fertilizer additions.

The inter-plant interactions that occur in field margins have been successfully modeled, using a range of species with differing life-history strategies (Schippers et al., 1999). The model builds on studies made within the project on establishment, competition and fate of plants, so that patterns of successional change in field margins are simulated. The model includes effects of fertilizer, herbicide, and cutting. Clearly, fertilizer additions lead to loss of subordinate species. Herbicides cause species loss and reduced diversity. Vegetation removal by cutting modifies competition and mortality, allowing species and therefore diversity to be maintained.

Introducing Vegetation Strips at Field Edges

successional changes, typically with annual weed plant species common in the first year which largely disappear in the second and subsequent years after sowing, except on regeneration plots. Successional patterns were similar between sites, indicating that soil fertility was a poor predictor of outcome. In general, few plants of the pre-existing margin colonize the new strips. The sown strips showed some convergence toward grass margins with colonization by more mobile species of grassland. In general, the species present showed low dispersal capabilities (Marshall and Moonen, 1997) and seedbanks were often impoverished. Colonisation by species not previously recorded in the above-ground flora was very low, indicating that the use of seed mixtures is likely to be essential in most farm situations.

The sown plots significantly reduced weed species populations. This effect was particularly apparent where the perennial *Cirsium arvense* achieved high populations where grasses were not sown in strips. The productivity of the strips was lower than might be expected on land taken out of arable production. Nevertheless, fertilizer use in the adjacent field affected plant growth, increasing biomass in the 10 to 20 cm nearest the crop (Kleijn, 1996). With the exception of some spread of the sown grasses, there was little evidence of the sown strips themselves causing weed problems in the adjacent crop (Marshall and Moonen, 1997; Smith et al., 1999; West et al., 1997).

In a survey of margins on two farms, it was shown that sown grass margins were associated with less disturbed and more species-rich field margin flora, indicating that such strips would give some protection to the pre-existing margin flora, with farming operations taking place further from the margins (Moonen and Marshall, 2001).

Invertebrates of Field Edges

Field margins were shown to increase the diversity and abundance of insects, especially if they were both botanically and structurally diverse (Thomas and Marshall, 1999). In experiments using pitfall trapping, a technique that measures activity-density, no significant differences were found between different field margin strips. However, it was demonstrated that the technique is unsuitable for comparing different vegetation structures and is unsuitable for small plots. Many invertebrates move significant distances, and studies should be made at the field scale, rather than the plot scale (Thomas et al., 1998). Alternative sampling methods showed that the more structurally complex and botanically rich plots supported the most abundance and diversity of invertebrates.

A novel field technique of marking individual ground beetles was developed in order to study the dispersal characteristics of these invertebrates (Thomas, 1995). These species are often polyphagous predators, implicated in the control of aphid and slug pest populations. Many species were particularly mobile, moving from the field margin into the crop over short intervals. Introduced field margin strips were

before dispersing into the crop in late summer/early autumn. Other species were part of a diverse community found within the arable crops. Mark-recapture studies provided quantitative results on dispersal for use in subsequent modeling, for example, maximum (c. 100 m day^{-1}) and average daily displacement. Other significant findings were that patterns of occurrence were not uniform. Some species were found in consistently high densities in certain locations, the reasons for which are unknown but which might be manipulated to reduce pest populations (Thomas et al., 2001). Another finding was that field margins, particularly a hedge, may act as a barrier to dispersal between fields.

Studies in Germany using mark-recapture methods demonstrated that ground beetles will move into adjacent crops in spring. A significant local influence of the field margin was recorded in the crop on populations of aphids. The reduction of aphids up to 5 m or more from the margin is likely to have resulted from a complex of predator species associated with the margin (Marshall, 1997).

The results demonstrate that field margins can support a diverse invertebrate fauna, some of which can contribute to the agricultural control of pest species. Suitable manipulation of the field margin may enhance the diversity and abundance of insects.

Managing Field Margins

Disturbance, particularly in the form of fertilizer and herbicide contamination, has adverse effects on the diversity of the perennial flora of field margins. Data indicates that significant herbicide drift will occur over 3 m from the crop and up to 4 m for fertilizer applied by spinning disk machinery (Rew et al., 1992; Tsiouris and Marshall, 1998). Sown margin strips 3 to 4 m wide, extending pre-existing boundaries, have the capability of reducing such disturbance by buffering drift. Such strips would be best managed by cutting and removal of clippings at a time of year when ground-nesting birds would be least disturbed.

In areas with a history of intensive agrochemical use, it is likely that the existing flora of field margins has adapted to high fertility conditions and is species-poor and unlikely to recover significant diversity without species introductions. Sown grass and wildflower strips would be suitable under such conditions. The seed used may need to be of local provenance.

Sown perennial vegetation strips located at field edges would not be suitable for the encouragement of rare arable weed species, many of which are threatened with extinction in modern arable farming. Such species are often found in soil seed banks at field edges. Where such species are known to exist, the technique of no-input crop edges (conservation headlands, ackerrandstreifen) is suitable. Studies indicate no significant increase in pest and disease problems and minor increases in weeds

CONCLUSIONS

At the plot scale, field margin floras are affected by adjacent management. Plants of the margin can forage into the arable area over short distances. Sowing seed mixtures of grasses and flowers as a field margin strip is a practical means of increasing local diversity and can protect adjacent pre-existing boundary flora. There is evidence that the technique can reduce the abundance of competitive annual hedgerow weeds, while increasing available habitat for invertebrates. There is evidence of limited seed spread from sown margins into crop areas, but this is not likely to be of economic significance.

At the field scale, impacts of field operations, particularly fertilizer application, are likely to be important for field margin flora. Contrasting results were shown in the development of margin flora: in sown strips, there was little evidence of farming system effect on the flora; in contrast, a farm with grass margins had significantly more diverse herbaceous hedge flora, compared with an adjacent farm without strips. Fauna studies show that species-specific spatio-temporal behavior occurs in Coleoptera. Margins are important as overwintering or aestivation sites. They may also be barriers to movement between fields for some species, which may have implications for recolonization and population recovery after field operations. Invertebrate abundance and diversity is encouraged by structural complexity and botanical diversity in margins.

At the landscape scale, boundary structure, land-use (farming type), location and the management of the boundary affect the shrub and herbaceous flora. Disturbance and eutrophication from adjacent farming are important in affecting margins. These various factors are in fact a result of many specific environmental and physical processes, which require further understanding. However, it is clear that local diversity of structure and land use favor diversity in margins. With a diversity of land use in Europe, and a range of overlying climatic differences, margins are variable features of agricultural landscapes. In those landscapes examined, field margins are an important contributor to biological diversity, but they are sensitive to the farming operations practiced.

RECOMMENDATIONS

1. Sown margin strips at least 3 m wide should be introduced at arable field edges, to buffer the effects of fertilizer and pesticide drift and to create new habitat on farmland. The technique will make significant contributions to the maintenance of biological diversity in agricultural systems.
2. The structural diversity of field margins should be maintained at farm, field and

ACKNOWLEDGMENTS

This project was funded by the European Commission, contract AIR3 CT-920476. We are grateful to the many researchers involved with the work, including Peter Schippers, Addi Kopp, Claudine Thenail, and many more. We are also grateful for associated funding from individual Member States that enhanced the overall program.

REFERENCES

Barr, C., Howard, D., Bunce, R., Gillespie, M., and Hallam, C. (1991) Changes in hedgerows in Britain between 1984 and 1990. Institute of Terrestrial Ecology.

Barr, C.J., Bunce, R.G.H., Clarke, R.T., Fuller, R.M., Furse, M.T., Gillespie, M.K., Groom, G.B., Hallam, C.J., Hornung, M., Howard, D.C., and Ness, M.J. (1993) Countryside Survey 1990. Main Report. Department of the Environment.

Baudry, J. (1988) Hedgerows and hedgerow networks as wildlife habitat in agricultural landscapes. In *Environmental Management in Agriculture*. European Perspectives. (Ed. J.R. Park), 111–124. Belhaven Press, London.

Burel, F. (1989) Landscape structure effects on carabid beetles spatial patterns in western France. *Landscape Ecol.*, 2, 215–226.

Burel, F. (1996) Hedgerows and their role in agricultural landscapes. *Critical Rev. in Plant Sci.*, 15, 169–190.

Burel, F. and Baudry, J. (1990) Hedgerow network patterns and processes in France. In *Changing Landscapes: An Ecological Perspective* (Eds. I.S. Zonneveld and R.T.T. Forman), 99–120. Springer-Verlag, New York.

Burel, F., Baudry, J., Butet, A., Clergeau, P., Delettre, Y., Le Coeur, D., Dubs, F., Morvan, N., Paillet, G., Petit, S., Thenail, C., Brunel, E., and Lefeuvre, J.-C. (1998) Comparative biodiversity along a gradient of agricultural landscapes. *Acta Oecologia*, 19, 47–60.

Denis, D., Thenail, C., Petit, S., Morvan, N., Moonen, N., Mérot, P., Le Coeur, D., Dubs, F., Burel, F., and Baudry, J. (1995) A method for analyzing field boundaries: Computerized management data. *INRA-SAD Armorique*, CNRS, URA ECOBIO, Rennes.

Fernandez Garcia, A., Griffiths, G.J.K. and Thomas, C.F.G. (2000) Density, distribution and dispersal of the carabid beetle *Nebria brevicollis* in two adjacent cereal fields. *Ann. Appl. Biol.*, 137, 89–97.

Fuller, R.J., Gregory, R.D., Gibbons, D.W., Marchant, J.H., Wilson, J.D., Baillie, S.R., and Carter, N. (1995) Population declines and range contractions among farmland birds in Britain. *Conserv. Biol.*, 9, 1425–1441.

Gerowitt, B. and Wildenhayn, M. (1997) Ecological and economic effects of intensifying arable farming systems: Results of the Göttingen INTEX-project 1990–94. Forschungs- und Studientzentrum Landwirtschaft und Umwelt der Universität Göttingen, Göttingen.

Greaves, M.P. and Marshall, E.J.P. (1987) Field margins: definitions and statistics. In *Field

Haines-Young, R.H., Barr, C.J., Black, H.I.J., Briggs, D.J., Bunce, R.G.H., Clarke, R.T.,
    Cooper, A., Dawson, F.H., Firbank, L.G., Fuller, R.M., Furse, M.T., Gillespie, M.K.,
    Hill, R., Hornung, M., Howard, D.C., McCann, T., Morecroft, M.D., Petit, S., Sier,
    A.R.J., Smart, S.M., Smith, G.M., Stott, A.P., Stuart, R.C., and Watkins, J.W. (2000)
    Accounting for Nature: Assessing Habitats in the U.K. Countryside, DETR, London.
INRA (1976) Les Bocages: histoire, écologie, economie. In *Compte Rendu de la Table Ronde*
    CNRS "Aspects physiques, biologiques et humaines des écosystemes bocagers des
    régions tempérées humides" INRA, CNRS-ENSA et Université de Rennes.
Jongman, R.H.G., Ter Braak, C.J.F., and Van Tongeren, O.F.R., Eds. (1995) *Data Analysis in
    Community and Landscape Ecology,* Cambridge University Press, Cambridge. 324 pp.
Jörg, E., Ed. (1994) *Field Margin-Strip Programmes.* Landesanstalt für Pflanzenbau und
    Pflanzenschutz, Mainz. 182 pp.
Kleijn, D. (1996) The use of nutrient resources from arable fields by plants in field boundaries.
    J. Appl. Ecol., 33, 1433–1440.
Kleijn, D. (1997) Species richness and weed abundance in the vegetation of arable field
    boundaries. Ph.D. thesis, Agricultural University, Wageningen.
Kleijn, D., Joenje, W., Le Coeur, D., and Marshall, E.J.P. (1998) Similarities in vegetation
    development of newly established herbaceous strips along contrasted European field
    boundaries. *Agric. Ecosys. Environ.,* 68, 13–26.
Kleijn, D. and Snoeijing, G.I.J. (1997) Field boundary vegetation and the effects of agro-
    chemical drift: botanical change caused by low levels of herbicide and fertilizer.
    J. Appl. Ecol., 34, 1413–1425.
Kleijn, D. and Van der Voort, L.A.C. (1997) Conservation headlands for rare arable weeds:
    the effects of fertilizer application and light penetration on plant growth. *Biol. Con-
    serv.,* 81, 57–67.
Kleijn, D. and Verbeek, M. (2000) Factors affecting the species composition of arable field
    boundary vegetation. *J. Appl. Ecol.,* 37, 256–266.
Le Coeur, D. (1996) La végégetation des élements linéaires non cultivés des paysages agricoles:
    identification à plusieurs échelles spatiales, des facteurs de la richesse et de la com-
    position floristiques des peuplements. Ph.D. thesis, Université de Rennes I, Rennes.
Le Coeur, D., Baudry, J., and Burel, F. (1997) Field margins plant assemblages: variation
    partitioning between local and landscape factors. *Landscape and Urban Planning,*
    37, 57–71.
Marrs, R.H. (1993) Soil fertility and nature conservation in Europe: theoretical considerations
    and practical management solutions. *Adv. Ecol. Res.,* 24, 241–300.
Marshall, E.J.P. (1988) The ecology and management of field margin floras in England.
    Outlook on Agric., 17, 178–182.
Marshall, E.J.P. (1989) Distribution patterns of plants associated with arable field edges.
    J. Appl. Ecol., 26, 247–257.
Marshall, E.J.P. (1993) Exploiting seminatural habitats as part of good agricultural practice.
    In *Scientific Basis for Codes of Good Agricultural Practice. Eur 14957.* (Ed. V.W.L.
    Jordan), 95–100. Commission for the European Communities, Luxembourg.
Marshall, E.J.P., Ed. (1997) *Field Boundary Habitats for Wildlife, Crop and Environmental
    Protection.* Final Project report of Contract AIR3-CT920476. Commission of the

Marshall, E.J.P. and Moonen, C. (1997) Patterns of plant colonization in extended field margin strips and their implications for nature conservation. In *Species Dispersal and Land Use Processes* (Eds. A. Cooper and J. Power), 221–228. IALE(U.K.), Coleraine.

Marshall, E.J.P. and Nowakowski, M. (1991) The use of herbicides in the creation of a herb-rich field margin. In *1991 Brighton Crop Protection Conference — Weeds,* 655–660. British Crop Protection Council, Farnham, U.K.

Marshall, E.J.P. and Smith, B.D. (1987) Field margin flora and fauna: interaction with agriculture. In *Field Margins* (Eds. J.M. Way and P.J. Greig-Smith), 23–33. British Crop Protection Council, Farnham, U.K.

Marshall, E.J.P., West, T.M., and Moonen, A.C. (1998) Creating and managing perennial field margin strips for botanical diversity in farmland. *Gibier Faune Sauvage/Game and Wildlife,* 15, 87–97.

Moonen, A.C. and Marshall, E.J.P. (2001) The influence of sown margin strips, management and boundary structure on herbaceous field margin vegetation in two neighbouring farms in southern England. *Agric, Ecosys. Environ.* 86, 187–202.

Moonen, C. (1995) The factors influencing field margin vegetation in two contrasted agricultural landscapes: bocage and polder. M.Sc. thesis, Agricultural University, Wageningen.

Morris, M.G. and Webb, N.R. (1987) The importance of field margins for the conservation of insects. In *Field Margins.* BCPC Monograph No. 35 (Eds. J.M. Way and P.W. Greig-Smith), 53–65. British Crop Protection Council, Farnham, U.K.

Paoletti M.G. (1999) Using bioindicators based on biodiversity to assess landscape sustainability. *Agric. Ecosys. Environ.,* 74, 1–18.

Paoletti M.G. (2001) Biodiversity in agroecosystems and bioindicators of environmental health. In *Structure and Function in Agroecosystems Design and Management — Advances in Agroecology* (Eds. M. Shiyomi and H. Koizumi), 11–44. CRC Press, Boca Raton, FL.

Paoletti M.G. and Lorenzoni, G.G. (1989) Agroecology patterns in northeastern Italy. *Agric. Ecosys. Environ.,* 27, 139–154.

Parish, T., Lakhani, K.H., and Sparks, T.H. (1995) Modelling the relationship between bird population variables and hedgerow, and other field margin attributes. II. Abundance of individual species and of groups of species. *J. Appl. Ecol.,* 32, 362–371.

Pollard, E., Hooper, M.D., and Moore, N.W. (1974) *Hedges.* Collins, London.

Rands, M.R.W. (1985) Pesticide use on cereals and the survival of grey partridge chicks: a field experiment. *J. Appl. Ecol.,* 22, 49–54.

Rew, L.J., Theaker, A.J., and Froud-Williams, R.J. (1992) Nitrogen fertilizer misplacement and field boundaries. In *Aspects of Applied Biology — Nitrate and Farming Systems,* Vol. 30, 203–206. Association of Applied Biologists, Wellesbourne, U.K.

Schippers, P., Snoeijing, I., and Kropff, M. (1999) Competition under high and low nutrient levels among three grassland species occupying different positions in a successional sequence. *New Phytologist,* 143, 547–559.

Schumacher, W. (1987) Measures taken to preserve arable weeds and their associated communities. In *Field Margins.* BCPC Monograph No. 35 (Eds. J.M. Way and P.W. Greig-Smith), 109–112. British Crop Protection Council, Farnham, U.K.

Smith, H., Firbank, L.G., and Macdonald, D.W. (1999) Uncropped edges of arable fields

ter Braak, C.J.F. (1987a) The analysis of vegetation-environment relationships by canonical correspondence analysis. *Vegetatio,* 69, 69–77.

ter Braak, C.J.F. (1987b) CANOCO — a program for canonical community ordination. Microcomputer Power, Ithaca, New York.

ter Braak, C.J.F. (1996) Unimodal models to relate species to environment DLO — Agricultural Mathematics Group, Wageningen.

Thomas, C.F.G. (1995) A rapid method for handling and marking carabids in the field. *Acta Jutlandica,* 70, 57–59.

Thomas, C.F.G., Cooke, H., Bauly, J., and Marshall, E.J.P. (1994) Invertebrate colonization of overwintering sites in different field boundary habitats. In *Arable Farming under CAP Reform — Aspects of Applied Biology* No. 40 (Eds. J. Clarke, A. Lane, A. Mitchell, M. Ramans and P. Ryan), 229–232. Association of Applied Biologists, Wellesbourne, U.K.

Thomas, C.F.G. and Marshall, E.J.P. (1999) Arthropod abundance and diversity in differently vegetated margins of arable fields. *Agric. Ecosystems and Environ.,* 72, 131–144.

Thomas, C.F.G., Parkinson, L., Griffiths, G.J.K., Fernandez Garcia, A., and Marshall, E.J.P. (2001) Aggregation and temporal stability of carabid beetle distributions in field and hedgerow habitats. *J. Appl. Ecol.,* 38, 100–109.

Thomas, C.F.G., Parkinson, L., and Marshall, E.J.P. (1998) Isolating the components of activity-density for the carabid beetle *Pterostichus melanarius* in farmland. *Oecologia,* 116, 103–112.

Tsiouris, S. and Marshall, E.J.P. (1998) Observations on patterns of granular fertilizer deposition beside hedges and its likely effects on the botanical composition of field margins. *Ann. Appl. Biol.,* 132, 115–127.

Vanruremonde, R.H.A.C. and Kalkhoven, J.T.R. (1991) Effects of woodlot isolation on the dispersion of plants with fleshy fruits. *J. Vegetation Sci.,* 2, 377–384.

West, T.M. and Marshall, E.J.P. (1996) Managing sown field margin strips on contrasted soil types in three Environmentally Sensitive Areas. In *Aspects of Applied Biology* (44) *Vegetation Management in Forestry, Amenity and Conservation Areas,* 269–276. Association of Applied Biologists, Wellesbourne, U.K.

West, T.M. and Marshall, E.J.P. (1997) Establishment and management of grass and wild flower strips at arable field edges in environmentally sensitive areas. In *Grassland Management in Environmentally Sensitive Areas.* BGS Occasional Symposium No. 32 (Ed. R.D. Sheldrick), 290–292. British Grassland Society.

West, T.M., Marshall, E.J.P., and Arnold, G.M. (1997) Can sown field boundary strips reduce the ingress of aggressive field margin weeds? In *1997 Brighton Crop Protection Conference — Weeds,* 985–990. BCPC, Farnham, U.K.

West, T.M., Marshall, E.J.P., Westbury, D.B., and Arnold, G.M. (1999) Vegetation development on sown and unsown field boundary strips established in three environmentally sensitive areas. In *Aspects of Applied Biology* (54): *Field Margins and Buffer Zones: Ecology, Management and Policy,* 257–262. AAB, Wellesbourne.

Whelon, D. (1994) The Hedgerow Incentive Scheme. In *Hedgerow Management and Nature Conservation* (Eds. T.A. Watt and G.P. Buckley), 137–145. Wye College Press, Wye.

Wilson, P.J. (1993) Conserving Britain's cornfield flowers. In *1993 Brighton Crop Protection*

CHAPTER 10

Changes in Landscape Diversity Patterns in the Province of Wielkopolska, Poland, Influenced by Agriculture

Andrzej Mizgajski

CONTENTS

INTRODUCTION

Wielkopolska is a historical region situated on the Central European Plain, in the drainage of the Warta, a tributary of the Odra. The main relief features of the region are the result of Pleistocene glaciations that moved in from the north. Consequently, Wielkopolska's landscape changes from the north to the south although it has no clear natural borders in the west and the east. In comparison to other regions of Poland, Wielkopolska has had a unique history that is reflected in its economic structure. This is especially true with regard to the last two centuries, when, because of the agricultural and industrial revolution, the modern economic structure of the region was taking shape. In the 19th century, the area roughly corresponding to historical Wielkopolska was a separate province under Prussian occupation. After

divided into smaller administrative units. Statistics quoted in this chapter refer to Wielkopolska within the historical administrative boundaries that existed at a given time.

A characteristic feature of the economy of today's Wielkopolska is highly intensive agriculture and a well-developed farm produce processing industry. The share of acreage of large farms (approximately 20%), in comparison to that of small peasant family holdings, is much higher here than in central Poland.

Throughout the region's history, agriculture exerted the strongest anthropogenic impact on ecosystems, which is reflected in the landscape. Man's impact on the environment has severely reduced the percentage of the land that is a natural or semi-natural ecosystem. For example, nature reserves cover only 0.18% of the present-day *voivodeship* of Wielkopolska.

NATURAL LANDSCAPE

The major natural characteristics of the region came into being during Pleistocene glaciations, which covered the area three times from 1 million to 11,000 years BP. They left behind a hilly, undulating, and plain terrain lying, for the most part, at 60 to 120 m above sea level. The lowest valley bottoms lie below this range, while end moraines rise above it. In terms of geological structure, boulder loams — a glacial sediment, and sands, both loamy and friable glacio-fluvial and fluvial sediments, in part reshaped by wind, dominate close to the surface. Medium-fertile and poor soils developed on them.

The latest glaciation (Würm) engulfed the northern part of the region, dividing it into two parts with different landscapes. The northern part shows greater relief energy and has glacial channels, on the bottom of which are many lakes. The southern part was molded by earlier glacier transgressions and certain processes occurring in the forefront of the last continental ice-sheet. The terrain here is less varied due to denudation, and there are no natural water reservoirs. Wielkopolska lies in the temperate zone. Its weather is characterized by the influx of polar-maritime air masses from the west, which dominate in the warm months. In winter and spring, a considerable impact on the weather is exerted by arctic and polar-continental air masses coming from the north and east (Woś 1994). Average temperatures for July, the warmest month, are 17.5 to 18°C, while in January, the coldest month, the average temperature is –2 to –3°C. Maximum rainfall is observable in summer (approximately 40%), while in the remaining three seasons the amount of precipitation is rather uniform. On average, the region receives 500 to 600 mm of rainfall annually.

There are two basic types of natural landscape: (1) the plain and undulating landscape of the lowlands and (2) the landscape of valleys and depressions (Figure 10.1). Within each type, there are several subtypes of peculiar morphogenetic

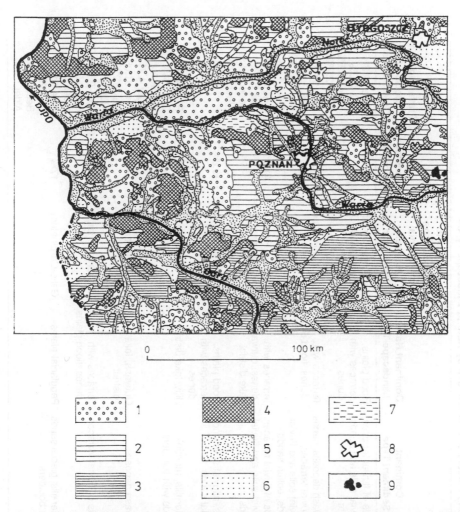

Figure 10.1   Natural landscapes of Wielkopolska. Legend: Flat and undulating landscapes: 1 — glacifluvial, 2 — glacial, 3 — periglacial. Landscapes of hills and heights: 4 — glacial and glacifluvial. Landscapes of valleys and depressions: 5 — flood plains, 6 — higher fluvial terraces, 7 — swamp plains. Landscapes strongly transformed anthropogenically: 8 — compact urban and industrial developments, 9 — strip-mining of lignite. (Adapted from Kondracki 1997.)

plateaus, dry-ground forests developed, whereas on higher valley terraces and out-wash deposits poor habitats of mixed and pine forests grew. Periodically flooded,

Potential Natural Vegetation and Forests of Wielkopolska by Natural Landscape Types

Regional Geographical	Dominant Sediment Type	Dominant Natural Landscape Type	Dominant Associations of Potential Natural Vegetation	Percentage of Forests	Prevailing Tree Species in Tree Stands
...rty [...of the Warta River]	Sands of fluvial terraces, alluvial soils and sands, peats	Flood plains, terraces with dunes	Leucobryo-Pinetum Ficario-Ulmetum	43%	Pine 92.8% Oak 2.0% Beech 0.5% Alder 2.2%
...ecka [...e Noteć River]	Outwash deposits, sands of fluvial terraces, alluvial soils and fluvial sands, eolian sands	Terraces with dunes, flood plains	Leucobryo-Pinetum	51.5%	Pine 93.6% Oak 1.1% Beech 0.3% Alder 1.5%
...otomysko-Kargowska [...wy Tomyśl and Kargowa]	Sands of fluvial terraces, alluvial soils and fluvial sands	Terraces with dunes, outwash, lake district	Leucobryo-Pinetum Pino-Quercetum	41.3%	Pine 88.1% Oak 3.6% Beech 0.6% Alder 4.0%
...owej Noteci [...e Middle-Noteć]	Peats, alluvial soils and fluvial sands	Flood plains, terraces with dunes	Fraxino-Alnetum	19.8%	
...elkopolskie [...ie Lake District]	Glacier tills, morainic sands with boulders	Hilly, lake district	Galio-Carpinetum	17.3%	Pine 80.8% Oak 7.4% Beech 0.4% Alder 4.7%
...nienskie [...ains of Gniezno]	Outwash deposits	Outwash, lake district	Pino-Quercetum	19.4%	Pine 81.0% Oak 5.7% Alder 4.8%
...ska [...w by Konin]	Sands of fluvial terraces	Terraces with dunes, flood plains	Leucobryo-Pinetum	31.9%	Pine 93.1% Oak 0.5% Alder 3.0%
...toszyńska [...rotoszyn]	Glacier tills, glacial sands with boulders	Periglacial plains	Galio-Carpinetum	17%	Pine 61.7% Oak 23.6% Beech 0.3% Alder 5.3%

...ted from Trampler et al. 1990.

self-regulating systems that were characterized by a long-term balance between the production of biomass and its conversion into energy (Pianka 1974). In those early days, similar to other organisms, man was totally dependent on the supply of food and on other environmental factors.

DEVELOPMENT OF AGRICULTURAL LANDSCAPE

Agriculture is the form of human activity that has had a decisive impact on the landscape throughout history. The origins of agriculture are related to the Neolithic revolution, or the process of transforming societies of hunters and gatherers into semi-settled and settled societies of food producers. It is widely accepted that in Wielkopolska the process began over 6,000 years ago, when the agricultural landscape began to emerge and later expand at the expense of the natural landscape over the next centuries. The earliest cultivated areas appeared in well-watered places where soil was light and medium-cohesive, next to lakes and in river valleys, and connected to the regulation of ground water.

According to Kurnatowski (1975), the share of the agricultural landscape did not exceed 10% in Wielkopolska in the 10th century. Later, settlement moved to higher areas, a tendency that became more conspicuous in the 13th century (Dunin-Wąsowicz 1974). The reason behind the move was a rapid growth of population and increased felling of forests, in particular, for constructing strongholds. Continued felling reduced the share of forests in Wielkopolska to 50 to 60% in the 14th century (Hładyłowicz 1932, Błaszyk 1976).

The consequences of forest clearings included a change of river regimes and more frequent overflooding which, in turn, made it necessary to move settlements to higher ground, leaving valley bottoms to renaturalize slowly. Larger portions of the natural landscape survived on the bottoms of large river valleys until the late phase of the self-supportive economy. The retreat of settlements from river valley bottoms meant a forced change in, but not abandonment of, the cultivation of hydromorphic soils. An important role in the expansion of the agricultural landscape in wet areas was played by Cistercian monks, who arrived in Wielkopolska in the 12th to 13th centuries. The second phase of the expansion took place in the 17th and 18th centuries, when settlers from western Europe, mainly from Holland, came to Poland. Their settlements (in Polish *olędry*) were established on wet wastelands that they subsequently cultivated. Cultivation works were most intensive in the last decades of the 18th century because of large-scale river regulations increasing the runoff rate.

A peculiar trait of self-supporting economy was the multiple use made of individual kinds of farmland that are clearly kept separate today. Animals, for instance,

A common trait of different agricultural systems throughout this period was a self-supporting type of economy. It was characterized by the absence of external, anthropogenic feeding of agroecosystems with matter and energy. The fertility of soils in self-supportive land cultivation systems was maintained by burning spontaneous vegetation (swidden farming), fertilizing with livestock excrements and natural regenerative processes (fallow farming) (Kostrowicki 1973). With growing population and developing farm produce market, the output of biomass from agroecosystems grew, while there was no external feeding with nutrients. In consequence, soil productivity fell, making it necessary to begin cultivating ever-newer land at the expense of the natural forest landscape. This process continued in Wielkopolska until the first decades of the 19th century, when farmlands reached their greatest acreage. According to different sources, in the early 19th century forests in Wielkopolska occupied as little as 20% of the land (Baur 1842, Janczak 1965). From 1801 to 1806, Meitzen (1868–1871) specified the structure of land use in what was then the Grand Duchy of Posen (an area roughly corresponding to Wielkopolska) in the following manner: meadows, pastures, used forested areas and wastelands made up 69.6%, 23% was taken up by arable land, and only 7.4% were covered by forests. It can be estimated that over half of the region's area was taken up by extensively used elements of the agricultural landscape, including forest pastures and land lying fallow.

In the 19th century, a new quality emerged in the human impact on the environment owing to the departure from the self-supportive economy and the growth of external feeding of agroecosystems with matter and energy. The main reasons were the introduction of artificial fertilizers and mechanical draught. The result was a stop to oligotrophication of agroecosystems and a more intensive land use. The main stimulus of the change in agriculture was the emancipation of peasants and a gradual drawing of them into the system of market economy. Wielkopolska was at that time within the borders of Prussia, where agrarian reforms were introduced in the 1820s and 1830s. Economic and ownership changes were reflected in the landscape by the shrinking of wastelands and the more intensive use of agricultural and forest ecosystems. In 1864, the share of wastelands in Wielkopolska was estimated at 16,500 ha (0.1% of the region) (Meitzen 1869–1871); however, among the very extensively used areas one has to consider lands classified as pastures (6% of the region). At the same time, the share of arable land reached 60%. However, it must be considered that some of this area lay fallow. Fallow farming was replaced by crop rotation in Wielkopolska in the course of the second part of the 19th century. This change was reflected in the landscape from which wastelands disappeared and were replaced by arable land, whose share increased by 25% (Klein 1973). Arable lands reached their greatest share in 1921 when they occupied two thirds of the region's area. The reason behind this increase was the food shortages during World War I and immediately afterward.

of independence in 1918 until World War II. As a result, the proportions were reversed and the share of small peasant holdings exceeded two thirds. The size structure of individual holdings that ensued then has not changed much until this day.

A peculiar trait of the evolution of Wielkopolska's agricultural landscape over the last 200 years is the elimination of infertile sandy lands from cultivation and instead afforesting them. The forests only slightly contracted in the 19th century, but profound structural changes did occur in their distribution. In the wake of the Napoleonic Wars, Prussia suffered from a deep economic crisis. A shortage of capital made many owners cut down their forests. This practice was further encouraged by the development of industry and a growing demand for timber. At the same time, the government encouraged afforestation, which became particularly intensive in the latter half of the century. What was afforested were heathland and the poorest, sandy cropland; they were planted with fast-growing conifers, particularly pine. Since the 1920s, the percentage of arable land has been on the decline, while the share of forests has been growing. According to statistics, from 1921 to 1973 forested areas grew by almost 170,000 ha in the *voivodeship* of Wielkopolska making the percentage of forests rise from 18.1 to 23.5%. Later, however, the share of forests remained stable. Afforestations carried out in the period between the two world wars were primarily a consequence of a difficult situation in agriculture and unprofitability of farming on the least fertile soils.

After World War II, a sizeable amount of afforestation was necessitated by errors in methods of farming on sandy soils belonging to state-owned or collective farms. In an effort to maximally increase crop yield at minimum outlays, it was a common practice to use large doses of fertilizers, in particular ones with nitrogen, and to abandon fertilizing with manure. Such a cultivation of light sandy soils led to a dramatic decline in organic content of soil and its rapid depletion. Consequently, afforestation was necessary to reclaim the soils.

A good illustration of this process is the change in the landscape structure in the *poviat*, or county, of Międzychód situated in the northwestern corner of Wielkopolska (Figure 10.2), where forested areas grew by as much as 60%. Such intensive afforestation was a result of the prevalence of poor soils in the *poviat*.

Next to tree felling, changes in water relations are a major influence exerted by agriculture on the landscape. The present state of the landscape owes much to the regulations of the Odra, Warta, Noteć, and Obra rivers, increasing the runoff rate. Vast regulation projects were undertaken in the last quarter of the 18th century making approximately 200,000 ha of valley bottom part of agroecosystems (Henning 1979, Falkowski and Karłowska 1961). Betterment drainages of land were continued until modern times; their intensity has noticeably decreased only recently. The area where betterment drainages have been carried out exceeds 1 million ha, i.e., 55% of farmland, in the *voivodeship* of Wielkopolska. The disappearance of water mills,

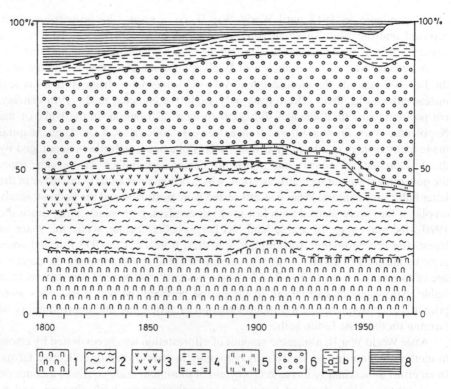

Figure 10.2 Development of land-use structure in the *poviat* of Międzychód since the 19th century. Legend: 1 — cereal crops, 2 — crops other than cereals, 3 — fallow land, 4 — meadows, 5 — pastures, 6 — forests, 7a — surface waters, 7b — developed areas, roads, 8 — heathland. (After Mizgajski 1990.)

Human activities aimed at increasing the rate of water runoff from ecosystems have led to profound, qualitative changes in the landscape of Wielkopolska (Kaniecki 1991). One such change is the lowering of subsurface water, by several dozen centimeters, to two meters. Another is the lowering of lake surfaces by 13% on the average, between 1890 and 1980, which is evident from cartographic analysis. However, the process was much more rapid in the case of small lakes (up to 20 ha), where the decline in surface area reached 50%. The smallest bodies of water (up to 1 ha) tended to disappear totally; from about 1890 to 1960, their numbers fell from over 11,000 to around 2,500. Because of drainages, the acreage of farmland has indeed grown, but problems related to the overdrying of land have appeared as well. Those problems become more acute with more intensive crop production.

The landscape and ecological consequences of anthropogenic changes in water

- Hydrological parameters — lowering of low water marks and ground-water table
- Soil properties — overdrying, gleying recession, and a quicker mineralization of organic substance causing a decrease in humus content
- The vegetation cover — elimination of swamp and riparian forests, spread of xerophytes and plant associations typical of dry habitats
- The fauna — decrease in the number of forest species in favor of species living in an open landscape; appearance of xerophilous mollusk species

The above list of consequences must be supplemented with a list of changes brought about by the elimination of surface waters, such as the disappearance of belts of spontaneous vegetation that serve as biogeochemical barriers deciding the buffer capacities of agroecosystems. (Ryszkowski 1999). An illustration of the extent to which this process took place in the second half of the 20th century may be the recorded changes in the landscape surrounding the village of Zamorze, located about 50 km west of Poznań (Mizgajski and Kafel-Głębowska 1990). In an area approximately 12 km², about 50 point and linear changes of topographical landscape elements that occurred in 1940–1982 were recorded. The most frequent type of change (27 places) was the drying up of ponds, in part accompanied by burying or removing of spontaneous vegetation communities surrounding them. Twenty cases of drainage ditch burying were recorded, frequently accompanied by the elimination of tree and bush belts growing along them. Furthermore, trees were felled along abandoned roads or ploughed field margins (six cases). A clear majority of changes consisting of the elimination of small landscape elements were direct or indirect consequences of drainage works carried out mostly in the 1970s. In this respect, the example of the village of Zamorze can be treated as representative of the whole of Wielkopolska.

PRESENT-DAY CHANGES IN AGRICULTURAL LANDSCAPE

Present-day intensive farming is characterized by extremely heavy inputs of matter and energy into agroecosystems by man. This is clearly seen in very high doses of artificial fertilizers. Moreover, the rise of large pig farms using industrial feeds resulted in breaking earlier systemic ties between crop and stock farming. Farmers ceased to rely on the supply of manure from their own animals for crop cultivation, replacing it with artificial fertilizers and catch crops. Owing to industrial feeds, the size of a stock farm ceased to be dependent on the amount of feed produced by the farm, i.e., on the acreage of cropland. Furthermore, farmers, who were earlier bound to their farms, which were the sources of subsistence, turned into workers supplying labor, while the farms ceased to be sources of produce for them. Thus, agroecosystems changed from closed systems dominated by stabilizing internal flows of matter and energy into open systems, where external relations are of primary

of large animal herds. This increased the significance of agricultural nonpoint pollution leading to the eutrophication of ecosystems and landscape effects such as the disappearance of lakes. This process continues despite the fact that agriculture became less intensive in the 1990s. The change of the political system in Poland initiated the process of profound socio-economic changes in rural areas. Subjecting agriculture to market pressures and abolishing subsidies to great state-owned farms resulted in an overall decrease in the intensity of land use.

Poland's transition to a market economy consisted mainly of the release of prices, withdrawal of subsidies to businesses, and deregulation of foreign trade. As a consequence, the ratio between produce prices and those of manufactured goods became less and less advantageous. This, in turn, resulted in the lowering of agricultural production intensity due to the shortage of funds for basic means of production, including fertilizers and pesticides. In due course, the consumption of fertilizers in central Wielkopolska fell from over 250 kg NPK/ha in the 1980s to approximately 100 kg NPK/ha in the early 1990s. The intensity of fertilizer use began to climb again to approximately 130 kg NPK/ha in the late 1990s.

Decreased use of fertilizers was not reflected, however, in less eutrophication pressure exerted by agriculture on the environment. Gołdyn and Grabia (1998) investigated the drainage of the Cybina, one of the smaller tributaries of the river Warta. They estimated that from the catchment area of 15,600 ha, the river annually receives almost 120 tons of N and over 5 tons of P as agricultural nonpoint pollution. From the point of view of matter flow, the landscape of Wielkopolska is dominated by autonomous areas (plains) with a significant share of allochthonous areas (no-drainage depressions). A smaller share is taken up by sloping areas, known as transit ones. As a result, the ecosystems of the region tend to retain substances introduced to them. Consequently, one should not expect that the decline in the use of fertilizers would be reflected soon in lower numbers of biogens in ecosystems. Thus, it can justifiably be claimed that present-day changes in Wielkopolska's landscape consisting of the disappearance of lakes are largely caused by eutrophication brought about by agricultural nonpoint pollution. The landscape changes aggravate the consequences of earlier draining projects that induced irreversible changes in water conditions.

The high price of the means of agricultural production and the low price of farm produce make farmers more interested in afforestation of the poorest lands. Such lands are usually found in large complexes, in areas that already now have a high percentage of forests. New afforestations will simplify the landscape by combining forested areas into large forest complexes, whereas good soils will be even more intensively farmed since their afforestation is practically out of the question for economic reasons. To counteract the simplification of the landscape-ecological structure of such areas, it is necessary to increase the share of biotopes of seminatural vegetation, specifically by planting trees along field margins and roads. Such actions have already been taken in Wielkopolska, and there are plans to continue them

the southern and eastern parts of the region and up to 15 ha in the northern part. Usually, land belonging to one farm is spread over several places, which means that individual fields are only a few hectares in size. Fields belonging to large farms, as a rule, make up compact complexes, with the size of individual fragments exceeding 100 ha. The present agrarian structure may not be considered stable.

In the future, one can expect the concentration of land ownership and field size to grow. Such a trend poses a danger of eliminating spontaneous vegetation growing in uncultivated habitats, in particular trees and shrubs. Structural changes may also result in a temporary increase in the share of fallow land in the landscape. In many parts of Poland, vast tracts of less fertile land are already left to lie fallow. This is especially true for former large state-owned farms. Meanwhile, in Wielkopolska, such lands are only marginal and, as such, virtually cannot be seen in the landscape.

An important aspect of the agricultural landscape is rural settlement. Its most common form is a compact village consisting of a dozen to several dozen home-steads. The spatial arrangement of homesteads varies depending on the age and origins of a settlement (Szulc 1972). In total, there are 3,746 rural settlement units in the *voivodeship* of Wielkopolska, which gives on average 5 km^2 of farmland per one village. Few and far between are individual homesteads surrounded by fields. Such sites came into being as a result of the parcellation of great landed properties carried out at the turn of the 19th century and in the 1920s and 1930s. Another type of settlement is made up of granges with adjoining workers' quarters. This type of building arrangement underwent the greatest transformations when farms were state-owned (1950–1990). It was then that many shabby farm buildings and farmhouses were built; poor quality workmanship and architecture and wasteful use of space characterize these constructions.

In the 1990s peasant homesteads were adapted for recreational purposes. The major force behind this process was a low return on farming on mediocre lands, which are often found in recreationally attractive areas, near lakes, surrounded by forests and varied land relief. After their use is changed, such buildings stand out favorably against poor homesteads of farmers. In suburban areas, the process of urbanization of the agricultural landscape is rapidly gaining momentum. One con-sequence of this process is a change in the character of suburban villages, where fewer and fewer peasant homesteads are found, while residential housing is mush-rooming, sometimes giving shelter to small businesses as well. In suburban areas, in particular along main roads, agricultural landscape is replaced by single-family housing estates or, in Poznań's vicinity, by industrial parks. One can also speak of the urbanization of rural areas in Wielkopolska in a qualitative sense because of the improving infrastructure in rural areas. The road network is getting more dense (presently over 80 km of roads with hard surface per 100 km^2) and about 70% of homesteads have access to a water supply network. Unfortunately, the construction

CONCLUSIONS

The present-day landscape is a combination of natural features and tendencies and the human impact on the environment. Agriculture played the most important role in landscape transformation. Historically speaking, the agricultural impact on the landscape of Wielkopolska can be divided into three qualitatively different phases:

- Domination of the process of expansion of the agricultural landscape at the expense of the natural one; this phase was related to self-supportive economic systems
- Domination of qualitative changes in the agricultural landscape related to increased inputs of matter and energy to agroecosystems, beginning with the industrial-agricultural revolution of the 19th century
- Decrease of the agricultural landscape due to afforestation and urbanization, processes that have become more intensive since the mid-20th century

The gradual expansion of the agricultural landscape continued until the beginning of the 19th century, bringing the area occupied by natural forest ecosystems to a minimum. Present-day forests are almost entirely the effects of plantings that are now in various phases of re-naturalization.

Agriculture is also responsible for changes in the landscape that has been shaped earlier by farming. Examples of qualitative changes in the landscape induced by agriculture include:

- Overdrying of landscape due to a quicker rate of river runoff, draining of land and elimination of small retention
- Eutrophication of ecosystems as a result of the irrational use of artificial fertilizers and slurry leakage
- Decline in biodiversity and landscape diversity because of the elimination of spontaneous vegetation from the agricultural landscape

For about 50 years, a tendency has been observed whereby the agriculture landscape has been shrinking in favor of forests and urban areas. These changes lead to the simplification of the landscape and diminishment of its diversity because of the increasingly greater share of forest complexes in areas of poor soil, more intensive farming in areas of fertile soil, and increasing urbanization in suburban areas.

Further development of the agricultural landscape will be conditioned by economic and cultural factors as well as legal regulations. Today's tendencies are related to the ever-closer integration of Poland with the European Union, and they make the agricultural landscape more contrastive. Agroecosystems based on fertile soils are subject to the increasing pressure of agriculture, while poor soils are taken out of crop or stock production and set aside for afforestation. Of vital importance for

REFERENCES

Baur K. F. 1842. *Forststatistik der deutschen Bundesstaaten.* Brockhaus, Leipzig. Teil I, 272 pp., Teil II, 280 pp.

Błaszyk H. 1976. Zmiany lesistości Wielkopolski [*Changes in Forest Cover of Wielkopolska*]. *Roczniki Akademii Rolniczej w Poznańiu,* Poznań, 73, 1–47.

Dunin–Wąsowicz T. 1974. Zmiany w topografii osadnictwa Wielkich Dolin na Niżu Środkowoeuropejskim w XIII wieku [*Changes in Topography of Settlement of Great Valleys on the Central European Plain in the 13th Century*]. Zakład Narodowy Imienia Ossolińskich, PAN, Wrocław-Warszawa, 178 pp.

Falkowski M., Karłowska G. 1961. Rozwój łąkarstwa w Wielkopolsce [*Development of Meadow Cultivation in Wielkopolska*]. Poznńskie Towarzystwo Przyjaciół Nauk, Prace Komisji Nauk Leśnych, IX, 2, Poznań, 198 pp.

Geografia, 5, 1964, Zeszyty Naukowe Uniwersytetu im. Adama Mickiewicza w Poznńiu, Nr 52.

Gołaski J. 1980, 1988. *Atlas rozmieszczenia młynów wodnych w dorzeczach Warty, Brdy i części Baryczy w okresie 1790–1960* [*Atlas of Water Mills in the Drainages of the Warta, Brda and Part the Barycz in 1790–1960*]. Akademia Rolnicza, Poznań, Part 1, 107 pp., Part 2, 100 pp.

Gołdyn R. and Grabia J. 1998. *Program ochrony wód rzeki Cybiny* [*The Cybina River Protection Program*]. Urząd Miasta, Poznań, 101 pp.

Henning F.-W. 1979. *Landwirtschaft und ländliche Gesellschaft in Deutschland. Part 2, 1750–1976.* Ferdynand Schönigh, Padeborn, 287 pp.

Hładyłowicz K. 1932. *Zmiany krajobrazu i rozwój osadnictwa w Wielkopolsce od XIV do XIX wieku* [*Changes in the Landscape and the Growth of Settlement in Wielkopolska from the 14th to the 19th c.*]. Instytut Popierania Polskiej Twórczosci Naukowej Lwów, 256 pp.

Janczak J. 1969. Okręgi Rolnicze Polski Zachodniej i Północnej w Pierwszej Połowie XIX Wieku [*Agricultural Centers of Western and Northern Poland in the First Half of the 19th Century*]. Zakład Narodowy Imienia Ossolińskich, PAN, Wrocław-Warszawa-KrakówWrocław, 226 pp.

Kaniecki A. 1991. Problem odwodnienia Niziny Wielkopolskiej w ciągu ostatnich 200 lat i zmiany stosunków wodnych [*The Problem of Dehydration of Wielkopolska Plain during the Last 200 Years and Changes of Water Conditions*]. The Proceedings of the Conference: Ochrona i racjonalne wykorzystanie zasobów wodnych na obszarach rolniczych w regionie Wielkopolski [*The Conservation and the Proper Use of Water Resources in the Agricultural Areas of Wielkopolska Region*]. Urzą Wojewódski, Poznań, pp. 73–80.

Klein E. 1973. *Geschichte der deutschen Landwirtschaft im Industriezeitalter.* Steiner, Wiesbaden, 192 pp.

Kondracki J. 1991. Krajobrazy naturalne (*Natural landscapes*). In Starkel L. (Ed.) *Geografia Polski. Środowisko przyrodnicze* [*Geography of Poland. Natural Environment*], Państwowe Wydawnictwo Naukowe, Warszawa, 670 pp.

Kostrowicki J. 1973. *Zarys Geografii Rolnictwa* [*An Outline of Agricultural Geography*], Państwowe Wydawnictwo Naukowe, Warszawa, 631 pp.

Kurnatowski S. 1975. Wczesnośredniowieczny przełom gospodarczy w Wielkopolsce oraz

262   LANDSCAPE ECOLOGY IN AGROECOSYSTEMS MANAGEMENT

Mizgajski A. 1990. Entwicklung von Agrarlandschaften im Mitteleuropaischen Tiefland seit
    dem 19. Jahrhundert in energetischer Sicht. Beispiele aus dem Emsland und Wielko-
    polska [The development of the agricultural landscape on the Central European Plain
    since the 19th century: an energy-based approach. Examples from Emsland and
    Wielkopolska], Münstersche Geographische Arbeiten, 33, 110 pp.
Mizgajski A., Kafel-Głębowska K. 1990. Przeobrażenia krajobrazu rolniczego od XIX wieku
    na przykładzie wsi Zamorze koło Pniew Wielkopolskich [Changes of the agricultural
    landscape since the 19th century with reference to the Zamorze village near Pniewy
    Wielkołolskie], Badania Fizjograficzne nad Polską Zachodnią, 41, A. Poznań, 75–85.
Pianka E.-R. 1974. *Evolutionary Ecology,* Harper & Row, New York, 356 pp.
Ryszkowski L. 1999. Ecological guidelines for development of sustainable agricuture in
    Poland. In Wicherek S. (Ed.) *Paysages agraires et environnement. Principies
    ecologiques de gestion en Europe et au Canada,* CNRS Éditions, Paris, 49–59.
Ryszkowski L., Bałazy St. 1999. Land-use change in the agricultural region of Wielkopolska,
    Poland. In Kroenert R., Baudry J., Bowler I.R., and Reenberg (Eds.) *Land-Use
    Changes and Their Environmental Impact in Rural Areas in Europe,* Man and the
    Biosphere Series, 24, UNESCO and Parthenon Publishing Group, Paris, 189–204.
Ryszkowski L., Bałazy St., Jankowiak J. 2000. Program zwiększania zadrzewień śródpolnych
    (Shelterbelt Planting Program), Postępy Nauk Rolniczych, 47(5), 83–107.
Szulc H. 1972. The development of the agricultural landscape of Poland. *Geographia Polon-
    ica,* 22, Warszawa, 85–103.
Szwed W., Ratyńska H., Danielewicz W., Mizgajski A. 1999. Przyrodnicze podstawy kształ-
    towania marginesów ekołogicznych w Wielkopolsce [Natural Foundations of Eco-
    logical Margin Modeling in Wielkopolska], Akademia Rolnicza, Poznań, 144 pp.
Trampler T., Kliczkowska A., Dmyterko E., Sierpińska A. 1990. *Regionalizacja przyrodniczo-
    leśna na podstawach ekołogiczno-fizjograficznych [Regionalization of the Natural
    Forests according to the Ecological and Physiographical Conditions],* Państwowe
    Wydawnictwo Naukowe, Warszawa, 159 pp.
Wodziczko A. 1947. Wielkopolska stepowieje [*Wielkopolska turns into Steppe*] In (Wodziczko
    A. et al. Ed.) Stepowienie Wielkopolski [*Wielkopolska turns into Steppe*]. Part 1,
    Poznańskie towarzystwo Przjaciół Nauk, Poznań, 141–152.
Wojterski T., Wojterska H., Wojterska M. 1978. Potencjalna roślinność naturalna Środkowej
    Wielkopolski [Potential Natural Vegetation of Central Wielkopolska], Badania Fiz-
    jograficzne nad Polską Zachodnią, 32, B, Poznań, 7–35.
Wojterski T., Wojterska H., Wojterska M. 1981. Potencjalna roślinność naturalna dorzecza
    Baryczy [*Potential Natural Vegetation of the Barycz River Drainage*] (mapa 1:200.000)
    Państwowe Przedsiębiorstwo Wydawnictw Kartograficznych, Wrocław.
Woś A. 1994. *Klimat Niziny Wielkołolskiej [The Climate of the Wielkopolska Plain]* [in Polish].
    Wydawnictwo Naukowe Uniwersytetu im. Adama Mickiewicza, Poznań, 192 pp.

CHAPTER **11**

The Impact of Agriculture on the Environment of the Northern Parisian Basin

Stanislas Wicherek

CONTENTS

INTRODUCTION

In terms of global production, France is the leading European agricultural nation. French agricultural production accounts for 24% of all European Union production, making France (after the U.S.) the second largest agricultural producer worldwide. Most of this great agriculture is situated in the Parisian basin, and the northeast region of Picardy in particular (Figure 11.1).

Photo 11.1   Regions of study in France.

to ensure a permanent framework for farmers. Technical and agronomic means of production are increasingly intensive. However, does this modern, high-performance form of agriculture contribute to environmental management and sustainable development of the environment? Or does it, instead, contribute to the disruption of ecosystems, causing an impoverishment in the diversity of cultivations, flora, and fauna, by raising the risks of soil and water pollution? Finally, is its recent evolution responsible for ecological imbalances (Wicherek, 2000)?

The major cultivations on the silty plains of northern Europe have, for several decades, experienced an intensification of soil erosion by water, giving rise to loss of soil fertility and harmful muddy floods. Yet neither the shallow typographic relief nor the limited erosion of the rains, which characterize these temperate regions of plains and hills, can explain the scale of these phenomena that one observed and measured in sloping agricultural basins. Moreover, according to the scientific literature, it would seem that no significant augmentation of the total rainfall or average intensity of precipitation (the common explanation for greater attacks upon the soils by rains) has been noted in these regions. However, most sloping agricultural basins affected by erosion problems have experienced a profound modification of cultivation systems and agrarian practices.

Numerous studies show that the evolution of cultivation practices, due to the mechanization of agriculture, has led to diminished water infiltratability as well as augmented soil erosion (Boiffin et al., 1986; Wicherek, 1993; Le Bissonnais and Le Souder, 1995). In addition, agrarian practices, adapted to the modernization of agricultural production, are inadequate in relation to the morphologic soil characteristics (Lechevallier, 1992), thus favoring runoff and a loss of further possibilities for infiltration (Larue, 1992).

If the changes in agrarian landscapes (in particular with the evolution from vegetation and utilization of soils) plays an undeniable role in the processes and harmful manifestations of soil degradation (Papy and Douyer, 1991), a historic approach is sometimes necessary to demonstrate and measure such evolutions over the long term. Documents produced by different state organizations, whether they be historic or current, effectively provide a record of agrarian landscapes over two centuries.

PICARDY — AN AGRICULTURAL REGION

Amongst its forests, wooded valleys, stone and brick villages, and cathedral cities, Picardy possesses an open-field landscape that stretches over the chalky plateaus of the Parisian Basin, "*la plaine*" ("the plain"), as farmers call it. The SAU (*surface agricole utilisée*; agricultural area in use) covers nearly 71% of the territory. This

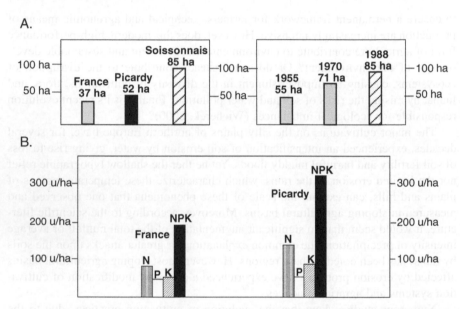

Figure 11.2 (A) (left) Average French agricultural area (SAU) in Picardy and Soissonnais (Agreste 1991); (right) Increase in average SAU in Soissonnais farmlands — 1955, 1970, 1988 (RGA, 1955, 1977, 1988). (B) Consumption of N, P, and K in France and Picardy.

Picardy, particularly Soissonnais, is also distinguished by its large area farms. The average farm size is 37 ha for all of France, 52 ha for Picardy, and 85 ha for Soissonnais. In Picardy more than 45% of farms cover more than 50 ha. In Soissonnais, the large size of farms is not a new phenomenon, since the average SAU per farm exceeded 55 ha in 1955, was 71 ha in 1977, and 85 ha in 1988 (Figure 11.2). Farms are frequently larger than 300 ha (with a record area of 5,000 ha) although the average is lowered by smaller market farms, animal farms, and the mixed farming of the valley floor.

The region is also distinguished by a higher consumption of fertilizers than the national average, whether this be for the separate elements N, P, and K, or for the total NPK: 200 u/ha in France vs. 308 u/ha in Picardy (Figure 11.2). Added to this is an imposing fleet of agricultural machines. After a strong increase in the 1970s, the number of tractors today stands at an average of 2.7 per farm. In 1988, 60% of farms had at least two tractors. From 1970 to 1988, the number of 110 CV tractors has tripled; on average there is one tractor for every 28 ha and one agricultural worker for every 100 ha.

The combination of this mechanization and sizeable consumption of fertilizers is associated with yields higher than the national average (Figures 11.3 and 11.4). Since the beginning of the 20th century, the average yield of wheat in France has

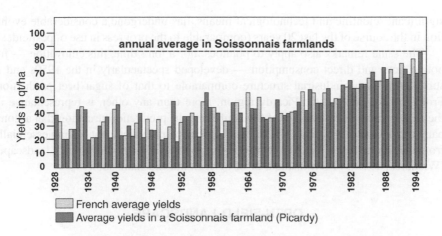

Figure 11.3   Growth in wheat yields. Comparison of France and Picardy. (From SNIE, 1993; farmer archives.)

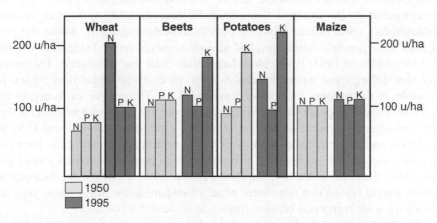

Figure 11.4   Comparison of NPK consumption in 1950 and 1995 (from investigations).

Agricultural crops and related industries are widely developed locally (production of sugar, potato flour, dehydrated mashed potatoes, potato chips, frozen French fries, etc.). They also include other vegetable production, such as high protein plants (peas, flageolet beans, etc.), carrots, and onions. In the Picardy region there are also numerous sugar refineries (Beghin Say, Générale Sucrière, CFS, etc.) and Vico (at Vic-sur-Aisne), D'Aucy, Bonduel, Findus, Saupiquet, of Flodor, Unichips and Mc-Cain, to cite only the best-known examples.

Sugar agriculture — "a world apart" (Lefèvre, 1993) — represents a colossal

significant scientific and technological means, has undergone a considerable evolution in the course of the last 20 years (comparable to the progress in use of pesticides).

The same remarks also apply to potatoes. For a long time, this cultivation — for potato flour and direct consumption — developed spectacularly in the 1950s and is surrounded by an industrial structure comparable to that of sugar beet production. From this perspective, the Picardy region, more than any other, is representative of the development of cultivations of foodstuffs and the production structures that accompany them. One may wonder whether this spectacular development (dating essentially from the second half of the 20th century) is responsible for a modification of landscapes (Wicherek, 1999), of fertilization, or of agricultural mechanization?

OPEN-FIELD LANDSCAPE

Yields have increased considerably, potato cultivation has greatly developed, and the sugar beet industry has perfected itself, but one observation stands out: the changes in the agrarian landscape and the size and distribution of plots of land are not extraordinary (for example, Soissonnais Plateau, according to the study of aerial photographs). Observations about the plateau indicate that, on the whole, the plots of land and open-field landscapes had already occurred under François I. Whether in 1936, 1958, or 1991, it was plots larger than 5 ha that dominated. The overall plot size enlargement naturally benefits large plots, those larger than 20 ha, for example, at the expense of those of less than 5 ha. This can be explained in two ways: first, the disappearance of small gardens given to agricultural workers for their personal vegetable production (sources of weeds — they disappeared around 1950, the producers preferring to compensate workers by distributing a part of the farm production); second, the enlargement of border plateau plots or converted former pastures. Over the whole of the township, from 1936 to 1993, not one hedge disappeared, for the simple reason that there were none. Elsewhere in the Soissonnais, very few townships were regrouped because there was no reason to do so.

The large farms at the beginning of the 20th century provided a means of living for about 100 people, people who often lived in housing estates adjoining the farms. Production was also more varied than it is today. At the beginning of the 20th century, most farms included a flock of sheep (sheltered in a sheepfold during the winter, they were moved to pastures in the spring and after the grain harvest, where they grazed on stubble). Therefore, the removal of stubble occurred rarely or not at all. An element conducive to the low erosion of bare soils in autumn but not favorable to the cleanliness of the land, direct ploughing of the stubble occurred later and lasted longer because of the slowness of animal haulage (draft horses and oxen). The oxen consumed the beet pulp, straw, and hay; the horses consumed oats and alfalfa. The disappearance of

needs for local consumption. From the 1960s, potatoes were grown for industrial uses (either for food or for potato flour production). The same is true of peas and legumes.

The transition from use of draft animals to mechanization and from fodder cultivation to industrial crop cultivation was rather slow since it took place over a period of about 20 years. However, the division of the land remained constant from the beginning of the 20th century until today. Where there is no evolution of the plots of land, one can trace the division back to the heritage of the Soissonnais Plateau where the farm origins are very old. In the Middle Ages, numerous forms of farms already existed under the supervision of the Prémontrés and the Cisterciens — such as the farm of Montramboeuf, or the farm of the Château, which still exists today in the commune of Vierzy. The farms of this township seem even to have descended from Gallo-Roman *villae* (archaeological vestiges), linked to the presence of the Brunehaut Way, a road joining Lutecia to Soissons. In addition, at the time of the Revolution the limits of property and of plots of land were spared. Another historical factor is pertinent — over the course of the centuries this region was crossed by waves of successive invasions by foreign armies. Clearly, it never presented a farmland intersected by hedges and trees as in the west of France. The great bare plateaus forced the enemy to expose itself. Very little wooded area existed; what wooded areas were there totally disappeared, as did rabbit warrens and groves of trees, with the onset of mechanization, during the 20th century.

FLOODS AND MUDSLIDES

Few studies on soil erosion have been carried out in the departments bordering the Seine valley (Wicherek, 1993; Papy and Douyer, 1991). This region is less affected by water erosion than is the rest of France. A preliminary understanding can be based on the analysis of the number of townships affected by natural disasters, such as floods and mudslides, which in agricultural areas are the principal consequences of the waterproofing of different pedological horizons and soil losses (Meyer et al., 1999). Several hundred reports of natural disasters were published in the Journal Officiel, following the law of the 13th of July, 1982. After reviewing the declarations of natural disasters occurring in the departments (Yvelines, Eure, Eure-et-Loir, Val d'Oise, but also Seine-Maritime, Oise, and Aisne), it is possible to establish certain statistics. Among these departments, about 95% of the townships became disaster stricken following floods and mudslides, only 3% of which following floods of the Oise or of the Seine (1993–1995). Furthermore, 63% of the townships are situated in rural areas, but 5% are located in urban zones where the impermeability of the land and the insufficiency of the rainwater purification network constitute the most probable explanatory factors for the damage incurred (Figure 11.5).

Figure 11.5   Example of damage resulting from erosion and pollution in a beet field in Erlon, northern Parisian basin (a station of measures belonging to the Center BIOGEO, CNRS/ENS). (From F. Tellier, June 1999.)

nonetheless reaches 75% even if the two floods of 1993 and 1995 are eliminated from the chronological series.

During this low-water period, floods by sheet overflows or water runoff have a much lower frequency than muddy floods or those not provoked by runoff rainwater pouring along slopes following storms. Nevertheless, these same slopes show well-developed vegetal cover, a factor that inhibits runoff. It is probable, as a result, that the intensity and the localization of storm-type summer rains are determining factors. So it would seem that the events related to hydric surplus are due as much to runoff as to overflow of water courses.

Alternatively one can imagine asking about a possible correlation between the number of disasters occurring during a summer period and those occurring during the preceding winter. A winter marked by a great number of natural disasters probably corresponds, in fact, to high pluviometric totals having favored the genesis of floods by the extension of areas contributing to the outflow of the flood. And the high winter pluviometry combined with little vegetation during "inter-cultivations" frequently provokes a rapid degradation of soil surfaces and the appearance of sedimentary crust (Boiffin et al. 1985). Although a succession of cycles of desiccation and wetting during the spring is likely to destroy the crust and thus reestablish

summers were preceded by winters without harmful events. So it would seem that the frequency of more or less muddy floods occurring in the summer is not unlike, in the northwest Parisian basin, the number of flood and mudslide disasters occurring during the preceding winter. Could erodability due to rain erosion have an effect on the formation of disastrous runoff? If these results are given a more in-depth analysis, however, they reveal a conceivable step.

The chronological series also permits an interpretation of possible spatial heterogeneities in the instances of natural disasters linked to hydric surplus. The departments of Eure-et-Loir and Yvelines are two times less affected than are Eure, Oise, or the Val d'Oise and six times less affected than Seine-Maritime, where the Pays de Caux and the Seine valley appear to be the most affected. If the events had been measured, analyzed, and then explained in the Pays de Caux, Soissonnais Plateau, would it not be interesting to try to understand the "non-events" in the nearby departments of Eure-et-Loir and Yvelines?

The spatial-temporal analyses of the non-event are often ignored in the study of morphogenic processes. In the framework of station measures to an outlet of the primary sloping basin, it can be observed frequently that the conditions for the formation of runoff — or lack thereof — although conditions presumed are filled, are not explained. Is there not, in this case, proof of the omission, or underestimation, of one of the explanatory factors of runoff formation? Thus, the analysis of the "non-events" permits a readjustment in hypotheses and factors initially considered.

In the seven regional departments studied, vegetation and soil use do not seem to influence the occurrence of mudslides (Wicherek, 1988). Indeed, of those seven departments, the Pays de Caux constitutes the agricultural region that was most affected by mudslides and that also has the greatest percentage of area grass covered in relation to the SAU. Inversely, the two least-affected departments — Eure-et-Loir and Yvelines — are the departments with the highest percentage of arable land. Consequently, it is probable that the frequency of damage caused by mudslides is independent of the state of the surface and soil use at the level of the department.

However, would not the Pays de Caux have experienced, over the course of recent decades, more profound modifications of its agrarian landscapes than the cultivated plateaus of Yvelines and western Eure-et-Loir? The difference in the effect of floods and mudslides on northwest Parisian basin departments may be explained not only by the differences in geomorphologic, pedologic, and climatic conditions but also by the rhythms of evolution of farming practices and agrarian structures, hence the interest in historical research that provides analysis of the influence of agrarian landscape transformations on the imbalances of the physical milieu. Mudslides and water erosion are expressions of these imbalances.

A FUNCTIONAL ECOLOGICAL AGRICULTURE

of agricultural yields, but overall they had reached the same degree of productivity: fertilization, pesticides, and yields were very similar to those of other major agricultural regions. The 20th century was a century of changes toward a commercially oriented agriculture and toward agrarian capitalism (Fiette, 1995). The man who works the land today has a farm that is 4.5 times more productive than that of his father in the 1960s; at the same time, the number of people so employed has been halved. Demonstrating a desire for competitiveness, farmers eliminated animal rearing and its accompanying crops, such as oats, and specialized in the food processing industry. Today's farmer often has no hesitation about designating his farm as a business farm from which he makes a profit.

Fertilizer and Pesticide Improvements and Their Impact on the Environment

Although the agricultural landscape has evolved very little, mechanization has, by contrast, made great leaps. Trucks were used for sugar beet harvesting beginning in 1930, but new machines for making pesticide treatments easier, such as the herbicide pulverizer, came into use around 1913. During World War I, steam locomotion was introduced to compensate for the lack of manpower. It was a system of winches and pullies connected to a steam-powered machine. This system was replaced first by the caterpillar tractors and then by the pneumatic tractors. Beginning in 1944, pneumatic tractors became widespread due to the arrival of American surplus machines.

Soil fertilization using manure is a longstanding practice in the region (Brunet, 1960). Farmers in the olden days were well known for their savoir-faire, for their means, and their example-setting; the Cistercians were already improving the manuring of land. Following the serious financial crisis of 1880, when the price of wheat did not rise, the only solution was to increase crop yields. This began in 1900 with the use of mineral fertilizers, which had appeared several decades earlier. For example, bills preserved at the Parcy Farm in Soissonnais show that in 1926 sodium nitrate and super phosphate mineral were used. By 1950, the use and nature of fertilizers had evolved little. Soluble mineral compounds (widespread today) were being used, but so too were solid compounds such as scoria, already used during the 19th century. The 1970s were a turning point, marked by both a great increase in use of mineral fertilization and the quasi-disappearance of other types of fertilization. Today, wheat, potatoes, and beets are more fertilized than they were in 1950 (Figure 11.6).

The situation is more or less the same for pesticides. At the beginning of the 20th century, they were used widely in Picardy but the types were totally different. For example, for wheat in 1926, copper nitrate (rather expensive), cuproazotine, or sulfuric acid could be applied. Beginning in the 1970s and continuing today, synthetic products, with numerous and frequent applications, are used. Examples are

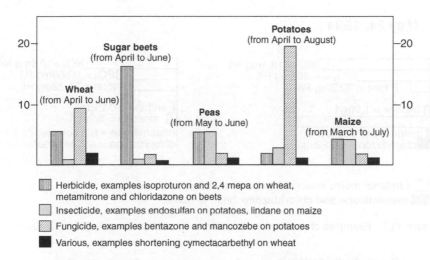

Figure 11.6   Present number of pesticide treatments (from investigations).

Environmental Risks

In considering their farms as businesses, farmers (agricultural businessmen), adopt cultural practices that create risks for the environment. Indeed, the coexistence of environment and agriculture is a difficult one. Years of productivity are emphasized as the basis of use of marked environmentally unfriendly practices. For several years, farmers have been attacked on all sides; they are called polluters or reactionaries, and they are widely accused of attacking the environment. Reasons given are the intensive monoculture on large plots and the excessive use of fertilizers and pesticides.

In Picardy, soil erosion comes from a very real situation, the same as the degradation of surface water and ground water from the chalk layer (Angéliaume, 1996). Cultivated soils suffer from repeated water erosion (encouraged by the packing of soils by agricultural machines and by spring crops such as beets, corn and potatoes). Chronic winter erosion, or occasional spring erosion, damages crops and the soil (impoverishment of fertile silt and organic matter). It also contributes to the pollution of waterways that receive the runoff waters filled with suspended matter, fertilizers, and pesticides, all dangerous for piscicultural life (Wicherek, 1999, 2000.)

Fertilizers and pesticides may be found in strong concentrations in surface waters, as indicated by results taken from a study of sloped basins (Figure 11.7). Water runoff samples taken during the course of two storms register concentrations of nitrates (10.6 to 13.1 mg N/l) close to the authorized limit for drinking water (11.3 mg N/l). A more critical fact may be observed regarding pesticides. Lindane and atrazine, whose use is currently regulated, are found in very high concentrations

May 24, 1994

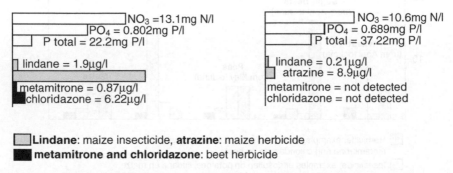

Lindane: maize insecticide, **atrazine**: maize herbicide
metamitrone and chloridazone: beet herbicide

Figure 11.7   Examples of pollutant concentrations in runoff waters. (Angéliaume 1996.)

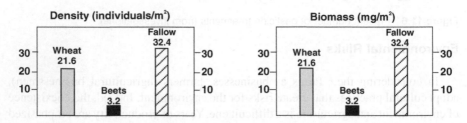

Dominant families (Carabidae, Lassidae, Cypselidae, Cantharidae, Muscidae, Staphylinidae)

Figure 11.8   Examples of impacts on cultural practices concerning biodiversity. (J. Karg 1996, in Angéliaume 1996.)

metamitrone and chloridazone) and are thus easily dissolved and washed toward waterways.

Pesticides can also influence the microfauna present in cultivated plots. The nature of these pesticides in relation to the cultivation is very important. Whether measured in microbial density or in biomass, biodiversity is markedly lower in beet fields (with repeated and apparently more noxious treatments) than in wheat and fallow fields (Figure 11.8).

Picardy agriculture is a high-performance (yields, rank in French production, growing practices, technical experimentation, etc.), modern (high technological means), export (brasserie barley, bread-quality flour, etc.) agriculture, but can it be qualified as clean and environmentally friendly? Can it be used as a model? It is difficult to imagine that it is when one considers problems such as erosion, water quality degradation, and the attack upon insect biodiversity. A change toward a cleaner agriculture seems even more difficult because of the current political and

the consequences of runoff and water erosion. This situation does not seem to occur north of the Parisian basin, where the occurrence of mudslides can be explained (among other reasons) by the expansion of crops favorable to runoff and erosion, such as sugar beets, potatoes, and endive. Even if the disappearance of hedges, woods, ponds, and meadows has greatly contributed to the worsening of the effects of runoff and water erosion, which is verifiable for former regions, it cannot, however, be applied blindly to all of France's agricultural regions, in particular those marked by a relative stability in the principal elements of the agrarian landscape. It thus seems essential to take into account the long-term evolution of agrarian landscapes in studies treating the links between agriculture and the environment.

The last 20 years have been a period of stress for French agriculture: 1975, the crisis of *"les trente glorieuses,"* then the PAC, then PAC reform. In Picardy, the agricultural situation remains full of contrasts: the coexistence of a small farm in the Aisne valley with its few cows, each named after one of the farm's creditors, and the immense farms on the plateau near Chemin des Dames, of several hundred hectares, which are villages within themselves. Large-scale cultivation, although affected by problems (drop in land prices since 1980, terres gelées, after years of productivity, difficulty of accepting fallow fields and other brakes on production) has been to a great degree spared by the crisis shaking the agricultural world today. The tendency is toward the stabilization of fertilizer and pesticide consumption, but not toward reduction. These products have brought about a spectacular rise in yields, growth that was even more pronounced on the large farms of the plateau. It is difficult to win acceptance for a drop in consumption, which is considered a step backward. Yet at the present time, with the set-aside, the notion of yields is often discussed. In the best of cases, growers can treat their fields with the objective of equalizing the cost of production and yields. They practice what is called *reasonable* treatment (prevention and treatment), but only of what is necessary, when it is necessary. It is difficult for them to reduce treatments because preservation of farm revenue depends upon maintaining the productivity acquired in past decades. By this very fact, this agriculture, still high performing, remains a source of damage to the environment and remains relatively reserved toward new agro-environmental developments.

ACKNOWLEDGMENTS

The author acknowledges the collaboration of M.O. Boissier, A. Sevestre, and A. Angéliaume.

REFERENCES

Angéliaume, A. (1999) Paysages et pratiques agraires dans le nord du Bassin parisien. Evolution du début du siècle à nos jours et impacts sur l'environnement (exemple du Soissonnais) pp. 101–114. In *Paysages agraires et Environnement*, S. Wicherek, Ed. CNRS Editions, 412 pp.

Boiffin, J., Papy, F., Peyre, Y. (1986) *Systèmes de production, systèmes de culture et risques d'érosion dans le Pays de Caux*. Rapport INA-PG/INRA/Min. Agri., 154 p. + appendices.

Brunet, P. (1960) Structure agraire et économie rurale des plateaux tertiaires entre la Seine et l'Oise. Thèse d'Etat, Université de Caen, 552 pp.

Fiette, A. (1995) *L'Aisne des terroirs aux territoires*. Comité d'expansion de l'Aisne, Compagnie Européenne de reportage et d'édition, 286 pp.

Karg, J. and Ryszkowski, L. (1996) Animals in arable land. In *Dynamics of an Agricultural Landscape*. Eds. Ryszkowski, N. French, A. Kędziora. PWRL Poznań, 138–172.

Larue, J.P. (1992) L'érosion des sols cultivés dans la région de La flèche, le rôle de l'évolution du paysage rural et des façons culturales, in *Bull. Ass. Géogr. Fr.*, 2, 127–134.

Le Bissonnais, Y., Le Souder, C. (1995) Mesurer la stabilité structurale des sols pour évaluer leur sensibilité à la battance et à l'érosion, in *Etude et Gestion des Sols*, 2, 1, 43–56.

Lechevallier, C. (1992) Evolution des structures agraires et érosion des sols en Pays de Caux, in *Bull. Ass. Géogr. Fr.*, 2, 101–106.

Lefevre, D. (1993) *Le retour des paysans*. Le cherche Midi Ed., 334 pp.

Meyer, E., Wicherek, S. and Peulvast, J.P. (1999) Processus de dégradation des sols et évolution des paysages agraires depuis le XVIIIe siècle. Exemple du nord — ouest du Bassin parisien. In *Paysages Agraires et Environnement*, S. Wicherek, Ed. CNRS Editions, 412 pp.

Papy, F. and Douyer, C. (1991) Influence des états de surface du territoire agricole sur le déclenchement des inondations catastrophiques. *Agronomie*, 11, 201–215.

Wicherek, S. (1998) Les relations entre le couvert végétal et l'érosion en climat tempéré de plaine, exemple de Cessières (Aisne, France). *Z. für Geomorphologie N.F.*, 32 (3), 339–350.

Wicherek, S., Ed. (1993) *Farm Land Erosion in Temperate Plains Environments and Hills*. Elsevier, Amsterdam, 584 pp.

Wicherek, S., Ed. (1999) — *Paysages agraires et environnement. Principes écologiques de gestion en Europe et Canada* [*Agrarian Landscapes and Environment. Ecological Principles of Management in Europe and in Canada*]. CNRS Editions, 412 pp.

Wicherek, S. Ed. (2000) *L'eau de la cellule au paysage* [*Water, from Cell to Landscape*]. Elsevier Editions, UNESCO, Paris, 424 pp.

CHAPTER **12**

Models Assessing the Impact of Land-Use Change in Rural Areas on Development of Environmental Threats and Their Use for Agricultural Politics

Armin Werner and Peter Zander

CONTENTS

278   LANDSCAPE ECOLOGY IN AGROECOSYSTEMS MANAGEMENT

INTRODUCTION AND OBJECTIVES

Especially in industrialized nations, changing economic conditions lead to large structural transformations in the businesses of land and its use (agriculture, forestry, in-land fish production, etc.). These transformations stem from changed politics of land use resulting from the diminishing importance of the land-use business within overall business and from economic and possibly climatic global change. Public perception of land use is also changing. More and more often not only is the supply function (food, fiber, wood, etc.) addressed, but continuously increasing ecological goals are proposed for land use and land-use planning. The changes in these driving forces will lead to changes in land use and thus in the impact of land-use systems on rural areas (economy, social aspects) as well as on the ecology (abiotic sources, nature). Due to the complex interactions of land-use systems with the relevant economic, cultural and ecological indicators, it is necessary to address the resulting problems of current or future land-use changes in an integrative way — *integrated land development* (Thöne, 2000). Integrated rural planning therefore can serve as an example of methods and approaches that help to achieve a sustainable development in human activities (Werner and Haberstock, 2001). The sustainable development of rural areas is a major goal of the national and international politics that are related to nonurban areas.

Most rural areas are dominated by agricultural land use. Therefore, the impact of agricultural land-use systems on the environment can be substantial. Assessing that impact on the environment is crucial for sustainable rural development. But in order to understand and control that impact and other relevant indicators, it is necessary to develop new methods for assessing the impact of land use and technologies and therefore of land-use politics or other relevant driving forces. These methods have to deliver general answers, that are scientifically sound and that can be generalized but still are transferable to the specific conditions of the studied area. The cooperative character of studies on future land use requires expert knowledge and innovative methods to analyze complex empirical data and to allow for the analysis of possible futures.

Many difficulties in the management of complex systems can be overcome with the use of decision-making support systems. This chapter summarizes and discusses methods to assess the impact of changing land use with models. An extensive example of the multi-criteria optimization approach shows the possibilities of using such a system in decision making for economic-ecological problems in rural areas.

SUSTAINABLE DEVELOPMENT OF RURAL LAND USE

several processes and values when the impact of technologies or agricultural politics has to be assessed. A common ground for defining a model for future land use is the sustainable development of land use and thus of rural areas which is *sustainable land development* (Werner et al., 1997). Economic and ecologic, as well as social and cultural, goals should be fulfilled (Barbier, 1987, Goodland, 1996).

With these goals in mind, integrative tools are necessary that support the decision-making process in rural areas and the related land-use planning (Maxwell et al., 1999). It is especially difficult to assess the impact of land-use systems on the environment of agrolandscapes (the complex of abiotic compartments and the nature). First, those indicators (values or objects) that should be addressed in the evaluation process are not yet sufficiently defined and agreed to (Maxwell et al., 1999). These indicators have to be selected by the affected stakeholders, the groups that participate in the decision-making process, or by society in general. Which indicators are suitable can be suggested by scientific evaluation through joint processes of different disciplines (Mansvelt, 1997). Second, it is necessary to have methods for deriving the values these indicators will have under specific land-use conditions. With this information, decision makers can select the feasible options in land use and landscape planning.

Many national or transnational (e.g., from the European Union) concepts related to the development of rural areas have similar approaches (Bosshard, 2000). They include:

- Strengthening regional marketing and food processing
- Enhancing the competitiveness of regional business
- Improving the social or cultural activities within a region
- Attempting to close regional matter cycles
- Preventing pollution of abiotic compartments
- Protecting and developing sensible biotopes as regional habitats for typical species

The character of these elements represents remarkably the necessary ingredients of the general concept of a sustainable development (Thierstein and Walser, 1997). For the groups and people who participate in the planning and running of a region, the common base is the available space and the natural resources of that region. There are no accepted and standardized methods available for defining all relevant groups for this participatory process of decision making and planning. Even intuition can be a crucial element of identifying the relevant stakeholders or actors (Baeriswyl et al., 1999). Successful decision making for rural development needs a systematic procedure and needs to be restricted to the relevant processes or compartments of the respective system, the rural area. One approach is to identify the necessary and important functions that have to be fulfilled within the rural area. For an integrated region-oriented policy, Kolk et al. (1999) identified 12 main categories of relevant

DECISION MAKING FOR SOLVING COMPLEX PROBLEMS
OF RURAL AREAS

The process for establishing sustainable rural development cannot be handled today by the traditional steps of rural land management: design and reconstruction of the area. Today it is mainly a problem of communicating with the relevant people or groups within the region and deciding how to let them participate in the decision-making process: "Implementation of policy objectives and targets is not likely to happen, without serious local participation and commitment" (Volker, 1997). This leads to the conclusion that integrated rural development is possible only with a democratic approach, from the bottom up (Meyer, 1997). For this purpose, round tables (Müller et al., 2000), environmental cooperatives (Glasbergen, 2000) and innovation groups (Horlings et al., 1997) are relevant instruments to determine and manage conflicting goals for the development of rural areas. In terms of ecological goals, recently the main focus of land-use development was on minimizing the impact of agriculture and forestry on the abiotic compartments of the environment (water, soil, atmosphere). Today increasing attention is directed toward goals that include, to a large extent, the living parts of agricultural landscapes (Harms et al., 1998).

Rural areas have a good chance to achieve sustainable development when

- instead of being driven mainly by exogenous business processes, endogenous forces dominate
- the perspective of business and administration changes from sole consideration of static-site factors to chance and possible developmental processes
- the perspectives in the region switch from economic forces to human action and initiative (Thierstein and Walser, 1997)

The last point in particular leads to the necessity of specific decision-making methods for rural areas. These methods should help stakeholders define possible solutions or pathways in the development (scenarios) and analyze the outcome of these solutions with respect to the views of all involved disciplines and groups (economy, ecology/nature conservation, social and cultural aspects). The process of defining objectives, assessing the impact of different land-use strategies, and eventually redefining those objectives will be repetitive and cyclic. Decision making for the development of rural land use has to be participative and iterative (Werner and Bork, 1998). The most promising way to achieve all these goals is to jointly define scenarios and assess the impact on the economy and ecology with simulation models (Figure 12.1). This leads to a change in the paradigms of land-use planning and land development with severe impacts on the related sciences, rural politics, consultation business, and regional or national administration (Magel, 2001).

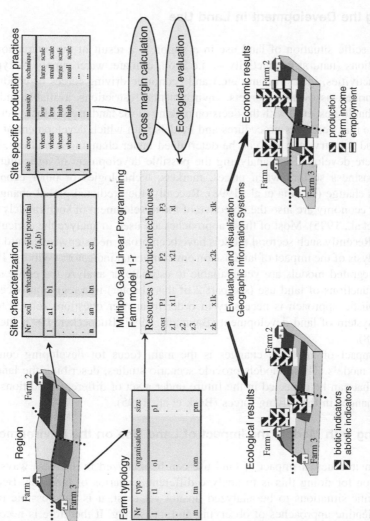

Figure 12.1 From site and farm typology to a regional evaluation of land-use scenarios (From Zander and Kächele, 1999.)

accepted moral axioms or principles" and therefore represent a nonauthoritative approach (Bosshard, 2000).

MODELS FOR ASSESSING THE IMPACT OF LAND USE AND LAND-USE CHANGES

Modeling the Development in Land Use

The specific situation of land use in a region is a result of the interactions of site conditions (natural productivity — i.e., soil, climate, water supply — type of business activities, infrastructure, etc.) and economic driving forces or conditions (politics, prices, assessable markets, environmental restrictions, available and permitted technologies, etc.) with the decision making of the land users. Much literature is available that describes procedures and models with which development of agricultural and forestry land use can be determined under changing conditions. Most models were developed for analyzing the possible development of agricultural or forestry business sectors when prices, markets, technologies or other economic conditions change (Bouma et al., 1998). Recently, the predicted global changes of climate or economy are also the objective of the development of such models (e.g., Mirschel et al., 1995). Most of these approaches are used to analyze the agricultural business. Recently such sectoral models have been broadened stepwise toward additional analysis of the impact of land use on environmental indicators (Wiborg, 1998).

No integrated models are yet available to estimate or analyze the entire set of different functions of land use (see part 2 of this chapter) in a region at once. But such a holistic approach is necessary in order to answer questions related to the complex system of land development (Baumann, 1997; Buchecker, 1997; Ittersum et al., 1998).

The impact of land use changes is the main focus for developing complex landscape models. These models provide scenario studies, describing the land-use situation that can be expected in the future under a set of different conditions with defined changes in the driving forces (Bork et al., 1995).

Assessing with Models the Impact of Land Use on the Environment

The environmental impact of land use can be analyzed in different ways. The main reason for doing this is to analyze different scenarios of land use. Because most specific situations to be analyzed do not yet exist, it is not possible to use classic scientific approaches of observing and measuring. It therefore is necessary to estimate or predict the relevant situations of land use with tools that are general

level of detail for the causal or indirect description of compartments or processes. Depending on the goals to be achieved, more or less sophisticated tools are developed: simple approaches that balance relevant flows of matter, nutrients, or energy (e.g., Bach, 1987); models that consist mainly of conditional clauses, rules, or table functions (e.g., Bork et al., 1995); models that describe the relevant processes in depth (e.g., Worral and Burt, 1999); or very complex models that look for a large set of indicators in detail (e.g., O'Callaghan, 1995).

Models that should help to analyze the complex impact of land use on relevant ecological indicators have a special demand for spatial data or they even have a spatial design (Costanza and Maxwell, 1991) because many ecologically relevant processes do occur as spatial interactions between compartments of a landscape (e.g., lateral flow of water, nutrients, matter, energy or migration of organisms). But the spatial interaction of business structures or human beings also has to be regarded when the whole landscape and its land use are analyzed (Bockstael, 1996, Werner and Bork, 1998). To accomplish this goal, most landscape or land-use models distinguish different processes in the landscape and describe relevant components separately. For the abiotic components, many models or sub-models exist for the local or the regional perspective (e.g., Addiscott and Mirza, 1998; Bass et al. 1998; Dunn et al., 1996). Only a few models are available now that try to assess the impact of land use on biotic components, species, or their habitats (e.g., Schultz and Wieland, 1995; Lutze et al., 1999).

A crucial point in using complex land-use models is the availability of data that describe the specific region in terms of the site conditions, the actual land-use structure and the noncultural biotopes of a landscape (Briassoulis, 2001). When assessing the impact of land-use changes for a large area, in most cases only data with a low level of spatial detail are available for site conditions, noncultural biotopes or structure of the land-use business. To collect data of actual land use, remote sensing can very easily provide a set of data that can cover completely the entire region (Wadsworth and Downey, 1996). Other data, especially those for actual management, are generally available with a high level of detail only for selected farms or other land-use businesses. In most cases, only data sets without a detailed spatial resolution are available as statistical data for an entire region.

Approaches of Modeling Land Use and the Effects of Land Use

After indicators to be addressed have been defined (Moxey et al., 1998), models are selected or developed that describe explicitly the processes or relate the values of the indicators to the specific land-use situation (e.g., Johnes, 1996). All these different models should be linked either physically within a software-framework (Dunn et al., 1996, Lutze et al., 2000, Tufford et al., 1998) or run separately, exchang-

processes for regions or rural areas (Belcher and Boehm, 1998). The different concepts for designing landscape models can, according to Lutze et al. (2000), be distinguished into two groups:

- *Integrated models* — All relevant processes are described with linked equations or algorithms; the entire model is developed in one set, and all parts must fit to each other from the internal logic and structure of the model as well as from the software development.
- *Modular models* — All relevant processes are modeled separately in different modules, which are linked by software calls; this approach allows different internal concepts for the submodels and even different programming languages.

Depending on the perspective on the landscape, it is possible to distinguish between different model types (Antrop, 2000) according to approach followed:

- *Thematic approach* — All relevant landscape components or compartments are described with the model and can be analyzed with regard to the effects of land use.
- *Regional or spatial approach* — The landscape is divided into hierarchical units that are described separately with the model and can be analyzed with regard to the effects of land use.

Modeling Spatial Aspects

Modeling the spatial aspects of interacting processes and driving forces is still a challenge for landscape modeling. In most cases, one-dimensional models (the dimension of time may always be added) are used to determine the impact of different land-use systems for defined points or homogeneous areas within the landscape (e.g., Kersebaum et al., 1995; Priya and Shibasaki, 2001; Wegehenkel 1999). These points or land units are selected from the entire landscape, so that (1) they are representative of a surrounding part of the landscape (in doing so, this piece of landscape is thought to be homologous for the relevant landscape properties; Verburg et al., 1999) or (2) sometimes the landscape is divided into cells of the same or different sizes, providing a full cover of the landscape with a grid (for each grid cell the one-dimensional model will be calculated; Børgesen et al., 2001). In both approaches, sometimes not all possible points or cells within a landscape are simulated; only a selected number of typical combinations of sites and land use are defined and simulated.

Stepwise models are also developed that take lateral processes into account or are valid for a complete portion of the space in the landscape. This modeling is mainly for water and matter flow (e.g., Ilyas and Effendy, 1996, Johnes and Heathwaite, 1997). Modeling biotic components and processes that are related to organisms requires spatial-explicit approaches. Only a few models are available that deal

(e.g., water-dynamic) can have different spatial pattern on different scales (Wenkel and Schultz, 1999).

Applying Models for Optimization of Land Use

In many cases of decision making for rural areas, it is necessary to provide suggestions that help find solutions for the specific problem. In order to enhance a goal-oriented selection of land-use combinations or land-use systems, models are applied with optimization procedures (e.g., Keith et al., 1999).

Zander and Kächele (1999) developed a complex model that allows estimation of the situation of predefined farms within large regions under different economic conditions. The model also includes estimations for ecological indicators for all possible combinations of crop management. With such an approach, simultaneous economic and ecological evaluations are possible. With a given set of preferences, the best feasible combination of ecological objectives and economic constraints can be found for a region (Meyer-Aurich et al., 1998).

DECISION MAKING FOR LAND-USE PLANNING IN RURAL AREAS THROUGH MULTI-OBJECTIVE OPTIMIZATION — AN EXAMPLE

Decision making in rural planning, as within other complex systems, requires having (1) different options, (2) sufficient information about these options, and (3) the power and other necessary resources to put a decision into action (Steffen and Born, 1987). Rural planning comprises agriculture, forestry, fishery, tourism, and infrastructure, among other areas. Simulation and ecological optimization of the management of each of these sectors requires specific models. Because agriculture dominates rural areas, assessing the impact of such land use is most relevant in analyzing the environmental threats of changing land-use systems for rural areas. Modeling agricultural decision making under consideration of ecological objectives is the focus of this section.

The classical decision-making process in agricultural land use is related to mainly one goal: maximizing the economic profit of the farm. Often stability of income as well as cash flow are defined as economic goals. Introducing additional goals, such as environmental objectives or those of nature protection, to the decision making of the agricultural business leads to a more complex situation. It is then necessary to optimize the activities and the production process toward more than a single goal (maximum economic output). Now selection of appropriate measures in land use must be done with several goals in mind. However, it is barely possible to maximize the outcome of all objectives simultaneously. In most cases, a compromise among

Table 12.1   Priorities for Environmental Quality Goals Specific for Single Fields (Principle Pattern, Examples)

Field-no. (example)	Abiotic Goals				Biotopes, Protected					Goals Related to Single Species				
	Protection of Ground Water	Ground Water-Recharge	Preventing Wind Erosion	Preventing Water Erosion	Valuable Ponds	Oligotrophic Biocoenoses	Dry Meadows	Valuable Biotopes	Ruderal Vegetation	Partridge	Gray Bunting	Barn Owl	Amphibia	Cranes
1	+	+	+	+	+	R	R	++	++	+	R	(+)	–	–
2	+	+	R	++	+	R	R	++	++	+	+	(+)	R	+
3	R	+	+	++	++	R	R	++	++	R	–	(+)	++	+
4	++	++	R	++	R	++	++	++	++	+	+	(+)	R	+
5	++	++	++	++	R	++	++	++	++	+	R	(+)	R	–

Relevance of the distinct goal for the corresponding areal unit: ++ = very important, + = important, R = rare, – = irrelevant, (+) = important if species does occur.
Source: Adapted from Plachter and Korbun (2001).

alternatives that can be considered in order to achieve the desired ecological goals. The decision makers on the production side need information about the effects these different land-use options have on the economy and the structure of the farms. The representatives for environmental protection in this decision-making process need information about the ecological impacts that would be caused by some specific options of the agricultural production systems. Representatives of both the farmers and those for the environmental protection need to consider the best solution for both sides.

The Multi-Optimization Model

The predefined cropping practices (Table 12.1) and their economic as well as their ecological effects (Figure 12.2) are a partial evaluation information of the crop management systems and the base for a complex simulation model (Figure 12.3) (Zander and Kächele, 1999). With this model one can find the most suitable combination of crop production systems for a given farm situation or region according to predefined ecological and economic goals. The search steps in MODAM are performed through a farm model by optimizing total gross margin with ecological objectives as restrictions (multi-objective-optimization). A series of consecutive runs of the model is conducted. In each run, the model is forced to use a 10% higher achievement rate for one specific environmental goal, going from the chosen reference situation and to 100% in the achievement of the goal (Figures 12.4 and 12.5).

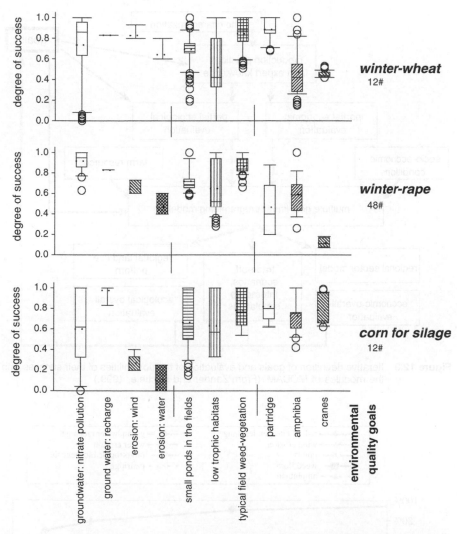

Figure 12.2 Distribution characteristics (box plots) in the success of environmental quality goals with different crop production systems for winter-wheat, winter-rape, and corn for silage. (Estimations for 192, 48, 132 standardized crop production systems, including four different soil fertility classes; production region: northeast German hill area; derived by estimating algorithms from Meyer-Aurich et al. (1998) in the multi-objective decision-making tool MODAM.)

predefined economic (agro-political) conditions on the developing land-use situation

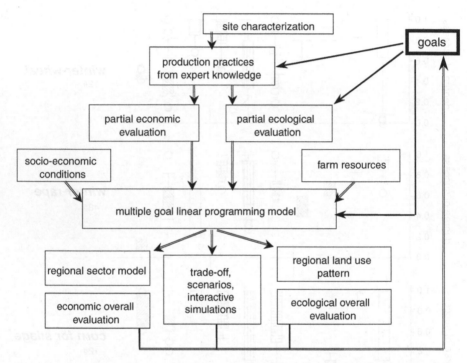

Figure 12.3 Iterative definition of goals and evaluation of the possibilities of their success with the modules of MODAM. (From Zander and Kächele, 1999.)

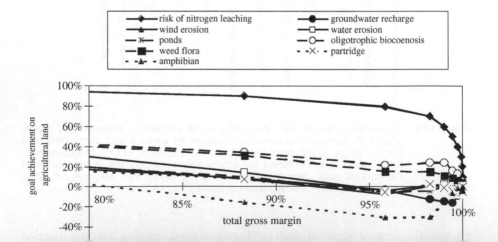

MODELS ASSESSING THE IMPACT OF LAND-USE CHANGE IN RURAL AREAS   289

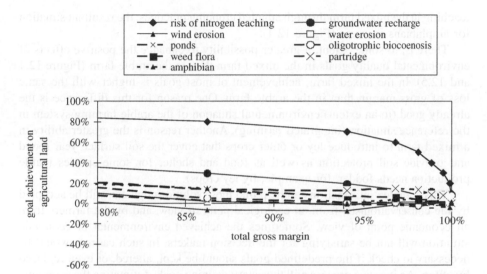

Figure 12.5 Trade-off between the reduction of nitrate leaching and the gross margin of the cash crop farm in the research area Wilmersdorf for selected environmental quality goals. Results from simulation studies with MODAM for an area of 1.821 ha in the southern Uckermark, Brandenburg, Germany. (From Zander, 1998.)

Simulations for the Study Area

For the study area, the environmental goals were available for each single field (see above section). These goals were given highest priorities (Table 12.1). When applying the MODAM system to these predefined goals, several solutions, fulfilling the goals to different extents, can be found.

It is possible to have substantial achievement in some environmental goals without losing too much gross margin for the farmers. For the mixed farm (Figure 12.4), it is possible to reduce nitrate leaching up to 85% by a 4% reduction in the gross margin (100% is the reference situation: integrated farming). This reduction is achieved by selecting cropping systems that will have the respective reduction of nitrate leaching.

As with all situations in real life when several goals should be reached at the same time, it is rarely possible that all goals can be fulfilled with the same approach. Therefore, the possible solutions concerning the given set of environmental quality goals for the analyzed region show different degrees in fulfilling the expected goals (Figures 12.4 and 12.5).

In addition, some environmental goals are congruent (their goal achievement heads in the same direction when cropping practices are changed). As can be seen

leaching in the mixed farm through adapted cropping practices, the resultant situation for amphibians is worse (Figure 12.4).

There is, to some extent, a greater possibility to enhance the positive effects of environmental quality goals in the mixed farm than in the arable farm (Figure 12.4 and 12.5). In the mixed farm, achievement of most goals is higher with the same loss of gross margin than in the arable farm. One reason for this difference is the already good (to an extent) environmental situation of the arable farming system in the reference situation (integrated farming). Another reason is the greater ability in a mixed farm to introduce ley or other crops that cover the soil surface year round and provide soil protection as well as food and shelter for some species (cattle production needs fodder, for example, by ley crops).

This complex information can help find those solutions that can still be accepted by the conservationists, from an ecological point of view, and by the farmers from an economic point of view. Sometimes the achieved environmental or economic situation will not be satisfying for the decision makers. In such cases it would be necessary to check if the predefined goals should be kept, altered, or even replaced by others. An iterative process will then start: defining goals, estimating the economic and ecological impacts, redefining goals, estimating the economic and ecological impacts, etc.

The developed method also can help define the necessary additional compensation that would be necessary to encourage the farmer to adopt the cropping practices that would lead to attaining a set of environmental quality goals. This compensation payment could come from groups that have interest in achieving the nonbusiness goals of the farmers (private organization in environmental protection, nature conservation, or society in general). Also, the appropriate compensation could be derived with the described method when farmers would be forced to attain the desired environmental quality.

In the process of finding suitable cropping systems for sustainable land-use development, it is necessary to have alternatives for the joint and participatory decision-making process. For the study area, two possible priorities (abiotic and biotic) were given as an example. With the MODAM tool possible land-use solutions were found (Figure 12.6). When a primarily abiotic set of goals is given, the resulting cropping practices can enhance the achievement of the abiotic goals by up to 20% (limited by too high economic losses with higher achievement levels). But a 6% or 11% loss in gross margin for the farm would result, depending on the farm structure (Figure 12.6). Greater losses in gross margin would occur when primarily biotic goals would be followed (Figure 12.6).

An Outlook with Scenario Studies of Agricultural Politics

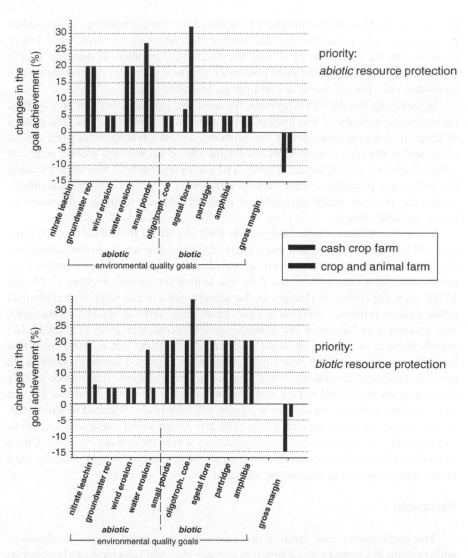

Figure 12.6 Success of selected environmental quality goals and the gross margin of simulated farms (cash crop and mixed farm) in the research area Wilmersdorf with
setting priorities on either primarily abiotic (top) or biotic (bottom) goals when
looking for economically and ecologically optimized crop production. Results from
simulation studies with MODAM for an area of 1.821 ha in the southern Uckermark, Brandenburg, Germany. (From Zander, 1998.)

environmental and nature protection, it is necessary to estimate the economic and

per ha of arable land and per number of employed persons for farming); (3) subsidies per land use area (subsidies per ha, amount dependent on regional yield level); (4) subsidies for protection of abiotic resourcesand (limitation of nitrogen supply dependent on the site specific yield potential); (5) continuation of the current system (subsidies only for specific crops and per ha of arable land).

By applying the MODAM system, the land-use situation that can be expected under specific conditions was estimated for a region of about 1.800 ha. The pattern of crops in a 6-year rotation and the economic optimal management practices on each field in the region were analyzed for the effects on selected abiotic goals, on selected species, and on habitat quality. The ecological analysis was done by using the estimation procedures from Meyer-Aurich et al. (1998). Economic conditions served the possible future agropolitical programs that currently are discussed as future political options.

These five scenarios were analyzed in their impact on four abiotic goals, three goals of habitat protection, and three goals of species protection. In the simulation, two different farming systems were assumed: a farm solely based on arable crop production and a farm with mixed crop and animal production. Figures 12.7A and 12.7B show the results as changes in the achievements of the respective ecological goals with the reference situation as base. In an arable farming system, all scenarios will produce a reduction in the achievements of ecological goals (Figure 12.7A) mainly because of reduced set-aside fields and the resulting loss in those crops that have permanent, or at least year-round, vegetation. In mixed farming systems, the possible economic conditions would lead to cropping systems that reduce mainly the impact on ponds and reduce soil erosion by water or wind (Figure 12.7B); an improvement in the ecological quality goals will take place. Reduced attainment of all other environmental quality goals, will also occur under the given scenarios. Consequently, all discussed economic conditions within the coming European Union agropolitics would deteriorate the environmental conditions in agriculturally used landscapes compared to the current situation.

Perspective

The participatory and iterative procedure was developed for the northeastern arable region in Germany. An adaptation to other sites and regions is easily possible. In order to make such an adaptation, the regional environmental quality goals have to be determined, the yield expectations (site-specific potentials) have to be estimated, and the sets of different production technologies (e.g., cropping practices) have to be defined. The latter can be done using local expert judgment.

The described procedure has to be developed continuously in the future. The MODAM system does not yet include the evaluation of nonagriculturally used areas

A.

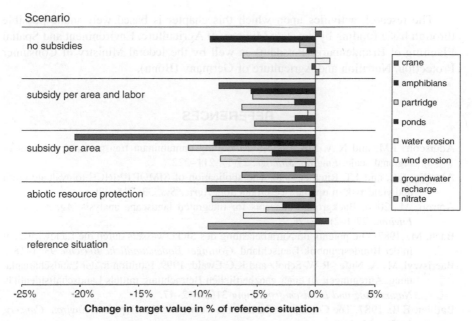

B.

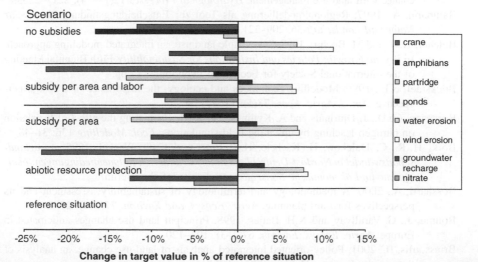

ACKNOWLEDGMENTS

The research activities upon which this chapter is based were made possible through basic funding by the state Ministry of Agriculture, Environment and Spatial Planning of Brandenburg (Potsdam) as well by the federal Ministry of Consumer Protection, Nutrition and Agriculture of Germany (Bonn).

REFERENCES

Addiscott, T.M. and N.A. Mirza, 1998. Modelling contaminant transport at catchment or regional scale. *Environ. Manage.* 27(1): 211–222.

Alocilja, E.C. and J.T. Ritchie, 1990. The application of SIMOPT2:RICE to evaluate profit and yield-risk in upland rice production. *Agric. Sys.* 33: 315–326.

Antrop, M., 2000. Background concepts for integrated landscape analysis. *Agric. Ecosys. Environ.* 77(1–2): 17–28.

Bach, M., 1987. Die potentielle Nitratbelastung des Sickerwassers durch die Landwirtschaft in der Bundesrepublik Deutschland. *Göttinger Bodenkundliche Berichte* 93: 1–186.

Baeriswyl, M., A. Nufer, R.W. Scholz and K.C Ewald, 1999. Intuition in der Landschaftsplanung. Anregungen zu einer ganzheitlichen Betrachtung mittels Landschaftsidentität. *Naturschutz und Landschaftsplanung* 31(2): 42–47.

Barbier, E.B., 1987. The Concept of Sustainable Economic Development. *Environ. Conserv.* 14(2): 101–110.

Bass, B., R.E. Byers and N.M. Lister, 1998. Integrating research on ecohydrology and land use change with land use management. *Hydrological Processes* 12(13–14): 2217–2233.

Baumann, A., 1997. Regionalmodellierung als Tool zur Entscheidungsfindung. *Archiv für Natur und Landschaft* 36: 209–221.

Belcher, K.W. and M. Boehm, 1998. Sustainable land use: an integrated modelling approach. In *Beyond Growth: Policies and Institutions for Sustainability.* Fifth Biennial Meeting of the International Society for Ecological Economics, 17 pp.

Bockstael, N.E., 1996. Modelling economics and ecology: the importance of a spatial perspective. *Am. J. Agric. Econ.* 78(5): 1168–1180.

Børgesen, Ch.D., J. Djurhuus and A. Kyllingsbæk, 2001. Estimating the effect of legislation on nitrogen leaching by upscaling field simulations. *Ecol. Modelling* 136: 31–48.

Bork, H.-R., C. Dalchow, H. Kächele, H.-P. Piorr und K.-O. Wenkel, 1995. *Agrarlandschaftswandel in Nordost-Deutschland unter veränderten Rahmenbedingungen: ökologische und ökonomische Konsequenzen.* Ernst & Sohn, Berlin.

Bosshard, A., 2000. A methodology and terminology of sustainability assessment and its perspectives for rural planning. *Agric. Ecosys. and Environ.* 77(1–2): 29–41.

Bouma, J., G. Varallyay and N.H. Batjes, 1998. Principal land use changes anticipated in Europe. *Agric. Ecosys. Environ.* 67(2–3): 103–120.

Briassoulis, H., 2001. Policy-oriented integrated analysis of land-use change: an analysis of data needs. *Environ. Manage.* 27(1): 1–11.

MODELS ASSESSING THE IMPACT OF LAND-USE CHANGE IN RURAL AREAS   295

Glasbergen, P., 2000. The environmental cooperative: self-governance in sustainable rural development. *J. Environ. Development* 9(3): 240–259.

Goodland, R., 1996. Environmental sustainability: universal and non-negotiable. *Ecol. Appl.* 6: 1002–1017.

Harms, W.B., P.J.A.M Smeets and A. Werner, 1998. Landscape Restoration and Policy in NW Europe. Nature and Landscape Planning and Policy in NW Europe; Dutch and German examples. In Dover, J.W., Bunce, R.G.H. (Eds.). *Key Concepts in Landscape Ecology.* 355–382.

Horlings, I., H.de Haan, B. Kasimis and M. Redcliff, 1997. Agricultural change and innovation groups in the Netherlands. *Sustainable Rural Development* 159–175.

Ilyas, M.A. and R. Effendy, 1996. Use of ANSWERS spatial distribution model for reviewing the impact of land use change on hydrographic water system, erosion, flood and sedimentation. *Jurnal Penelitian dan Pengembangan Pengairan* (Indonesia) 11(36): 10–21.

Ittersum, M.K., R. Rabbinge and H.C. van Latesteijn, 1998. Exploratory land use studies and their role in strategic policy making. *Agric. Sys.* 58(3): 309–330.

Johnes, P.J., 1996. Evaluation and management of the impact of land use change on the nitrogen and phosphorus load delivered to surface waters: the export coefficient modelling approach. *J. Hydrol.* 183: 323–349.

Johnes, P.J. and A.L. Heathwaite, 1997. Modelling the impact of land use change on water quality in agricultural catchments. *Hydrol. Processes* 11: 269–286.

Keith, B.M., A.R. Sibbald and S. Craw, 1999. Implementation of a spatial decision support system for rural land use planning: integrating geographic information system and environmental models with search and optimization algorithms. *Comput. and Electron. Agric.* 23: 9–26.

Kersebaum, K.C., W. Mirschel and K.-O.Wenkel, 1995. Makroskalige Abschätzung der Auswirkungen von Landnutzungsänderungen in Nordost-Deutschland auf den Stickstoffhaushalt mit Hilfe eines Simulationsmodells. *Mitteilungen der Deutschen Bodenkundlichen Gesellschaft* 76 (Heft II): 823–826.

Kolk, A.J. van der and J.N.M. Dekker, 1999. Functions in integrated region-oriented environmental policy: a classification system. *Land Use Policy* 16(2): 107–119.

Lutze, G., R. Assmann, R. Wieland, M. Voß und K.-O. Wenkel, 2000. ELANUS — Prototyp für ein Entscheidungsunterstützungssystem zur Landschaftsanalyse und zur integrativen Bewertung alternativer Landnutzungsstrategien. *Zeitschrift für Agrarinformatik* 2: 28–35.

Lutze, G., R. Wieland und A. Schultz, 1999. Habitatmodelle für Tiere und Pflanzen-Instrumente zur integrativen Abbildung und Analyse von Lebensraumansprüchen mit direktem Bezug zur Landschaftsstruktur und zur Landnutzung. In Blaschke, T. (Ed.), *Umweltmonitoring und Umweltmodellierung: GIS und Fernerkundung als Werkzeug einer* nachhaltigen Entwicklung. H. Wichmann Verlag, Heidelberg, 223–236.

Magel, H., 2001. Paradigmenwechsel in der Landentwicklung und Flurbereinigung Europas. *Landnutzung und Landentwicklung* 42: 4–9.

Mansvelt, J.D., 1997. An interdisciplinary approach to integrate a range of agrolandscape values as proposed by representatives of various disciplines. *Ecosys. Environ.* 63: 233–250.

Maxwell, T.J., G.W. Hill, and K.B Mathews, 1999. Sustainable rural land use. *J. Royal Agric.*

Meyer-Aurich, A., P. Zander, R. Roth, and A. Werner, 1998. Design of agricultural land strategies appropriate to nature conservation goals and environmental protection in North East Germany. *Landscape and Urban Planning* 41(2): 119–127.

Mirschel, W., A. Schultz, and J. Pommerening, 1993. Modellierung des Wachstums und der Ertragsbildung in komplexen Agroökosystemmodellen, dargestellt am Beispiel Winterweizen und Winterroggen. *Agrarinformatik* 24: 183–204.

Mirschel, W., K.-O. Wenkel, J. Pommerening und F. Reining, 1995. Grundlagen und Modelle zur Abschätzung von Klimaänderungseffekten auf Boden und Pflanze. In *Schriftenreihe des Bundesministeriums für Ernährung, Landwirtschaft und Forsten,* Reihe A: Angewandte Wissenschaft, H. 442 (Klimawirkungsforschung im Geschäftsbereich des BML, 248.

Moxey, A., M. Whitby, and P. Lowe, 1998. Agri-environmental indicators: issues and choices. *Land Use Policy* 15(4): 265–269.

Müller, K., H.-R Bork, A. Dosch, K. Hagedorn, J. Kern, J. Peters, H.-G. Petersen, U.J. Nagel, T. Schatz, R. Schmidt, V. Toussaint, T. Weith und A. Wotke (Eds.), 2000. Nachhaltige Landnutzung im Konsens — Ansätze für eine dauerhaft- umweltgerechte Nutzung der Agrarlandschaften in Nordostdeutschland. 190 S., Focus-Verlag, Gießen.

O'Callaghan, H.R., 1995. NELUP: An introduction. *J. Environ. Planning and Manage.* 38: 5–20.

Plachter, H. and T. Korbun. A methodological primer for the determination of nature conservation targets in agricultural landscapes. In Flade et al. (Eds.). *Nature Protection in Agricultural Landscapes. Methods for Analysis, Goal Definition and Evaluation.* Meyer-Verlag, Heidelberg, in press.

Priya, S. and R. Shibasaki, 2001. National spatial crop yield simulation using GIS-based crop production model. *Ecol. Modelling* 136: 113–129.

Schultz, A. and R. Wieland, 1995. Die Modellierung von biotischer Komponenten im Rahmen von Agrarlandschaften. *Archives of Nat. Conserv. and Landscape Res.* 34: 79–98.

Steffen, G. and D. Born, 1987. Betriebs- und Unternehmensführung in der Landwirtschaft. Uni-Taschenbücher 1423, 455 pp.

Thierstein, A. and M. Walser, 1997. Hoffnung am Horizont? — Nachhaltige Entwicklung im ländlichen Raum. *Zeitschrift für Kulturtechnik und Landentwicklung* 38(5): 198–202.

Thöne, K.F., 2000. Effects on the development of rural areas: 30 years of regional planning and land law in Germany. *Entwicklung und Ländlicher Raum* 34(1): 11–15.

Tufford, D.L., H.N. McKellar, Jr. and J.R. Hussey, 1998. In-stream no point source nutrient prediction with land-use proximity and seasonality. *J. Environ. Qual.* 27: 100–111.

Verburg, P.H., G.H.J. de Koning and K. Kok, 1999. A spatial explicit allocation procedure for modelling the pattern of land use change based upon actual land use. *Ecol. Modelling* 116: 45–61.

Volker, K., 1997. Local commitment for sustainable rural landscape development. *Agric. Ecosys. Environ.* 63: 107–120.

Wadsworth, R. and I. Downey, 1996. The role of remote sensing in a land use planning decision support system. In C.H. Power and L.J. Rosenberg (Eds.), *Remote Sensing and GIS for Natural Resource Management,* 6–13.

Wegehenkel, M., 1999. Möglichkeiten und Grenzen der GIS gestützten Wasser und Stoffhaushaltsmodellierung als ein Beispiel der integrierten Landschaftsmodellierung. In

Werner, A. and H.-R. Bork, 1998. Integrating diverging orientors: sustainable agriculture: ecological targets and future land use changes. In Müller, F. and Leupelt, M. (Eds.), *Eco Targets, Goal Functions and Orientors.* Springer-Verlag, Berlin, 566–584.

Werner, A. and W. Haberstock, 2001. Landeskultur — quo vadis? *Landnutzung und Landentwicklung* 42, 97–99.

Werner, A., K. Müller, K.-O. Wenkel, H.-R. Bork, 1997. Partizipative und iterative Planung als Voraussetzung für die Integration ökologischer Ziele in die Landschaftsplanung des ländlichen Raumes. *Zeitschrift für Kulturtechnik und Landentwicklung* 38: 209–217.

Wiborg, T., 1998. KRAM — a sector model of Danish agriculture. Background and framework development. CARD-Working Paper No. 98-WP193, 37 pp.

Worrall, F. and T.P. Burt, 1999. The impact of land-use change on water quality at the catchment scale: the use of export coefficient and structural models. *J. Hydrol.* 221: 75–90.

Wurbs, A., C. Kersebaum and C. Merz, 1999. Quantification of leached pollutants into the groundwater caused by agricultural land use — model based scenario studies as a method for quantitative risk assessment of groundwater pollution. In Lal, R. (Ed.) *Integrated Watershed Management in the Global Ecosystem,* CRC Press, Boca Raton, FL, 239–250.

Zander, P. and H. Kächele, 1999. Modelling multiple objectives of land use for sustainable development. *Agric. Sys.* 59: 1–15.

CHAPTER 13

Ecological Guidelines
for the Management of
Afforestations in Rural Areas

Stanisław Bałazy

CONTENTS

INTRODUCTION

The widespread transformations of natural landscapes reflect thousands of years of human impact on natural ecosystems. Pastures, cultivated fields, plantations, villages, towns, industrial complexes, and huge metropolies replaced forests, grasslands, and swamps. Removal of permanent plant cover resulted in intensification of water and wind erosion, higher leaching rates of chemical compounds into underground and surface water reservoirs, impoverishment of biological diversity, and other deteriorations of the biosphere.

In areas dominated by arable fields, protective functions against destructive climatic effects on settlements, livestock, and plants were performed by the remaining afforestation fragments and bushes. Their protective role has long been recognized and appreciated, and is best witnessed by the continuously supported networks

the northwestern parts of France (Meynier 1976, Pape 1976, Stanners and Bourdeau 1995, J. P. Marchand, personal communication). Until now, these structures were mainly composed of native plant species, despite growing tendencies to introduce foreign tree and shrub species in the reafforestation of areas excluded from agricultural production (Hooper 1976, Delelis-Dusollier 1986, Alexander et al. 1986).

For similar reasons, the hedgerows are permanently maintained on agriculturally utilized areas in mountains and uplands of strongly diversified relief along the wide central European belt from Cantabrian to the Carpatian Mountains. In the Council of Europe paper, Biber (1988) used the French term *bocage* for this kind of afforestation. Stanners and Bourdeau (1995) called them *semi-bocage*, whereas in the majority of papers the terms *haies*, *shelterbelts*, *hedgerows* and *windbreaks* were commonly used as synonyms.

As early as the 1820s Chłapowski (1835) introduced and recommended midfield afforestation systems in the Wielkopolska region of Poland modeled on English hedgerow patterns, pointing out a number of their advantageous effects for agriculture. He considered the most important benefits to be timber, forest fruits, herbs and honey, protection of fields and cattle against wind, and elimination of shepherd employment on pastures surrounded with thorny hedgerows, which simultaneously provided shade for animals. Among tree species particularly useful in shelterbelts Chłapowski recommended the black locust (*Robinia pseudoacacia*) and among shrubs he recommended the hawthorns (*Crataegus monogyna* and *C. oxyacantha*), but he also planted more than 20 species — including some exotic ones, such as honey locust (*Gleditschia triacanthos*) and catalpa (*C. bignonioides*). Chłapowski also stressed the role of shelterbelts for their scenic values and their shaping of field contours. As an educated military officer and participant in the Napoleonic campaigns, Chłapowski also appreciated their military significance. The afforestations that he had established on an area of about 10,000 ha survived both World Wars and gave rise to a large-scale research program started in the 1950s by the Research Centre for Agricultural and Forest Environment of the Polish Academy of Sciences in Poznań (RCAFE).

In forest-steppe and steppe areas of southwestern Russia, strongly deforested in the 18th and 19th centuries, the local population started to plant the edges of fields and settlements with afforestations to counteract damages caused by dry winds and soil erosion (Sobolev 1948, Petrov and Skachkov 1987).

The farmland abandonment in the set-aside program recently proposed by the Common Agricultural Policy of the European Union produced a new problem of management of areas permanently or temporarily excluded from tillage. It coincided with the recognition of new environmental threats, such as water shortages and water contamination, caused by intensified plant production, large-scale livestock farming, and industrial emissions. A simultaneous strong decline of biological resources in farmland has been ascertained. It became necessary to counteract diffuse pollution

habitats for rare or endangered species protection and other forms of management supporting nature conservancy.

SHELTERBELTS AND WOODLOTS IN THE STRUCTURE OF BIOGEOCHEMICAL BARRIERS

Studies on the limitation of environmentally harmful side effects of intensive farming, carried out over many years as part of a broad research program on energy flow and matter cycling in agricultural landscapes (Ryszkowski 1975, Ryszkowski and Bartoszewicz 1987, Ryszkowski et al. 1997, 1999, Prusinkiewicz et al. 1996, Kędziora and Olejnik 1996) confirmed the pertinence of biogeochemical barriers elaborated by Ryszkowski and co-workers (1990). Under the moderate climatic conditions of Europe, the main structural landscape elements of such barriers are mid-field woodlots, shelterbelts, wetlands, communities of perennial vegetation, and open (surface) water networks.

The following processes determine the functions of biogeochemical barriers:

- The kind of vegetation cover modifies solar energy fluxes used for evapotranspiration, soil, and air heating, thus influencing water circulation and biochemical reaction intensity, as well as local air mass movement (wind, heat advection, and convection processes) (Ryszkowski and Kędziora 1987, Kędziora 1996, Kędziora and Olejnik 1996). Plant cover structure also determines the potential for wildlife survival (Karg and Ryszkowski 1996, Ryszkowski et al. 1999).
- The main methods of matter cycling in agricultural landscapes are migration by underground and open waters, transportation by air movements, and transformations brought about by living organisms (primary production, decomposition, mineralization and migrations — Ryszkowski et al. 1990).
- Spatial structure of the plant cover modifying speed and direction of air masses also influences airborne pollution. Chemistry of ground and surface water is controlled by vegetation and animal migrations.

More than 20 different functions favorable to the environment, human economy, health, and culture have been listed in papers concerning the role of shelterbelts and woodlots (Zajączkowski 1997). One can distinguish, however, the following more general categories of function that should be considered the most important for agricultural landscapes:

- Increase of water retention by the restricting surface runoff and drying effect of wind
- Purification of ground waters and counteraction of nonpoint pollution spread in the landscape
- Prevention and restriction of wind and water erosion effects

Water protection, both of its resources and purity, is the paramount problem for most regions. Afforestations affect water retention because of the greater water storage capacity of their litter and soil than that in arable fields. Plant cover increases infiltration rates by slowing runoff. These phenomena, including soil penetration by tree and shrub roots which enables deeper percolation through impermeable layers of loess or argillaceous soil profiles, were recognized before the end of the 19th century. This recognition provided some theoretical basis for regional reafforestation programs aimed primarily to ameliorate soil humidity conditions and to restrict erosion in steppes or deforested agricultural areas. Such efforts started during the first half of the 20th century in the chernozem areas of Ukraine and the European part of Russia, and then broadened to include some Asiatic parts of the former Soviet Union (Sobolev 1948, Gorshenin 1949, Aderichin et al. 1983, Lamin 1983, Petrov and Skachkov 1987). Shelterbelt networks further augment water retention because they reduce wind velocity that would otherwise diminish evapotranspiration from the fields. Moreover, recent meteorological studies have shown that the indices of horizontal precipitation (dew, fog) are greater in densely afforested fields than in open ones (Kędziora 1993). According to estimates by Ryszkowski and Kędziora (1987, 1993), water deficits in rural areas may be reduced by 40 to 60 mm by well-managed shelterbelt networks, which is especially important in low precipitation seasons.

Beneficial effects of afforestations on water purity result from phytosorption of chemical compounds from the soil solution. However, the trees have a greater ability to intercept these compounds than do herbaceous plants because of the extensive root systems and significantly greater evapotranspiration by aboveground structures. Moreover, some amount of contaminants is subjected to additional transformation by the floor vegetation and the superficial layer of soil and litter, whereas a considerable part of absorbed chemical compounds remains immobilized for dozens or hundreds of years in accumulated timber biomass. They may be released from local cycling after burning or microbiological decay. In view of the growing problems of nonpoint pollution control, these processes, overviewed historically by Correll (1997), have become extraordinarily important for intensive agriculture areas. Investigations of water purification efficiency in the Central Povolzhe chernozem region of Russia by the 12.5-m wide deciduous shelterbelts (composed of 40% birch, oak, and linden and 60% maple) showed that pollution was reduced 20–30% in the case of shelterbelts alone compared with open fields, and by over 50% in the case of shelterbelts with an embanked ditch (Antonov 1987). Fertilizers not absorbed by cultivated plants in the shelterbelt-only decreased 9 to 42.5% and herbicides decreased 91.4%, whereas for the shelterbelt with embanked ditch the reductions were 44.5 to 76.3 and 96%, respectively. Detailed studies in RCAFE (Ryszkowski 1987, Ryszkowski and Bartoszewicz 1987, Ryszkowski et al. 1997, 1999) showed

of nitrogen leached from sub-watershed covered 28% by biogeochemical barriers (forests, shelterbelts, meadows, and wetlands) was only 1.5 kg/ha, whereas other areas practically devoid of such barriers (99% of arable fields) had 20.4 kg/ha.

Trees have long been widely applied as anti-erosive measures on arable fields. Their rather obvious suitability for this purpose results from relatively great height and flexibility of stems and branches as well as deep and strong root systems that anchor in the soil. These root systems together with the roots of plants growing under the canopy form a dense stabilizing network resistant to erosion. In order to prevent water erosion processes, the application of arborescent plants must be combined with other, often rather complicated, protective activities, especially in mountain and upland regions of diversified relief. Methods for counteracting different kinds of erosion have been described and are currently being improved (Instruction 1952, Sobolev 1948, Tibke 1988, Ticknor 1988), but it would be difficult to imagine effective erosion control without arborescent plants as part of the complex of preventive measures.

Afforestations are able to counteract or minimize the effects of extreme climatic or weather phenomena (particularly low and high temperatures, droughts, and storms) that have a detrimental influence on plant cultures, open animal breedings, living conditions, and human health; for example, they have an obvious anti-erosive function. The peri-Atlantic region hedgerows (mentioned above) and numerous other naturally woodless or otherwise deforested areas have been protected mostly for these reasons. Different and often expensive efforts aiming at reafforestation of such areas have been undertaken in many regions. Among a wide range of problems that it addresses, the activity of the Canadian Prairie Farm Rehabilitation Administration, Shelterbelt Centre (PFRA) is directed at the management of hedgerow systems in woodless North American areas, especially those with a relatively humid climate and severe winters (Schroeder 1988). It cooperates with some divisions of the USDA and with scientific and experimental institutions of Russia, where similar conditions dominate great areas (Lamin 1983, Gabeev 1983, Stadnik 1984). Annual reports of the Shelterbelt Centre, published by the PFRA, contain useful information on progress in nursery techniques, results of scientific research, establishment of shelterbelts, pest and disease protection in afforestations, improvement of planted tree value, and use of their resources for human needs and environmental protection.

A diverse range of activities linked with tree or shrub planting is applied in arid zones for the protection of soil humidity and to increase soil fertility (e.g., by mulching with bush twigs), counteract torrid heat, shade pastures, and improve living conditions of livestock. Recently, dense shelterbelts with rich bushy undergrowth have been planted on some Polish farms aimed at qualified seed production, in order to restrict the possibility of undesired cross pollination between cultivars (Bałazy et al. 1988).

All types of afforestations favor wildlife resource preservation in agricultural landscapes and maintenance of rich biodiversity. Until the past few decades, agricultural areas were regarded as having little significance for nature protection because of the well-known destructive effects of tillage and agricultural chemicals. However, investigations of ecological effects of intensive agriculture, carried out by the RCAFE as part of broad programs, showed that in the differentiated landscape structure of arable fields, abounding in small nonproductive natural elements (woodlots, bushes, wetlands, seminatural plant associations, water bodies, and others), the richness and diversity of living organisms are comparable to those of other nonutilized ecosystems. This point of view was expressed in the Polish strategy of biological resource protection (Ryszkowski and Bałazy 1991) and later accepted as a general guideline of the European Community landscape and biological diversity protection Biological and Landscape Diversity Strategy (Pan-European strategy 1996, Ryszkowski et al. 1996).

Thus, biogeochemical barriers enable landscapes to diminish loads and to restrict dispersal of diffuse (nonpoint) pollution, transforming them by biochemical processes into gaseous emissions, absorption by soil-inhabiting biota, and accumulation in timber biomass and organic deposits. The patches and strips covered by woody plants usually form basic networks for biogeochemical barriers that

- reduce various forms of soil erosion and improve solar energy partitioning, water regime and local agroclimatic conditions because of biological structure of trees
- stimulate self-purification processes because of networks of roots, under-canopy vegetation, and organic deposits
- protect rich associations of biota due to persistent, multi-storied habitats, functioning as refuges and horizontal corridors

Apart from the functions listed, afforestations significantly augment aesthetic and recreational landscape values, often inspiring artists, and are subjects of ecological education and scientific research. They simultaneously produce a range of directly utilizable goods, such as wood, honey, fruits, mushrooms, and pharmacological plants, and improve habitats for game. Hence, they are always multifunctional elements in any landscape structure. Their efficiency has been determined mostly by the types, share, distribution in the landscape, species composition, and state of management (i.e., quality). All those elements should be considered when new shelterbelts or their functional networks are laid out.

PRINCIPLES OF MID-FIELD AFFORESTATION MANAGEMENT

Past efforts to protect or introduce shelterbelts that embraced ecological premises

Vistula River outlet in Żuławy of Poland, planting of fast-growing trees in riparian sites of the Hannovermünde area in Germany). The recent progress in agro-ecological research made it possible to predict relatively accurately the impact of shelterbelts on agricultural environments.

Afforestation problems have usually been underestimated in comparison to other factors determining socioeconomic development of rural areas. Their importance depends largely on climatic conditions, relief, and land-use structure, as well as on applied tillage systems, general level of agriculture development, environmental hazard intensity, and material resources of the society. The last factor most fully determines the possibilities and extent of environmental protection programs that also involve afforestation tasks. After the period of political, social, and economic changes in central and east European countries in the 1980s, programs for agricultural landscape protection and management were developed for Hungary, Poland, the Czech and Slovak Republics, Lithuania, and the Ukraine (Bałazy and Ryszkowski 1999). Only in the first three of these countries was the income per person greater than $5000 USD, an amount, according to some economists, below which efficient citizen participation in costs of environmental protection is impossible. For steppe areas of Moldova, the Ukraine, and Russia, antierosive and water-protection programs of afforestations were not realized until the 1980s. It seems that despite proprietary, structural, and administrative changes in the agriculture of countries of the former Soviet Union, these programs will continue to be developed and implemented because of the highly significant impact of windbreak networks on crop yields in the chernozem soil regions. In the early 1980s, a modification of the principles for agricultural land use in the Ukraine was presented (Tarariko 1992a, 1992b, Medvedev and Sokołovaki 1992). It aims at regionalizing row and cereal crops on soils in two situations with erosive hazards because of slope inclinations 0 to 3° and 3 to 7°. The third soils group with sloping over 7° would not be subjected to cultivation except for sustaining grassland cover or reafforestation. To obtain a relative balance of the factors destabilizing and stabilizing landscape resistance, it would be necessary to introduce approximately 50,000 ha of forest and shelterbelts in the forest zone, 500,000 ha in the forest-steppe zone, and 750,000 ha in the steppe zone, independent of a range of other changes in the land-use structure and agrotechnologies implemented thus far.

The Hungarian and south Romanian plains are very similar to east European forest-steppe areas and are threatened by the same processes although of a slightly lower intensity because of a more favorable seasonal distribution of yearly precipitation. However, reafforestation or shelterbelt planting programs may be difficult to introduce, especially in Romania, because of denationalization of large state and cooperative farms that were divided among countryside inhabitants as individual or family farms (Von Hirschhausen et al. 1999). The creation of rationally functioning

border. However, Kort's (1988) review on worldwide references concerning the effects of shelterbelts on different crop yields showed only three instances of crop output decline out of the 97 considered. Two of them were large oats fields and the third was spring wheat in a field strip located leeward of a shelterbelt. In 93 cases, the yield increase was 3 to 35% for most cereal crops and 65% for millet, and 70% to over 100% for hay and some papilionaceous perennial crops (in one case of alfalfa the increase was 203%). As a result of this analysis, the author concluded that yield increase from shelterbelt was usually more conspicuous in drier regions. Some tree or shrub species inappropriately included in shelterbelts may stimulate appearance of pest insects, mostly aphids (Voegtlin et al. 1988) or plant diseases caused particularly by rust fungi or viruses (Bałazy 1997b). However, such undesirable phenomena may be easily eliminated or reduced by careful selection of species, appropriate spatial distribution, and intervention in shelterbelt managements, such as crown shapening, root pruning, and opening of shelterwoods. Positive effects of rationally managed mid-field afforestations always prevail over the negative, even though they are not conspicuously reflected in yields (Kort 1988, Knauer 1990, Fitzpatrick 1994, Ryszkowski et al. 2000).

Despite the benefits of shelterbelts for the environment and their increasing implementation, often performed by specialized institutions, analyses of landscape structure frequently show a decreased afforestation in some countries. For example, the estimate of shelterbelts in England (Hooper 1984) showed that in the two decades after World War II approximately 70,000 km of hedgerows were removed either to augment arable areas or to install electrical supply lines. Mannion (1995) reported declines as high as 22% until the mid-1980s. French legal mandates of the 1940s regarding mergence of scattered fields caused great losses in the hedgerow networks of the northwestern part of this country (Baudry and Burel 1984, Augé 1999). The analysis of satellite pictures of different regions of France (Bras and Piveteau 1999) showed an approximately 1% annual decline in hedgerows, woodlots, and single trees in agricultural areas. In 1991–1996, the greatest decreases occurred in Calvados, the Brittany Peninsula, and the central parts of the Parisian Basin. A 1.3% increase in afforested areas was observed in only a few small but highly afforested districts mainly in the south and in the vicinity of Calais. Full or partial adjustment of the former DRG agriculture to EC standards would likely result in a decline of linear elements differentiating the landscape of arable fields (shelterbelts, balks, roadsides, etc.). Full adjustment would decrease those elements from 920 m/100 ha to 910 m/100 ha; partial adjustments would result in a decrease to 700 m/100 ha. However, a simultaneous increase of afforested areas would also occur, which should lengthen forest-field borders from 670 m to 720 m and 740 m, respectively (Stachow and Haberstock 1995).

Estimations in recent decades show decreased shelterbelts in farmland in most

beginning of the 19th century, as well as poplar afforestations, planted in the 1950s as part of the country's plan to augment timber resources. Felling afforestation trees to increase timber yields, which also occurred in other countries of eastern and central Europe, is connected with the impoverishment of the rural population because of the associated agricultural transformation processes (Hirschhausen et al. 1999, Bałazy and Ryszkowski 1999).

The degree of shelterbelt participation in the farmland of Poland varies greatly. Quantitative estimations carried out by the RCAFE in 33 districts along the country's western border showed the differentiation range from about 100 m to 1500 m/100 ha and was usually similar in adjacent communes. In D. Chłapowski Agroecological Landscape Park, which has a surface area of 17,200 ha and is considered to be one of the country's best afforested agricultural areas, mean shelterbelt length/100 ha amounts to approximately 1740 m (Ryszkowski et al. 2000), whereas outside its borders, although in the same district, it amounts to 150 to 550 m/100 ha.

In terms of the farmland afforestation program (Ryszkowski et al. 2000), the set-aside areas should be qualified according to the following principle. When the total area of soil shows low fertility, large-scale reafforestation gains priority. In the case of a significant amount of good quality soils, the acreage with shelterbelt planting should predominate over reafforestated areas. However, local situations and individual landowners' attitudes or intentions will considerably modify possibilities for both kinds of afforestations. The total area for the optimal variant of shelterbelt network to be planted in Poland was estimated at 350,000 ha, and for the minimum variant 250,000 ha. The most pressing needs occur in the central Polish lowlands with the greatest water deficits, an area of approximately 130,000 km^2, and in the foothills and mountains of southern Poland that are severely or moderately threatened with erosion.

Afforestations in rural areas should be designed and managed according to the following six guidelines.

1. Adjustment to the functions afforestations should perform in the environment

Because afforestations are by nature multifunctional, one or more main functions should be ascribed to each of them. To fulfill that function they should be optimally situated and given proper species composition and tree-cover structure. It is therefore necessary to recognize the types and scale of environmental hazards significant for a given area and to assess which shelterbelt forms would most effectively counteract them. Afforestations perform a relatively full range of anticipated functions only if they form a rationally distributed spatial network. Thus it is absolutely necessary that they be harmoniously embedded into the structure of biogeochemical barriers, usually, and that embedding be, their most essential element. Water protection functions are most effectively performed by shelterbelts 10 to 20 m wide, the equivalent of three to seven tree rows. Since the interception of chemical substances

wind. It is advisable not to plant in outer rows those species with extending root systems, especially when they tend to produce sprouts (e.g., poplars). The belts should be localized transversally in relation to the direction of ground water flow, which in water-bearing layers is not always compatible with slope-surface angle.

In agricultural landscapes with diversified structure, purification effects of shelterbelts and other elements of biogeochemical barriers are of basic significance for control of eutrophication processes in open water reservoirs. Multiyear monitoring of water contamination in the Agricultural Landscape Park in Turew showed, for example, that the spring content of nitrates, 45 mg/dm^3, in ground waters under arable fields decreases to 2 to 5 mg/dm^3 in waters outflowing to the drainage canal of the Park in a relatively dense and well-managed afforestation network.

Two-row roadside afforestations of the allee type, if compact, well managed, and enriched with an under-canopy shrub layer, and distributed approximately along the contour lines of slopes, and distanced 400 to 600 m, may sufficiently reduce water contamination at a slope inclination up to 3%. With slopes of over 5%, afforested belts should be widened to 10 to 20 m and distributed alternately with strips of grasslands, arable fields, and berry-bearing shrub plantations in transversal arrangement.

On excessively wet sites, increased use of trees that absorb large amounts of water for biomass unit production (birch, oaks, larch, poplars) is recommended to stimulate drying and purifying effects. The depth of the ground water table is an important determinant of shelterbelt significance because the purification effect may be insignificant if tree root systems do not reach the water-bearing horizon or at least the capillary soaking layer in the soil profile. In the protection zones of water intakes, as well as in the screening green around polluting landscape elements, the share of afforestations — together with compact forests — should be increased to 30% or even more.

Typical wind-breaking shelterbelts should be rather narrow (one to three tree rows), not compact, and situated in general for Polish conditions, approximately north to south to break the predominant western wind, in distances of 350 to 500 m depending on soil and relief characteristics. Such structure, density could be attained in most cases by planting one or two tree rows along linear elements of farmland infrastructure (roadsides, ditches, streams, and other marginal patches). Species with shallow root-systems are of little value for these shelterbelts. Of course, the wind-breaking function also is performed by all other forms of afforestations.

Active landscape management for nature protection and biodiversity preservation requires maintenance and introduction of woodlots and bushes to function as wildlife refuges. They should be connected with greater areas of natural or seminatural ecosystems, such as marshes, peri-aquatic rushes, xerotermophilic bushy communities, greater afforested areas, and river valleys. Shelterbelts established or adapted for this purpose should generally be wider than the common wind-breaking strips and

2. Adjustment of the species composition to local site conditions

The use of native species and local ecotypes of trees and shrubs best adapted to the regional climate and soil should be a general principle in shelterbelt planting. The recommended components for Polish and central European areas have been listed in many scientific papers, books, and manuals, but because of economic reasons and insufficient coordination, nursery production does not meet the actual demand for seedlings (too narrow a scope of native wild species and simultaneously an overabundance of ornamental cultivars). However, situations or requirements such as the following may call for the use of nonnative or even exotic species:

- Atmospheric and soil pollution forcing the use of more resistant species (black pine, plane tree, horse chestnut, red oak)
- Good increment and favorable wood characteristics (Euro-American poplar cultivars, Douglas fir, red oak)
- Fruit and melliferous species that, apart from direct use by man, are also attractive for beneficial arthropods and birds, stimulating biodiversity (many species of trees and shrubs)
- Some ornamental forms traditionally planted

Purposeful dissemination of highly expansive species that threaten to dominate native associations in Poland, as for example, the black cherry or ash-leaved maple, should not be allowed, nor should native or exotic species indirectly posing hazards of spreading plant diseases or significant pests (e.g., berberry, bird's cherry, alder buckthorn, white pine).

3. Continuity of shelterbelt existence in landscape structure

To properly fulfill their protective functions, afforestations have to be persistent elements in the landscape. Hence it is necessary to secure their existence legally. It is also essential to recognize local residents' attitudes toward shelterbelt plans and to establish valid legal records precluding groundless afforestation removal. Composition and management of afforestation species are also of great importance for ensuring their long-lasting benefits. Biogroups should be differentiated according to fast-growing and long-lived species in proportions adequate to site conditions. Tree generation exchange should be carried out over several dozen-year periods. On the basis of the RCAFE research, the systems of afforestation barriers proposed for wet habitats in Poland (alder and riparian plain forest sites) should contain 60 to 70% species of 30- to 70-yr felling cycles (black and white poplars and their selected hydrids, black alder, birch and willows), and 30 to 40% long-lived species (mostly European ash, oaks, larch, elms), ensuring persistence of shelterbelt structure for long time spans (150 to 250 or more years). In drier areas of moraine plateau, the components typical for horn beam-oak associations should dominate (oaks, linden, beech, horn beam, elms, birch, rowan, sycamore) with small admixtures of larch,

beginning from 60 to about 100 years. In all these cases, the component exchange should be carried out by single-tree selection cutting or small group felling. Neither one-row and allee-type afforestations nor many-row belts and small-area woodlots can be cut by means of clear cutting. Principles of management and utilization of afforestation should be mandated in space management plans of communes.

4. Economical management of farmland

Spatial arrangement of shelterbelts should be to the maximum degree connected with linear elements of the fields' infrastructure, locating them in one- to four-row strips along roads, watercourses and on the borders of different land use forms (grassland-field ecotones, riparian lines of water bodies, etc.). In those cases when they must be run across arable fields, it is advisable to use low-utility areas. In areas of intensive farming, shelterbelt participation should not exceed 2.5 to 4% of arable fields. Areas with poor soils should also be cautiously qualified for persistent reafforestation because reserves of agriculturally nonutilized fields may facilitate combining farming with other forms of economical activities, such as agrotourism, small forms of services, and crafts. This aspect is particularly important for regions with long agricultural traditions and high unemployment levels. In such cases, planting of shelterbelts would be a noncontroversial and environmentally advantageous form of management of selected set-aside fields. It should also meet the tendency to manage nature conservation and landscape protection resorts in the areas withdrawn from the agricultural use, which are more and more popular in the European Community countries (Bastian 1994, Latesteijn 1994, Bacharel and Pinto-Correia 1999).

5. Nature protection support and biological control enhancement

The functions involving biodiversity preservation in rural areas are best performed by woodlots of rich species composition, growing on differentiated sites, and linked by shelterbelts with greater areas of natural or seminatural ecosystems (forest, swamps, and rushes). Rationally managed networks of biogeochemical barriers usually contain well-functioning refuges of many species and ecological corridors for their migration, as confirmed by extensive studies on biological resources in the Agro-ecological Landscape Park near Turew (Karg and Ryszkowski 1996, Karg 1997, Ryszkowski et al. 1998, 1999). Their differentiation by biocenotical admixtures of fruit or melliferous plants, as well as by limited removal of dead trees, favors creation of nesting habitats for birds and increased attractiveness for invertebrates beneficial for agriculture (spiders; pollinating, parasitoid and predacious insects and mites; different saprophagous forms) and entomopathogenic microorganisms (Bałazy 1994, 1997, Hemmati et al. 1997, Bałazy et al. 1998).

Crop protection development prospects (Lipa 1998, 1999), considered among nonchemical methods, widened utilization of a habitat's natural potential to prevent

for natural populations of beneficial organisms with the primary aim to include them in integrated pest control methods, observing general principles of biodiversity preservation.

6. Increase of the economic value of shelterbelts

Although afforestations are mainly introduced in farmland for their protective functions, they make it possible to obtain various products, timber included, with no harm to their principal role. To fully utilize this potential, certain conditions must be fulfilled, of which the most important are:

- Planting tree species yielding high quality timber
- Proper stand management to obtain possibly large amounts of high quality timber
- Establishment of methods and quotas of timber yields to secure the continuity of shelterbelt functioning
- Indication of ways for tree-cover renewal which ought to be carried out immediately after timber removal

CONCLUSIONS

The problem of mid-field afforestation has not yet been ultimately settled by Polish law. No proper system of economic incentives has thus far been introduced to encourage farmers to establish shelterbelts. The adoption of such a system would comply with European Union Directive 2078/92 defining subsidies to compensate losses resulting from extension of economic activity involving proecological technologies and supporting nature protection. In the postwar period, several planting actions were undertaken, but only some were successful (Zajączkowski 1996, Budzyński 1997).

Fifty years of Polish practice has taught us it is extremely difficult to maintain a shelterbelt in a proper state unless it is supervised by a landowner or administrator, or if location generates conflicts with neighbors. For these reasons, a person or institution should be designated responsible for the maintenance of every afforestation.

Farmland afforestations are indispensable factors supplementing the "country's program of afforestation increase" since they fulfill the majority functions. Hence, it is necessary to coordinate closely mid-field afforestation and the country's reafforestation projects, both on the general scale and in particular watersheds.

The number of shelterbelts, the spatial structure of their networks, and their species composition should be adjusted to the kinds and intensity of environmental threats that they must counteract. They should be designed according to the principle of maximum high quality farmland economy.

REFERENCES

Aderichin P. G., Belgard A. L., Zonn S. V., Krupenikov I. A., Travteev A. P. 1983. The impact of forest vegetation on chernozem. In *The Russian Chernozem. 100 Years after Dokuchaev*. Eds. V. A. Kovda, E. M. Samojlova. Tzd. Nauka. Moskva. Part 1. Chapter 7: 117–125 [in Russian].

Alexander K. N. A., Kirby K. J., Watkins C. 1998. The links between forest history and biodiversity: the invertebrate fauna of ancient pasture-woodlands in Britain and their conservation. In *The Ecological History of European Forests: Based on Presentations Given at the International Conference on Advances in Forest and Woodland History*. Ed. K. J. Kirby. CAB International. Wallingford (U.K.): 73–80.

Antonov V. I. 1987. Water pollution control by shelterbelts. *Lesnoi Zhurnal* 3: 17–21 [in Russian].

Augé S. 1999. La bourse aux arbres dans la Manche. *Le Courrier de l'Environnement de l'INRA* 36: 25–38.

Bacharel F., Pinto-Correia T. 1999. Land use, nature conservation and regional policy in Alentejo, Portugal. In *Land-Use Changes and their Environmental Impact in Rural Areas in Europe*. Eds. R. Krönert, J. Baudry, I. R. Bowler, A. Reenberg. Man and Biosphere Series. Vol. 24. Paris: 65–79.

Bałazy S. 1994. Significance of ecotones in agricultural landscape for the distribution of insect pathogens. In *Functional Appraisal of Agricultural Landscape in Europe*. Eds. L. Ryszkowski, S. Bałazy. Research Center for Agricultural and Forest Environment, Pol. Acad. Sci. Poznań: 153–163.

Bałazy S. 1997. Afforestations in relation to diseases and pests of cultivated plants. In *Zadrzewienia na obszarach wiejskich*. Instytut Badawczy Leśnictwa. Warszawa: 11 pp. [in Polish].

Bałazy S. 1997. Diversity of entomopathogenic fungi in agricultural landscapes of Poland and France. In *Ecological Management of Countryside in Poland and France*. Eds. L. Ryszkowski, S. Wicherek. Research Centre for Agricultural and Forest Environment, Pol. Acad. Sci. Poznań: 101–111.

Bałazy S., Ryszkowski L. 1999. Protection de la diversité biologique et paysagère dans les paysages d'Europe Centrale et Orientale. Conseil de l'Europe. *Sauvegarde de la nature*, No. 94. Strasbourg. 58 pp.

Bałazy S., Ziomek K., Weyssenhof H., Wójcik A. 1998. Principles of mid-field afforestations management. In *Kształtowanie środowiska rolniczego na przykładzie Parku Krajobrazowego im. Gen. D. Chłapowskiego*. Eds. L. Ryszkowski, S. Bałazy. Zakład Badań Środowiska Rolniczego i Leśnego PAN. Poznań: 49–65 [in Polish].

Bastian O. 1994. Land withdrawal as a means to improve ecological conditions in rural landscapes. In *Functional Appraisal of Agricultural Landscape in Europe*. Eds. L. Ryszkowski, S. Bałazy. Research Centre for Agricultural and Forest Environment, Pol. Acad. Sci. Poznań: 291–307.

Baudry O., Bourgery C., Guyot G., Rieux R. 2000. Les haies composites reservoirs d'auxiliaires. CTIFL. Ser. Hortipratic. Paris. 116 pp.

Baudry J., Burel F. 1984. "Remembrement": landscape consolidation en France. *Landscape*

Chłapowski D. 1835. *Agriculture*. Poznań. 164 pp. [in Polish].

Correll D. L. 1997. Buffer zones and water quality protection: general principles. In *Buffer Zones: Their Processes and Potential in Water Protection*. Eds. N. Haycock, T. Burt, K. Goulding, G. Pinay. Quest Environmental. Harpenden: 7–20.

Delelis-Dusollier A. 1986. Histoire du paysage par l'analyse de la végétation: l'exemple des haies. *Actes du colloque: Du pollen au cadastre*. Lille. Hommes et Terres du Nord 2–3: 110–115.

Ferron P. 2000. Base écologique de la protection des cultures. *Currier de l'Environnement* INRA. 41: 33–41.

Fitzpatrick D. 1994. *Money Trees on Your Property: Profit Gained through Trees and How to Grow Them*. Incata Press Pty, Butterworth-Heinemann, Chastwood (Australia). 174 pp.

Gabeev V. N. (Ed.). 1983. *Regeneration and Stability of Western Siberia Forests*. Sukachev Institute of Forestry. Izdatelstwo Nauka. Moskva. 174 pp. [in Russian].

Gorschenin N. M. 1949. Shelterbelts and drought control. *Priroda* 38 (2): 3–23 [in Russian].

Hemmati F., Pell J. K., Mc Cartney H. A., Deadmann M. L. 1997. Aerial dispersal of the entomopathogenic fungus. *Erynia neoaphidis* from reservoirs in hedgerows. *SIP Banff* 97: 26–27.

Hooper M. D. 1976. Historical and biological studies on English hedges. In *Les bocages: histoire, écologie, économie*. CNRS-ENSA et Université de Rennes. Ed. M. J. Missonier. Rennes: 225–227.

Hooper M. D. 1984. What are the main recent impacts of agriculture wildlife? Could they have been predicted, and what can be predicted for the future? In *Agriculture and the Environment*. Ed. D. Jenkins. Institute of Terrestrial Ecology. Cambridge: 33–36.

Inglis I. R., Wright E., Lill J. 1994. The impact of hedges and farm woodlands on woodpigeon (*Columba palumbus*) nest densities. *Agric. Ecosys. Environ.* 48 (3): 257–262.

Instruction 1952. Instruction for designs and distribution of shelterbelts in steppe and forest-steppe regions of the European part of the SSSR. 1952. *Les i Step* 9: 3–18 [in Russian].

Ivanov A. F. 1984. Shelterbelts in the Kulunda steppe. *Lesnoe Khoziaistvo* 9: 41–43 [in Russian].

Karg J. 1997. Biocenotic role of afforestations in agricultural landscape. In *Zadrzewienia na obszarach wiejskich. Instytut Badawczy Leśnictwa*. Warszawa. 12 pp. [in Polish].

Karg J., Ryszkowski L. 1996. Animals in arable land. In *Dynamics of an Agricultural Landscape*. Eds. L. Ryszkowski, N. R. French, A. Kędziora. PWRiL. Poznań: 138–173.

Kędziora A. 1993. Estimation of heat and water balances for large areas. *Meteorologicheskie Issledovanija* 29: 33–47 [in Russian].

Kędziora A. 1994. Energy and matter fluxes in an agricultural landscape. In *Functional Appraisal of Agricultural Landscape in Europe*. Eds. L. Ryszkowski, S. Bałazy. Research Centre for Agricultural and Forest Environment, Pol. Acad. Sci. Poznań: 61–76.

Kędziora A. 1996. The hydrologic cycle in agricultural landscapes. In *Dynamics of an Agricultural Landscape*. Eds. L. Ryszkowski, N. R. French, A. Kędziora. PWRiL. Poznań: 65–75.

Kędziora A., Olejnik J. 1996. Heat balance structure in agroecosystems. In *Dynamics of an agricultural landscape*. Eds. L. Ryszkowski, N. R. French, A. Kędziora. PWRiL.

Lamin L. A. 1983. Persistence of shelterbelts in the Northern Kulunda. In *Vozobnovlenie i ustojchivost lesov Zapadnoj Sibirii*. Ed. V. N. Gabeev. Izdatelstwo Nauka. Moskva: 150–169. 173 pp. [in Russian].

Latesteijn H. 1994. Ground for choices: a policy oriented survey of land use changes in the European Community. In *Functional Appraisal of Agricultural Landscape in Europe*. Eds. L. Ryszkowski, S. Bałazy. Research Centre for Agricultural and Forest Environment, Pol. Acad. Sci. Poznań: 289–297.

Lipa J. J. 1998. Applied entomology and plant protection in 21st century. *Wiadomości Entomologiczne* 17 (Supplement): 51–77 [in Polish].

Lipa J. J. 1999. Does species diversity restrict outbreaks of agrophagous pests? In *Uwarunkowania ochrony różnorodności biologicznej i krajobrazowej*. Eds. L. Ryszkowski, S. Bałazy. Zakład Badań Środowiska Rolniczego i Leśnego PAN. Poznań: 81–98 [in Polish].

Mannion A. M. 1995. *Agriculture and Environmental Change*. Wiley. Chichester. 405 pp.

Medvedev V. V., Sokołovaki A. N. 1992. Concept on efficient use of soil cover in the Ukrainian. In *Comparisons of Landscape Pattern Dynamics in European Rural Areas*. EUROMAB Research Program "Land Use Changes in Europe and their Impact on the Environment." Seminars Rennes-Lieury-Kiev: 63–69.

Meynier A. 1976. Typologie et chronologie du bocage. In *Les bocages: histoire, écologie, économie*. CNRS-ENSA et Université de Rennes. Ed. M. J. Missionier. Rennes: 65–67.

Pan-European Biological and Landscape Diversity Strategy. 1996. Council of Europe. UNEP. European Centre for Nature Conservation. Amsterdam. 50 pp.

Pape L. 1976. Bocage et voies romaines dans le département des côtes-du-Nord. Essai de chronologie. In *Les bocages: histoire, économie, écologie*. CNRS- ENSA et Université de Rennes. Ed. M. J. Missionier. Rennes: 75–78.

Petrov P., Skachkov B. 1987. Forests and fields. *Science in the USSR* 4: 88–99.

Prusinkiewicz Z., Pokojska U., Józefkowicz-Kotlarz J., Kwiatkowska A. 1996. Studies on functioning of biogeochemical barriers. In *Dynamics of an Agricultural Landscape*. Eds. L. Ryszkowski, N. R. French, A. Kędziora. PWRiL. Poznań: 110–119.

Rauce K., Kyrstja S. 1986. The soil threats control. *Agropromizdat*. Moskva. 222 pp. [in Russian].

Read R. A. 1958. The Great Plains shelterbelt in 1954. Great Plains Agric. Council Publ. No. 16, 125 pp. [in Russian].

Read R. A. 1964. *Tree Windbreaks for the Central Great Plains*. USDA. Forestry Service, Agric. Handbook No. 250. 68 pp.

Ryszkowski L. 1975. Review of studies of the effects of afforestations on adjacent field habitats carried out in Turew. *Zeszyty Problemowe Postępów Nauk Rolniczych* 156: 71–82.

Ryszkowski L. 1987. Ecological agriculture. *Zeszyty Problemowe Postępów Nauk Rolniczych* 324: 15–42 [in Polish].

Ryszkowski L., Bałazy S. 1991. Strategy of living resources protection in Poland. *Zakład Badań Środowiska Rolniczego i Leśnego PAN*. Poznań. 95 pp. [in Polish].

Ryszkowski L., Bałazy S., Jankowiak J. 2000. Program of mid-field afforestation increase. *Postępy Nauk Rolniczych* 5/287: 83–106 [in Polish].

Ryszkowski L., Bartoszewicz A. 1987. Impact of agricultural landscape structure on cycling

Ryszkowski L., Bartoszewicz A., Kędziora A. 1999. Management of matter fluxes by biogeochemical barriers at the agricultural landscape level. *Landscape Ecol.* 14:
479–492.
Ryszkowski L., French N. R., Kędziora A. (Eds.). 1996. *Dynamics of an Agricultural Landscape.* PWRiL. Poznań. 223 pp.
Ryszkowski L., Gołdyn H., Arczyńska-Chudy E. 1998. Plant diversity in mosaic agricultural
landscapes: a case study from Poland. *Planta Europe.* Eds. H. Synge, J. Akeroyd.
Uppsala: 281–286.
Ryszkowski L., Karg J. Kujawa K. 1999. Protection and management of biological diversity
in agricultural landscape. In Uwarunkowania ochrony różnorodności biologicznej i
krajobrazowej. Eds. L. Ryszkowski, S. Bałazy. Zakład Badań Środowiska Rolniczego
i Leśnego PAN. Poznań: 59–80 [in Polish].
Ryszkowski L., Kędziora A. 1987. Impact of agricultural landscape structure on energy flow
and water cycling. *Landscape Ecol.* 1: 85–94.
Ryszkowski L., Kędziora A. 1993. Energy control of matter fluxes through land-water eco-
tones in an agricultural landscape. *Hydrobiologia* 251: 239–257.
Ryszkowski L., Kędziora A., Marcinek A. (Eds). 1990. *Water Cycling and Biogeochemical
Barriers in Agricultural Landscape.* Zakład Badań Środowiska Rolniczego i Leśnego
PAN. Wydawnictwo Uniwersytetu A. Mickiewicza. Poznań. 187 pp. [in Polish].
Ryszkowski L., Pearson G., Bałazy S. (Eds). 1996. *Landscape Diversity: A Chance for Rural
Community to Achieve a Sustainable Future.* Research Centre for Agricultural and
Forest Environment, Pol. Acad. Sci. Poznań. 223 pp.
Schroeder W. R. 1988. Planting and establishment of shelterbelts in humid severe-winter
regions. *Agric. Ecosys. Environ.* 22/23: 441–463.
Sobolev S. S. 1948. Development of erosion processes in the European part USSR and their
control. *Izdatelstvo Akademii Nauk SSSR.* Moskva. T. 1. 307 pp., T. 2. 248 pp. [in
Russian].
Soltner D. 1977. L'arbre et la haie. *Collection Sciences et Techniques Agricoles.* Angers. 112
pp.
Stadnik A. P. 1984. The means of the increase of afforestations in northern steppe area of
Western Ukraine. *Lesnoe Khoziaistvo* 9: 46–48 [in Russian].
Stanners D., Bourdeau P. 1995. *Europe's Environment.* European Environment Agency.
Copenhagen. 676 pp.
Tarariko A. G. 1992a. Optimization of agricultural landscapes in the Ukraine. In *Comparisons
of Landscape Pattern Dynamics in European Rural Areas.* EUROMB Research Pro-
gram: "Land Use Changes in Europe and their Impact on the Environment." 1991
Seminars. Rennes-Lieury-Kiev: 57–58.
Tarariko A. A. 1992 b. Optimization of the rural landscapes in the Ukraine using soil-protecting
systems of agriculture. In *Comparisons of Landscape Pattern Dynamics in European
Rural Areas. EUROMB Research Program.* "Land Use Changes in Europe and their
Impact on the Environment." 1991 Seminars. Rennes-Lieury-Kiev: 59–62.
Tibke G. 1988. Basic principles of wind erosion control. *Agric. Ecosys. Environ.* 22/23:
103–122.
Ticknor K. A. 1988. Design and use of field windbreaks in wind erosion control systems.

Zajączkowski K. 1996. State of afforestation in Poland and actual problems of their development. Forestry Research Institute. Scientific documentation. 15 pp. [in Polish, unpublished].

Zajączkowski K. 1997a. Afforestations in spatial planning. In *Znaczenie zadrzewień w krajobrazie rolniczym oraz aktualne problemy ich rozwoju w przyrodniczo-gospodarczych warunkach Polski*. Płock: 111–121 [in Polish].

CHAPTER **14**

Land Units and the Biodiversity of Forest Islets: From Satellite Images to Ground Analysis

Marc Galochet, Vincent Godard, and Micheline Hotyat

CONTENTS

INTRODUCTION

The landscape of extensively farmed countryside in Poland, as in France, is made up of a mosaic of agricultural plots linked to noncontiguous wooded areas of various sizes and shapes, sometimes connected to one another by hedges or wooded corridors. These wooded areas have been and still are greatly influenced by man and his methods of cultivation. The biodiversity of forest islets thus depends as much on human actions as on the environmental conditions.

What is the influence of forest management methods and forestry traditions on the biodiversity of forest islets? What kind of diversity, in terms of structure and function, characterizes forest islets that have found a niche in ecosystems dominated by intensive farming? To illustrate our remarks, we use a sector in Greater Poland, representative of extensively farmed countryside, and situated approximately 50 km to the south of Poznań. First, the land units are delimited with the help of satellite data. Second, an automatic extraction process pinpoints the forest islets. Finally, the biodiversity of these forest islets is analyzed using a frame sample.

LAND UNITS AND THE BIODIVERSITY OF FOREST ISLETS

Research on land cover units allows us to look at the forest islets in context and to understand the environmental conditions in which they are best situated. The biodiversity of the forest islets can be appreciated when their spatial distribution and the modifications that have taken place over time are considered at one and the same time. This leads to cross-referencing the environmental conditions as well as the uses of the land have had and still have an influence on the biodiversity of the islets.

Defining Land Units

The term *land unit* is difficult to pin down in that there is still a certain amount of uncertainty in its definition, indeed in its conception. Nevertheless, some aspects of the term can be fixed:

- Overall, a land unit is homogenous and constitutes a whole with its own identity,
- A land unit can be defined by the fact that there is sufficient difference between it and the units which surround it. This difference allows the spatial boundaries of the unit to be defined, and its characteristics to be discerned.
- This said, *homogenous* does not mean *monotonous* and does not imply that there is only one feature appearing repeatedly. On the contrary, a land unit is made up of diverse features, organized in such a way that they give the unit its originality and its identity. As a consequence, to delimit a land unit, it is necessary to list all

areas, woods, forests, and so on), as well as features, such as the size and shape of the plots (whether these are agricultural or forest plots), the types of land cover, and the topography (Papy, 1992). Each land unit is thus defined by a particular physiogomy and therefore a structure that expresses the uses to which it is put, that is to say, the use of the soil.

Forest Islets: From Land Units to Vegetal Diversity

Forest islets, when seen as land units, can be approached on two complementary levels (Hotyat et al., 1999):

- On a regional scale in order to determine the organization of land units in its totality, by distinguishing a mosaic of agricultural plots, which constitute the agrosystem, and the woods, which represent the sylvosystem (Figure 14.1). By thus looking at the wooded areas in their regional context among the other landscape elements, we can reveal their relationship toward each other in space. This scale allows us to appreciate the fragmentation of forest cover (Harris, 1984).
- On the woodland scale, we can appreciate structural and vegetal diversity in their totality, both of which express the internal functioning and structure of the forest islets.

The fragmentation of forest cover and its spatial organization have an effect on the plant biodiversity of the forest islets. In fact, these wooded areas of different sizes can be considered "islands" totally isolated in both coastal and continental areas. However, these continental islands or forest islets, like the summits of isolated mountain chains, that come within the boundaries of agrosystems have specific characteristics, which evoke the insular biogeographical theory (MacArthur and Wilson, 1967).

The theory of dynamic equilibrium elaborated by MacArthur and Wilson in 1963 and 1967 respectively, explains the constitution and evolution of insular ecosystems through the conjunction of two opposing phenomena: the immigration of new species and the extinction of species (Ramade, 1994). Forest islets find themselves all the more easily enriched by migrating species since the places from which they come are nearby. Inversely, the extinction rate is all the higher since the forest islets are isolated.

The insular theory which looks at the distance that separates islets from one another and from other source habitats, as well as their size, does not, on the other hand, take into account the respective density of species, their abundance-dominance, their capacity to produce seeds, and the possibility they have to colonize the environment, or the potential of said environment. This theory greatly informed the concepts of the evolution of biocenosis in the 1960s; nevertheless, it is appropriate to add to this theory environmental data and to take into consideration the influence of human action effected on the environment.

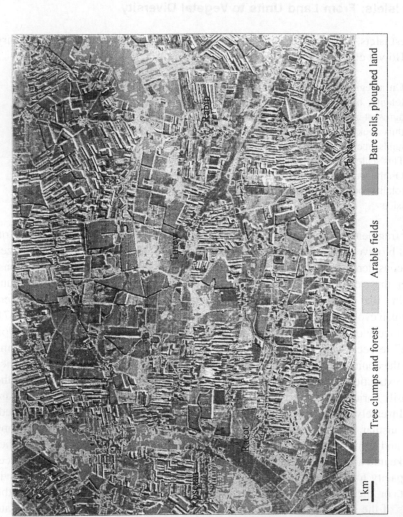

Figure 14.1 Shaded composition of General Chłapowski's Agro-ecological Park (Greater Poland), Satellite Image SPOT HRV-17, June 1997.

and the plant diversity of the forest islets (Galochet, 2001). All these forested plots
are largely influenced by human activity and local history, notably forestry traditions.

The wooded areas in Greater Poland are marked, as if in indelible ink, both by
a form of forestry inherited from the Prussian invasion and then by almost half a
century of collectivism. These are important events which marked the established
societies of the period by modifying their economic, social, and agricultural structures over several generations. Yet, as the historian Fernand Braudel wrote in his
Grammar of Civilisations, "the past of a civilisation is [...] only the history of
continual borrowings which they have made from one another, over the course of
centuries, without, however, losing their specificity, or their originalities."

Thus, after the third partition of Poland in 1795, which caused it to disappear
from the political map of Europe, Greater Poland was under Prussian rule until 1918,
the date when Poland became independent once more following the defeat of the
central empires, and the Russian Revolution (Le Breton, 1994). This occupation had
serious consequences for forestry traditions and the forest environment, something
that can still be seen today through the numerous mono-specific and even aged linear
formations most frequently composed of Scotch pines *(Pinus sylvestris)*.

Nevertheless, an exception is found in Greater Poland, a nobleman's estate —
General Dezydery Chłapowski (1788–1879), a prominent landowner, whose landscape improvements survive to this day. Recruited into the Napoleonic army at the
age of 18, after having studied at the Military School in Paris, he became the
Emperor's aide-de-camp and accomplished some great military campaigns at his
side, something that allowed him to travel across western Europe. On his return to
Poland around 1820, he devoted himself to his land (about 10,000 ha)* and made
some important developments in forestry and agriculture, influenced by various
observations he had made over the course of his numerous trips to England, Scotland,
Ireland, and France. He planted lines of trees along the edges of pathways, made
countless experiments with drainage, crop rotation, and seed testing, built hedges to
act as windbreaks and to limit erosion from wind, introduced exotic touches such
as the locust tree, and established forest islets.

The ideas of General Chłapowski were adopted by some of his noble friends
who thus spread his agroforestry model. But this model did not extend to territories
outside Greater Poland because of the proximity of the frontier between Prussia and
Russia.

These old forestry developments, which modified and shaped wooded areas, still
influence countryside diversity today, something that can be identified using satellite
images. A methodology capable of defining land units, identifying them on the
ground, analyzing them, and ensuring their cartographic representation must still be
finalized. Remote sensing, where it is adapted to an overall view of the principal
types of land cover seen from above, on a regional scale, provides these possibilities.

REMOTE SENSING TO DELIMIT LAND UNITS
AND ESTABLISH THE LOCATION OF FOREST ISLETS

The distribution, size, and shape of forest islets on extensively farmed plots, in Poland as in France, are determined by the environmental conditions and the intensity of their cultivation by man. The confrontation of natural and cultural factors (Hotyat et al., 1997) explains the specific vegetal, locational and countryside diversity that can be perceived through the analysis of remote sensing images.

Remote Sensing to Determine the Forest Islets

To what extent can forest islets in extensively farmed areas, considered geographical and radiometric land elements, be individualized in relation to other types of land cover?

Remote sensing uses the physical properties of the objects under observation to obtain information about their nature, through electromagnetic radiation that these objects reflect. In this way, objects can be differentiated from one another by their own individual spectra.

Using a specific color composition (Figure 14.1; vegetation index, Roberts filter on this index, and analysis of the principal components, coded respectively in blue, green and red) we can appreciate the spatial distribution of wooded areas (in shades of green) in comparison to other types of land cover (cultivated fields in orange, and bare ground in blue), and we can detect their structure and their dominant features.

Remote sensing creates a dual interest. On the one hand, it gives us numerical information by means of the different channels and their spectral resolution, and, on the other, it provides an exhaustive study, a sort of objectivity, about the areas studied as a whole. However, it must not be forgotten that objects of a smaller scale than that of the spatial resolution cannot be spotted (unless very strong radiometric contrasts are used) and that topographical variations, due to the effects of shade, can bring about confusion in the identification of the objects that make up the landscape. We must therefore be careful when it comes to interpreting data; in Laurence Le Du's words, "The use of close-range infrared, and of medium-wave infrared, in the study of landscapes is necessary, but obliges us, to an even greater extent than for visual observations, to make a good interpretation of the data which necessitates a good knowledge of the ground. We must not only link the information from the images to a specific reality, but also decide to what extent this reality can be considered to be a proper criterion for defining a land unit" (Le Du, 1995). This situation raises questions of automatization: to what extent are these steps possible when you take into account that certain features that are too small (hedges, plot boundaries, isolated trees and so on) are not detectable, while others, that are not visible in the landscape,

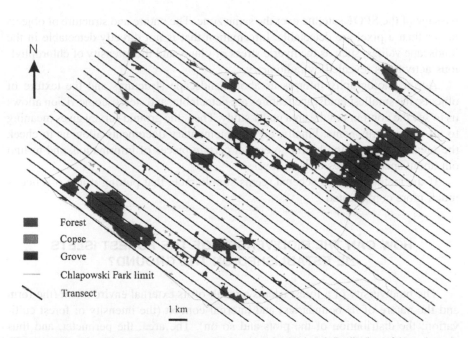

Figure 14.2 Extraction of forest islets by the masking method and distribution of the transects.

us to combine the rapidity of satellite data treatment, the precision of other sources of information, and the possibility of finally reaching some form of generalization. Moreover, this step means a trans-scalar approach to space is favorable.

To perceive the forest islets in their totality, a procedure of extraction by masking is carried out (Figure 14.2). This population of islets forms the base of the survey, from which individual islets are selected for study.

The Choice of Samples in the Land Units: From Medium to Large Scale

This multisource and multiscale approach to space integrates a maximum amount of information, mostly complementary. However, the documents have different scales, which obliges us to scale the different documents, then geo-reference them. However, the depth of this kind of approach which fits different scales together, favors the division of the land into units through the use of satellite images, from the midst of which representative samples are selected as necessary to the analysis of the next scalar level down. In this study, satellite data, aerial photographs and

passage of the SPOT satellite over the same zone. The nature and structure of objects larger than a pixel are detectable. Information that is not directly detectable in the landscape values, such as the hydric stress of plants and the intensity of chlorophyllous activity, is revealed by radiometry.

Aerial photography informs us about landscape structures and the texture of objects; the smallest landscape features are detectable, and stereoscopic vision allows us to see the contours of the area under study. Then, the ground survey gives meaning to all these components. Finally, the complementary documents allow us to check the meaning of certain objects, to extract certain pieces of information, and to find out about specific types of land cover.

If the methodological schema seems simple, note that putting it into practice is much more complicated.

HOW CAN THE BIODIVERSITY OF THE FOREST ISLETS BE ESTABLISHED ON THE GROUND?

The biodiversity of a forest islet is a result of its external environment (the form and the nature of its boundaries) and internal context (the intensity of forest cultivation, the distribution of the plots and so on). The area, the perimeter, and thus also the density of each forest islet are determining factors when it comes to the appreciation of their characteristics.

Only implementing a ground survey system can allow us to characterize the biodiversity of these islets. This step comprises several stages: first, a census of the population must made and is done with the help of isolation of the forest islets from their context in order to make their spatial distribution, shape, and size stand out. Then certain islets are selected before the ground survey procedure is launched. Finally, this step as a whole requires some adjustments, e.g., the choice of transect as a sampling unit, the direction, the spacing, the selection of the individual islets surveyed.

The Choice of Sampling Unit: The Transect

With regard to areal frame sampling, if one wishes to describe the totality of land-cover types, it is necessary to use the technique of nonaligned systematic sampling of the areas under survey in order to make up the sample (Godard, 1991; Eastman, 1995). This technique ensures a uniform distribution of the sample over the whole of the territory, while avoiding any bias caused by the regularities to be found in the landscape. On the other hand, when we are concerned with just one type of land cover, characterized by objects that are generally small, compact, and

in measuring the distance of each change in physiognomy, or even of facies, encountered in crossing a forest islet, from the starting point of a transect. The nature of the point of contact at the entrance to the wood is localized and then researched, as is also the case for the point of exit. Then, each transition inside the wood is measured in relation to the entrance point. The homogenous zones on the one hand, and on the other the zones of transition, are researched with regard to their structure, strata, dominant components, abundance-dominance of the main species, management method, and so on.

In addition to the qualification and quantification of the transitions, the transect method allows us to evaluate the different facies, their distribution and number, all in the absence of previous knowledge of the ground. It also allows us to carry out a few ratio calculations, and therefore also calculation of surface area for the different facies encountered. It is also acknowledged that the transect method ensures minimal but sufficient control for the nonsupervised remote sensing treatments (Fournier and Gilg, 1985).

On the other hand, still in the framework of remote sensing treatments, the transect method provides few informative pixels; the ground survey can, in fact, provide only information about a narrow strip of land. Because of this fact, it is difficult to allocate facies to pixels, although this allocation is necessary for the initialization of supervised treatments. Finally, with the transect method, it is particularly important to take into account problems of progression and the avoidance of obstacles in facies-pixel allocation. Recording the trajectory with the aid of a GPS (Global Positioning System) type satellite-positioning instrument is particularly recommended.

Although it is not easy to match the ground and satellite images with the transect method, the possibility of perceiving the transitions (lengths and shapes) and of indicating between which facies they occur means that we have retained this method of exploring the land. In order to put it into practice, it was necessary to determine the azimuth of the transects, their spacing and their number.

Orientation of the Transects

The risk of any systematic method is to fall foul of the regularity of the landscape which may not have been detected when sampling was carried out, leading to the over-representation of certain types of land cover in relation to others. To avoid this pitfall, we chose to test the main directions followed by the landscape. These directions are followed according to the way the forest is divided into plots, and to the felling areas, the tracks and so on.

This information is included in the satellite images of the zone under study. In order to extract it, we used Fourier's transformation analysis, which transposes the

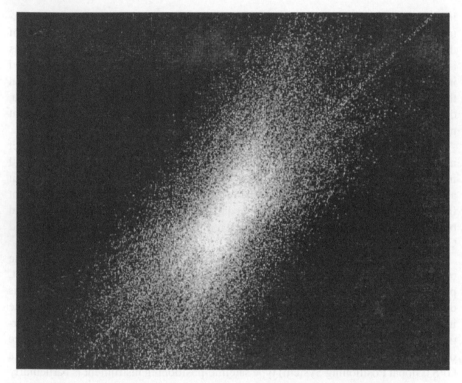

Figure 14.3 Fourier transformation applied to the forest islets of General Chłapowski's Agro-ecological Park (Greater Poland).

After having determined the direction of the transects, their spacing and their number remain to be defined before moving on to the ground phase.

Transect Number

We have previously stressed the relation that exists between biodiversity and the density of the forest islet. The latter is not easy to see, either on the ground or through images. It is therefore necessary to use other parameters, correlated to density, that are more easily perceived. Among these, the size of the forest islets is probably the simplest. It remains to be seen if, for General Chłapowski's Agro-ecological Park (our study zone), the size and the density of the forest islets are sufficiently correlated for it to be possible to substitute one for the other. If the correlation is judged to be sufficient, it will then be possible to stratify the population of the forest islets in three classes (small, medium, and large islets), then to sample

with a total surface area between 5 and 50 ares; groves, with a total surface area between 50 ares and 4 ha; and forests, with a surface area of above 4 ha (IFN, 1985). Moreover, the spatial resolution used, of 10 m width per pixel, does not gives us any information about sparse trees or any surface areas less than 5 ares.

In fact, the extraction of forest islets had to be carried out using the same satellite data as that used to calculate Fourier's transformation (see above). Because of the restrictions of time, these extractions were carried out on digitized topographical maps with a 1/50,000 scale. This situation calls for two comments:

- First, that there is a time difference — the last update of the maps (in 1992) precedes the satellite data (from 1997), which is practically contemporary with the ground data. Some differences are thus to be expected between the ground survey and the digitized data.
- Second, due to problems related to cartographic representation, the existence, the size and the localization of the forest islets are subordinated to the presence of other cartographic objects that are superimposed upon them, perhaps explaining why the extraction of the forest islets left only objects of 9 pixels width, in other words, 9 ares.

In that we worked with the Idrisi software, the density factor (I_c), taken from Claude Collet's book (Collet, 1992, p. 133), had to be adapted to the features of a squared grid. In our example, the following formula is used:

$$I_c = \sqrt{\frac{A_z}{\left(\frac{P_z}{4}\right)^2}}$$

with A_z, standing for the area of the islet, P_z for the perimeter of a square object of the same area of the islet A_z. Theoretically, it varies from the value 1, which represents a square object of maximal density, to the value 0, which is an object without surface area of an infinite length. In practice, certain wooden corridors can have a factor close to 0. Having made these comments, it is now possible to analyze the relationship existing between the size and the density of forest islets at the heart of General Chłapowski's Agro-ecological Park.

If forests (in the IFN's sense of the word) are excluded in order to keep only forest islets, a linear relation of a relatively strong intensity (R = 0.7) can be made clear between surface area and density, which is significant and carries a risk of error less than 1 in 1000. It therefore seems useful to try to stratify the islets according to whether they are small, medium, or large.

In order to do this categorization, the density of the islets was recoded according to their size, using the IFN criteria described above. This recoding proved to be very

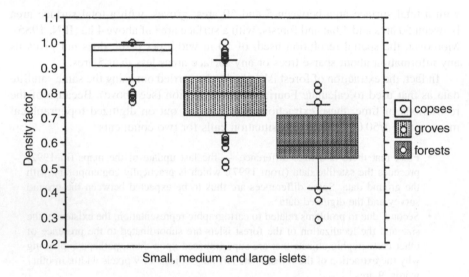

Figure 14.4 Density of the forest islets when divided into three size groups.

The stratification of the islets gives rise to the following distribution within the limits of General Chłapowski's Agro-ecological Park: 101 copses, 94 groves, and 34 forests. But the samples within each stratum must still be provided.

Distribution of the Transects

Where the spacing of the transects is concerned, satellite images can, once again, be useful. Fourier's transformation indicates the direction of the highest spatial recurrences (Figure 14.3), but does not give us a direct view of the rate (the distance) of the regularity relating to this axis. Methods coming under the heading of variographic analysis were successfully put into practice on a number of occasions on various study zones in order to determine the rate of sampling preceding the ground analysis (Godard, 1994; Fournier et al., 1997). They were applied over the entirety of survey zone and on the totality of the types of land cover, which was not the case for our study ground in Poland.

After overall analysis of the distance between islets, the transects are placed at regular intervals of 500 m, generating an aligned systematic sampling. In this way, 31 transects were marked (Figure 14.2) which allowed us to extract 93 transect segments (that is to say, a section of transect that crosses a forest islet). Their distribution is as follows: 22 segments for the copse stratum and 34 for the groves. Each segment crosses only one islet in its group. On the other hand, the case of the

moving from 500 m to 1000 m, exclusively for this stratum. In the end, 9 segments were examined out of the 34 islets of this stratum.

CONCLUSION

Remote sensing allows us to define the land units at the heart of which the main plant types are identifiable, and the forest islets stand out individually. Supplementing information concerning soil moisture and chlorophyllous activity, satellite images reveal variations that relate to the vertical and horizontal structures of the plant cover. The reliability of this information nevertheless depends on the quality of the ground survey and above all on the inventory protocol put into place for the collection of samples. The method of making an inventory using transects allows us to perceive the contacts between forest formations and their agricultural environment, but also the contacts and the transitions between the facies that make up the forest islets.

However, the allocation of the ground data, collected along a transect, to the pixels of satellite images is not simple, for they represent a narrow strip of information in the landscape. Yet it is necessary to make the most of it in order to establish a typology of the diversity of forest landscapes in the Chłapowski Agro-ecological Park and constitutes one of the axes of research of this project that must use ground survey and satellite images jointly. The methodology developed here can be reproduced and will be applied in France in the Gatinais region with a view to making a comparative study.

REFERENCES

Collet C. (1992). *Systèmes d'information géographique en mode image.* Presses polytechniques et universitaires romandes, coll. Gerer l'environnement, no. 7, Lausanne, 186 pp.

Eastman J.R. (1995). *Idrisi, un SIG en mode image* (Version 4.0 et 4.1). CRIF. Lausanne.

Fournier Ph., GIlg J.-P. (1985). *Télédétection et observation terrain.* Cours du GDTA. Toulouse. 22 pp.

Fournier Ph., Geroyannis H., Godard V. (1997). Analyse variographique de données satellitaires pour déterminer la taille des unités d'échantillonnage. Exemple d'un inventaire d'occupation du sol en milieu agricole et forestier. In *Annales Littéraires de l'Université de Franche-Comté. Cahiers de Géographie.* 35: 85–89.

Galochet M. (2001). Les îlots boisés, des lieux de diversité en terre de grande culture. Comparaison France et Pologne. Thèse de Doctorat en Géographie, Université de Paris IV — Sorbonne. 344 pp.

Godard V. (1991). *Utilisation conjointe de la télédétection et de l'enquete de terrain lors des inventaires d'occupation du sol. Recherche appliquée au Sahel sud-mauritanien.*

Hotyat M., Galochet M., and Liège F. (1997). Petits bois et leurs lisières dans les plaines de grande culture: « entre nature et culture ». Exemple pris dans le Gâtinais occidental, In *La dynamique des paysages protohistoriques, antiques, médiévaux et modernes.* Editions APDCA. Sophia Antipolis. 493–504.

Hotyat M., Galochet M., and Liège F. (1999). Structure et dynamique des ilots boisés: intérêt de l'étude multicapteurs et de l'analyse diachronique. In *Paysages agraires et environnement,* S. Wicherek (Ed.), CNRS éditions. Paris. 357–370.

IFN (1985). *But et méthodes de l'inventaire forestier national.* Ministère de l'agriculture, Service des forêts. Paris. 67 pp.

Le Breton J.-M. (1994). *L'Europe centrale et orientate de 1917 à 1990.* Nathan, coll. Fac. Histoire. Paris. 304 pp.

Le Du L. (1995). Images du paysage. Télédétection, intervisibilité et perception: l'exemple des Côtes d'Armor. Thèse de Doctorat, Universite de Rennes 2. 334 pp.

MacArthur R.H., Wilson E.O. (1967). *The Theory of Island Biogeography.* Princeton University Press. Princeton, New Jersey. 203 pp.

Papy M. (1992). Effets des structures agraires sur le ruissellement et l'érosion hydrique. *Bulletin de l'Association des Géographes Français* 2, 115–125.

Ramade F. (1994). *Elements d'écologie. Ecologie fondamentale.* 2nd edition. Edisciences International. Paris. 573 pp.

Robin M. (1995). *La télédétection: des satellites aux SIG.* Nathan, coll. Fac. Géographie. Paris. 318 pp.

CHAPTER **15**

Agrolandscape Ecology in the 21st Century

Gary W. Barrett and Laura E. Skelton

CONTENTS

A new century causes us to reflect on the accomplishments and problems of the past century and especially to address challenges regarding the future. In this chapter, we outline five topics (perhaps some would term them *problem areas*) that ecologists, agronomists, and resource planners need to address if sustainable agriculture is to become a reality during the 21st century. As Albert Einstein once stated, "The significant problems we face cannot be solved at the same level of thinking we were at when we created them." We feel that it is imperative that new approaches be implemented to address agricultural problems and to create opportunities at greater temporal/spatial scales. Barrett (in press) and Barrett and Odum (1998) term this new transdisciplinary approach *integrative science*.

Goodland (1995) defined sustainability as "maintaining natural capital." We sug-

to quantify the value of ecosystem services and natural capital on a large-scale basis (e.g., Costanza et al. 1997). These attempts lend evidence to our assertion that the time is right to consider and implement a new integrative approach to agriculture at the landscape scale. This chapter describes five guidelines to address this task.

NATURE AS A MODEL SYSTEM

Natural ecosystems have endured far longer than conventional agroecosystems have been in existence. Intensive-input agriculture, as currently practiced in much of the developed and developing world, represents a waste of scarce, finite resources. Intensive-input ecosystems (i.e., systems focused on a single crop or product with maximum yield as a goal) do not sustain themselves but instead rely on large amounts of labor and subsidies (fossil fuels, fertilizers, and pesticides) for production. Typically a single crop occupies a field during a single growing season (a monoculture approach to agriculture). Problems arise when subsidies are applied at one level (species or single crop) and then used without further study at another level (community, ecosystem, or landscape). Problems intensify when that single crop is planted and cultivated to maximize yield. Furthermore, traditional practices of crop rotation and allowing fields to lie fallow are not common in modern agriculture. Processes that are thought to sustain natural systems, such as natural means of pest control and detritus accumulation (Altieri and Nicholls 2000), are often absent or discouraged in modern agricultural practices.

Natural systems also contain a co-evolutionary system of checks and balances between herbivores and predators that aids in the regulation of potential pest species. Thus, insect and weed pests thrive in monoculture cropping systems where they are not out-competed by native species or consumed by predators. Lower abundance of pest populations in heterogeneous systems is most likely due to the presence of natural enemies (Karel 1991). Monocultures also attract specialist herbivores, thus providing low diversity of food sources for predators (Letourneau and Altieri 1983). Also, monocultures provide less refuge for beneficial insects, predators, and parasitoids than found in nature (Letourneau and Altieri 1983). A greater abundance of these desirable insects helps regulate populations of specialized herbivores.

Conventional tillage results in loss of topsoil, loss of wildlife habitat, and increased rates of soil erosion, among other consequences. Alternatives to conventional tillage include no-tillage, reduced tillage, and low-input sustainable agriculture (LISA). Surface litter in reduced or no-till systems contributes to water retention and cooler temperatures in topsoil (Coleman and Crossley 1996). Surface litter also provides home to soil micro- and macroinvertebrates, bacteria, and fungi that aid in decomposition of stubble left after harvesting.

found to be more abundant in no-till than in conventionally tilled cropping systems (Warburton and Klimstra 1984). Rates of decomposition are also slower in agricultural systems not subject to tillage (imitating decomposition processes in natural ecosystems); therefore, a constant supply of nutrients is available for mineralization. Arthropods and earthworms in no-till systems aid decomposition (House and Parmelee 1985). In addition to providing cover for invertebrates (House and All 1981), detritus helps to maintain cooler and moist soil conditions. Thus, there exists ample evidence that natural or polyculture systems not only require less subsidies but also have evolved regulatory mechanisms necessary to control insect pests, aid in nutrient cycling, and improve soil conditions. A challenge for the 21st century is to couple or integrate mechanisms found in natural systems with traditional cropping systems. Only an integrative approach will accomplish this goal.

AN AGROLANDSCAPE PERSPECTIVE

Agronomic research and planning have traditionally focused on the field or agroecosystem level. More recently, however, several studies have addressed interchanges of insects between agricultural and surrounding landscapes (Ekbom et al. 2000). The science of landscape ecology considers not only the development and dynamics of spatial heterogeneity but also the exchanges of biotic and abiotic resources across heterogeneous landscapes, including how this spatial heterogeneity influences biotic and abiotic processes (Risser et al. 1984). Traditionally, a single field (agroecosystem) approach was employed to address questions and to solve problems related to problems or concepts such as pest management, restoring or conserving biotic diversity, reducing subsidy input, or improving crop yield (Barrett 2000). Thus, a new landscape or regional perspective is warranted.

Just as societies learned that biotic diversity cannot be protected or conserved by a single-species approach (Salwasser 1991), we predict that resource managers and agronomists will learn during the 21st century that a single farm/field (agroecosystem) approach cannot, among other larger scale challenges, sustain agricultural productivity, reduce regional or watershed eutrophication, or regulate pest species. Rather, an agrolandscape approach is needed in which landscape elements (patches, corridors, and the landscape matrix) are patterned and managed to optimize factors such as insect pest control, biotic (genetic, species, and habitat) diversity, soil restoration, net primary productivity, nutrient retention, and landscape connectivity (Barrett 1992, 2000). This emerging field of study, based on the concepts of sustainability and linking ecological capital with economic capital (Barrett and Farina 2000), should provide solutions to such challenges as ecologically-based pest management (National Research Council 2000), ecosystem stress and crop yield relationships (Odum and Barrett 2000), role of corridors in helping to regulate arthropod popu-

nitrogen-fixing cover crops) have in many cases reduced pest damage, created habitat for wildlife, and decreased the use of subsidies (pesticides, fossil fuels, and commercial fertilizers). This landscape perspective based on transdisciplinary approaches is likely to continue, deriving from nature that mutualistic, rather than competitive, mechanisms increase as systems become more complex. As societies mature (i.e., reach the human carrying capacity) during the 21st century (Barrett and Odum 2000), it is anticipated that these mutualistic interactions will accompany the maturation process.

REDUCING EUTROPHICATION AT THE LANDSCAPE SCALE

Agricultural practices have modified the nitrogen cycle found in nature. Although nitrogen is typically cycled in a rather closed manner in nature, nitrogen fertilizers applied to crops in massive amounts, to stimulate plant growth thus maximizing crop yield, are now being lost to agricultural systems in great amounts. Fertilizers are often applied in excess, causing nitrogen from commercial fertilizers to be released to the environment.

A consequence of excess nitrogen at the landscape scale is contamination of watersheds and ground water. Nitrogen not consumed by plants in the form of nitrates seeps into ground water or is released into the atmosphere. This typically limiting resource (when combined with phosphorous in fresh water habitats) causes eutrophication (the growth of toxic algal blooms in lakes). Eutrophication is known to limit the survival of aquatic life and decrease biodiversity (Vitousek et al. 1997). Algal blooms create lakes that are uninhabitable to most forms of life. Excess nitrogen not only acidifies ground water, lakes, and streams but also acidifies soil, which drastically changes the microclimate for soil fauna, thus making plant survival difficult. Another concern is the possibility of decreased biotic diversity of plant species. Opportunistic species that respond well to increased nutrient input typically become dominant and suppress native species that do not grow well when exposed to excess nitrogen (Vitousek et al. 1997).

Unfortunately, mature ecosystems (e.g., forests, prairies, and wetlands) are sometimes converted to agricultural fields, thus altering ecosystem processes and functions. An important function of wetlands, for example, is their ability to denitrify nitrates. Water is released from wetlands very slowly so that less fixed nitrogen is passed to rivers and estuaries. Therefore, societies must preserve wetlands. Forested lands and riparian zones also must be maintained to divert excess nitrogen from cropland (Vitousek et al. 1997). Thus, the patterns and heterogeneity of ecosystems within a landscape are key to regulating the nitrogen cycle.

however, has been paid to academic fragmentation (see Barrett in press). Numerous new interfaced fields of study emerged during the latter half of the twentieth century, such as restoration ecology, ecological engineering, landscape ecology, ecological toxicology, and conservation biology. These fields of investigation, including agro-landscape ecology, have contributed to a clearer understanding of our natural world, including a deeper understanding of the relationships and emerging properties among various disciplines in the physical, biological, and social sciences.

Unfortunately, however, institutions of higher learning have failed to promote and establish mechanisms or structures to administer these interdisciplinary fields of study. Within this context Barrett (in press) has suggested a transdisciplinary scientific field of synthesis termed *integrative science*, based on the noosystem concept to integrate, rather than continue to fragment, this academic process.

Along those lines, we suggest a 2-1-2 (5-year) undergraduate degree — 2 years of liberal arts; 1 academic year internship; 2 years focused on either an applied or basic science major. The internship would permit undergraduates interested in areas such as agroecosystem or agrolandscape ecology to understand better the scale of and challenges associated with seeking solutions to these problems, as well as the opportunities afforded those with this holistic perspective. For example, Barrett et al. (1997) stressed the need to more fully understand processes (e.g., energetics, regu-lation, diversity, and evolution) that transcend all levels of organization. Unfortu-nately, most courses and disciplines focus more on reductionist science at the lower levels of organization (molecule, cell, and organism) rather than on holistic science at higher levels of organization (ecosystem, landscape, and world). An internship option should help elucidate the need to merge basic and applied science, to wed disciplinary and interdisciplinary approaches, and to appreciate how major processes transcend all levels of organization.

This approach will help to ensure that fields such as agronomy and ecology become integrated during the 21st century. It will also demonstrate why net energy and net economic currency will lead to sustainable agriculture (and a sustainable landscape) rather than to maximize crop yield as a societal goal. Barrett (1989) notes that a sustainable society is characterized by the virtues of preventive medicine, critical thinking, and problem solving on a landscape scale. A citizenry educated in this manner will focus on concepts such as net energy and net economic currency rather than on goals such as maximum growth and agricultural productivity.

LINKING URBAN-INDUSTRIAL AND NATURAL LIFE-SUPPORT SYSTEMS

One of the greatest challenges of the 21st century will be to link urban-industrial

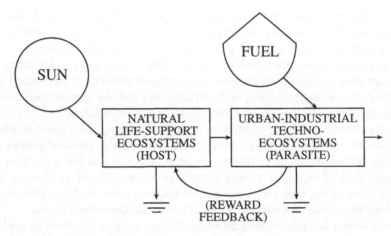

Figure 15.1   Model illustrating the need to link natural life-support ecosystems with urban-industrial ecosystems, including a reward-feedback loop. (Modified after Odum 2001.)

Systems can also be classified based on the ratio of energy produced by primary production (P) to energy used for respiration or system maintenance (R). Natural and agricultural systems are termed autotrophic where P/R > 1. In contrast, urban systems are heterotrophic systems where P/R < 1. Barrett et al. (1999) defined sustainable systems as those systems or landscapes where long-term P/R ratios equal 1. Thus, to meet this definition it is imperative to link urban-industrial (heterotrophic) systems with natural (including agricultural) life-support (autotrophic) systems at the landscape scale (Barrett et al. 1999 describe this developmental process).

Naveh (1982) and Odum (2001) refer to these fuel-powered urban-industrial systems as *techno-ecosystems*. A modern city, for example, is a major techno-ecosystem. Techno-ecosystems represent energetic *hot spots* on the landscape that require a large area of natural and agricultural countryside to support such systems. Wackernal and Rees (1996) note that techno-ecosystems have a very large "ecological footprint."

Figure 15.1, modified from Odum (2001), depicts the need to link urban-industrial and techno-ecosystems with natural life-support ecosystems. The analogy to a host-parasite relationship describes how these two entities, we hope, will co-evolve. It is important to note that if the parasite (city) takes too much from the host (life-support system), both will die. Cairns (1997) is optimiztic that natural and techno-ecosystems will co-evolve in a mutualistic manner. A landscape perspective, including the development of reward feedback mechanisms (Figure 15.1) between these two systems, should lead to the mutual linkage of urban and agricultural systems —

CONCLUSION

This chapter afforded us the opportunity to reflect on the agricultural enterprise during the past few decades, then to reflect upon agriculture during the 21st century. Crosson and Rosenberg (1989), in a special issue of *Scientific America* (September 1989, "Managing Planet Earth") noted that "agricultural research will probably yield many new technologies for expanding food production while preserving land, water and genetic diversity. The real trick will be getting farmers to use them."

Essentially, Crosson and Rosenberg were correct. For example, farmers, ecologists, policy makers, and resource managers now debate the costs and benefits of transgenic crops (Marvier 2001), while societies remain concerned about land use practices, water quality, and biotic diversity. Interestingly, it is not only the farmers and practitioners who need to modify human behavior, but also society as a whole. Society appears to know and understand the benefit derived from quality landscape health, protection of scarce biotic and abiotic resources, and adaptation of a sustainable approach to food production. A primary goal during the 21st century must be to use this knowledge and understanding to develop not only a sustainable approach to agriculture, but to become a mature and sustainable society as a whole. The time appears right to take a major step in that direction. Children and grandchildren will likely not forgive unless we use this knowledge and understanding on their behalf in the very near future.

ACKNOWLEDGMENTS

We thank Lech Ryszkowski for inviting us to contribute to this book. We especially thank Terry L. Barrett for her editorial comments and help with the preparation of this manuscript. The first author is indebted to the numerous graduate students and postdoctoral fellows at Miami University of Ohio who, over the years, encouraged him to reflect on sustainable agriculture at the ecosystem and landscape scales.

REFERENCES

Altieri, M. A. and Nicholls, C. I. (2000) Applying agroecological concepts to development of ecologically based pest management strategies, *Proceedings of a Workshop Board on Agriculture and Natural Resources Professional Societies and Ecologically Based Pest Management*, National Academy Press, Washington, D.C., 14–19.

Barrett, G. W. (1989) Viewpoint: a sustainable society. *BioScience*, 39, 754.

Barrett, G. W. (1992) Landscape ecology: designing sustainable agricultural landscapes, in Integrating Sustainable Agriculture, Ecology, and Environmental Policy, Olson, R. K.,

Barrett, G. W. Closing the ecological cycle: the emergence of integrative science, *Ecosys. Health,* in press.

Barrett, G. W., Barrett, T. L., and Peles, J. D. (1999) Managing agroecosystems as agrolandscapes: reconnecting agricultural and urban landscapes, in *Biodiversity in Agroecosystems,* Collins, W. W. and Qualset C. O., Eds., CRC Press, Boca Raton, FL, 197–213.

Barrett, G. W. and Farina, A. (2000) Integrating ecology and economics, *BioScience,* 50, 311–312.

Barrett, G. W., Peles, J. D., and Odum, E. P. (1997) Transcending processes and the levels-of- organization concept, *BioScience,* 47, 531–535.

Barrett, G. W. and Odum, E. P. (1998) From the President: Integrative Science. *BioScience,* 48, 980.

Barrett, G. W. and Odum, E. P. (2000) The twenty-first century: the world at carrying capacity, *BioScience,* 50, 363–368.

Cairns, J. (1997) Global coevolution and natural systems and human society, *Revisa Sociedad Mexicana de Historia Natural,* 47, 217.

Coleman, D. C. and Crossley, D. A., Jr. (1996) *Fundamentals of Soil Ecology,* Academic Press, San Diego.

Collins, W.W. and Qualset, C.O. (1999) *Biodiversity in Agroecosystems,* CRC Press, Boca Raton, FL.

Costanza, R., et al. (1997) The value of the world's ecosystem services and natural capital, *Nature,* 387, 253–260.

Crosson, P. R. and Rosenberg, N. J. (1989) Strategies for agriculture, *Sci. Am.,* New York.

Daily, G. C., Alexander, S., Ehrlich, P. R., Goulder, L., Lubchenco, J., Matson, P. A., Mooney, H. A., Postel, S., Sch Tilman, D., and Wodwell, G.M. (1997) Ecosystem services: benefits supplied to human societies by natural ecosystems, Issues in Ecology, 2, Ecological Society of America, Washington, D.C.

Ekbom, B., Irwin, M. E., and Robert Y., Eds., (2000) *Interchanges of Insects between Agricultural and Surrounding Landscapes,* Kluwer Academic Publishers, Dordrecht, the Netherlands.

Goodland, R. (1995) The concept of environmental sustainability, *Annu. Rev. Ecol. Systematics,* 26, 1–24.

Harris, L. D. (1984) *The Fragmented Forest,* University of Chicago Press, Chicago, IL.

Hendrix, P. F., Parmelle, R. W., Crossley, D. A., Jr., Coleman, D. C., Odum, E. P., and Groffman, P. M. (1986) Detritus food webs in conventional and no-tillage agroecosystems, *BioScience,* 36, 374–380.

Holmes, D. M. and Barrett, G. W. (1997) Japanese beetle (*Popillia japonica*) dispersal behavior in intercropped vs. monoculture soybean agroecosystems, *Am. Midland Natur.,* 137, 312–319.

House, G. J. and All, J. N. (1981) Carabid beetles in soybean agroecosystems, *Environ. Entomol.* 10, 194–196.

House, G. J. and Parmelee, R. W. (1985) Comparison of soil arthropods and earthworms from conventional and no-tillage agroecosystems, *Soil Tillage Res.,* 5, 351–360.

Karel, A. K. (1991) Effects of plant populations and intercropping on the population patterns

Letourneau, D. K. and Altieri, M. A. (1983) Abundance patterns of a predator, *Orius tristicolor* (Hemiptera: Anthocoridae), and its prey, *Frankliniella occidentalis* (Thysanoptera: Thripidae): habitat attraction in polycultures versus monocultures, *Environ. Entomol.* 12, 1464–1469.

Marvier, M. (2001) Ecology of transgenic crops, *Am. Scientist*, 89, 160–167.

National Research Council. (2000) *Professional Societies and Ecologically Based Pest Management*, National Academy Press, Washington, D.C.

Naveh, Z. (1982) Landscape ecology as an emerging branch of human ecosystem science, *Adv. Ecol. Res.*, 12, 189–237.

Odum, E. P. (1997) *Ecology: A Bridge between Science and Society*, Sinauer Associates, Sunderland, MA.

Odum, E. P. (2001) The techno-ecosystem, *Bull. Ecol. Soc. Am.*, 82, 137–138.

Odum, E. P. and Barrett, G. W. (2000) Pest management: an overview, in *National Research Council Report, Professional Societies and Ecologically Based Pest Management*, National Academy Press, Washington, D.C.

Risser, P. G., Karr, J. R., and Forman, R. T. T. (1984) Landscape ecology: directions and approaches, Special Publication 2 in Illinois Natural History Survey, Champaign, IL.

Salwasser, H. (1991) Roles and approaches of the USDA Forest Service, in *Landscape Linkages and Biodiversity*, Hudson, W. E., Ed., Island Press, Washington, D.C.

Thies, C. and Tscharntke, T. (1999) Landscape structure and biological control in agroecosystems, *Science*, 285, 893–895.

Vitousek, P. M., Aber, J. D., Howarth, R. W., Likens, G. E., Matson, P. A., Schindler, D. W., Schlesinger, W. H., and Tilman, D. G. (1997) Human alteration of the global nitrogen cycle: sources and consequences, *Ecol. Appl.*, 7, 737–750.

Wackernal, M. and Rees, W. (1996) *Our Ecological Footprint: Reducing Human Impact on the Earth*, New Society Publishers, Cabriola Island, British Columbia, Canada.

Warburton, D. B. and Klimstra, W. D. (1984) Wildlife use of no-till and conventionally tilled corn fields, *J. Soil Water Conserv.*, 39, 327–330.

CHAPTER **16**

Agriculture and Landscape Ecology

Lech Ryszkowski

CONTENTS

OVERVIEW

It can be argued that objectives for agriculture are inherently different from those for nature protection, and, therefore, it is unwise to rely on the agricultural management of land as the primary means for environmental protection. As agricultural productivity expanded, natural landscapes were converted to rural landscapes composed of cultivated fields, pastures, settlements and patches of nonarable habitats. Farmers, to a great extent, have been responsible for the development and stewardship of the landscape. Agricultural activity is focused on production; therefore, all available resources were used to maximize yields. This attitude led to neglecting the importance of all nonproductive elements of the rural landscape and, because of a lack of knowledge, resulted in the widespread treatment of small patches of wetlands, mid-field ponds, hedges, groves, and so on as idlelands, or habitats of no use.

There seems to be little question that, because of this attitude, as cultivated fields were developed, to increase production, many nonproductive landscape elements

the side effects from farming were overcome — which was facilitated by loss of landscape diversity — environmental threats started to appear. Accelerated rates of water and wind erosion, diffuse pollution of water reservoirs, lowered ground water levels, increased frequency of floods, broader variation in microclimatic factors, outbreaks of crop pests, and many other phenomena brought by farming intensification began to be observed worldwide.

Two reports on changes in the European environment (Stanners and Bourdeau 1995, European Environment Agency 1998), requested by the environment ministers for all of Europe, clearly documented environmental deterioration from agriculture, industry, urbanization and other human activities. The same situation in other parts of the world is evidenced by a variety of publications. Although there has been much more concern shown for environmental threats caused by industry or urbanization, agricultural activities have played an important role in overall human impact on the environment. Ground water quality can be decreased by high concentrations of nitrates brought by the use of artificial fertilizers and manure on cultivated fields. The spatial scale of these impacts can be enormous. Stanners and Bourdeau (1995), using the figures provided by model computations, reported that the nitrate concentrations levels in ground water from 87% of the agricultural land in Europe exceeded $25 \text{ mg} \cdot \text{l}^{-1}$ (the European Union target value) due to leaching from soils. The drinking water standard of $50 \text{ mg} \cdot \text{NO}_3^- \cdot \text{l}^{-1}$ was exceeded in 22% of the area. Light soils, high inputs of nitrogen fertilizers, and the loss of small nonarable habitats, such as permanent vegetation patches and mid-field wetlands and ponds, are the main factors leading to high nitrate passing loads into aquifers. Contamination of water by pesticides leached from soils, erosion, soil compaction, desertification, impoverishment of biological diversity, pest outbreaks, and other threats to agroecosystems have been recognized not only by scientists but more recently even by decision makers.

THE NEW COMMON AGRICULTURAL POLICY (CAP)

These environmental considerations became an important issue during the formulation of the new Common Agricultural Policy (CAP) in the 1990s by the European Union (EU) within the context of Agenda 2000, issued in 1997, which included a revised EU policy on development. The CAP, which makes up one of the sectorial policies of Agenda 2000, proposes that farmers should observe a minimum standard level of environmental protection measures defined in the codes of good agricultural practices and that any additional environmental services should be financed by society through agro-environment programs.

Enacting such administrative regulation is a good example of the formulation of

to environmental disturbances. Distortion of landscape functions that may be caused by farming intensification and lead to altered energy fluxes and organic matter cycling patterns should either (1) be counteracted by mechanisms (negative feedbacks) that diminish inputs that cause threats or (2) be controlled by rearranging the system structure in such a way that the distortion itself is arrested. It is important to note that because the landscape distortions are caused by farmers, the elaboration of negative feedback mechanisms should encompass both environmental and societal components of the landscape.

The development of the new CAP has relied mainly on the first method to counteract increased environmental deteriorations. The CAP reform marks an important step toward implementing a sustainable agriculture, thereby enabling the transition from identification of a crisis to its control. In other words, it is a transition from recognizing the causes of environmental threats and observing their impact on the economy or living conditions of the general public, which stimulates calls for the administration to take action, to undertake effective measures to control the actual threats. This transition was realized by involvement of administrative structures of society in the control of threats. The sooner we understand that scientific recognition of environmental impacts, public pressure on administration to eliminate such threats, and the forging of control regulations by decision makers are a normal part of learning to deal with crises, the easier it will be to design policies of sustainable development. The efficiency of such control mechanisms depends very much on the elaboration of the ways of opening channels of communication between those who understand the process in question and the general public and decision makers.

The environmental problems created by agriculture are handled within the context of the new CAP mainly by imposing restrictions on agricultural activities — for example, the Nitrates Directive issued by the Commission of the European Communities (CEC) in 1991 — or by protection of biota as well as habitats against human interferences — for example, the Birds Directive issued in 1979 or the Habitats Directive of 1992. Thus, the concerns are directed toward mechanisms constraining intensity of inputs that are causing the threat (elaboration of negative feedback loops concept).

Implementation of the Nitrates Directive has been rather unsuccessful (Com 1999). One reason is probably that the Nitrates Directive does not mandate compensation to farmers because of the "polluter pays" principle, which requires farmers to observe good agricultural practices without payment. This situation demonstrates that the efficiency of negative feedback loops that encompass societal structures is to some extent tied to the exchange of capital.

The main restrictions of the Nitrates Directive are connected with further elaboration of limits on amounts of fertilizers used and imposed requests concerning

SUSTAINABLE AGRICULTURE

The most important achievement of the new CAP is its recognition of landscape properties for the sustainable development of agriculture (Com 1999). The study of landscape properties leads to disclosure of links between natural and societal processes that change landscape structures and also provide opportunities for economic activities. A landscape has multifunctional characteristics that make it significant as a material object but also creates a sensitive relationship between humans and their environment. The sustainability concept is based on three main pillars — economic, social, and environmental — which should be balanced in such a way that change in one pillar does not undermine the whole system. Thus, production increases should complement a higher standard of life and also sustained improvement in the quality of the environment. Maintenance of such a situation could be achieved by two systems of management. The first system relies on development of more environmentally friendly technologies that can be used when production is increasing. The second system focuses on building buffer capacities in the system that could neutralize any negative side effects of the increased production.

The codes of good agricultural practices describe ways to introduce environmentally friendly technologies, but enforcement requires administrative regulations. Maintenance of good soil quality and soil fertility can provide long-term economic gains. Cultivation technologies that aim to increase storage capacity of soil and to preserve or improve its physical, chemical, and biological properties are of the utmost importance for successful development of sustainable agriculture.

In the arsenal of available technologies one can include diversified crop rotation patterns, minimizing exposure of bare soil, which constrains erosion rates and simultaneously prevents accumulation of hazardous biochemical compounds (e.g., phenolic acids) that inhibit growth of cultivated plants (Ryszkowski et al. 1998). Plowing green manure into soil as well as frequent application of farm manure restores and preserves the organic matter in soils that have the most importance for sustaining soil fertility over long periods. Increased integration of crop and animal production within the farm, leading to tight recycling of nutrients and therefore less loss of nutrients to ground and surface water reservoirs (Granstedt 2000), is another option for farm-level control of environmental problems. Integrated methods of pest and disease control can reduce the use of pesticides and no-till technologies and can restore the physical and chemical properties of soils as well as stimulate rich communities of soil microorganisms.

Nevertheless, as shown by Ryszkowski and Jankowiak in Chapter 2, all those technologies cannot eliminate problems caused by agriculture; they can only reduce them. Agroecosystems can be managed for simplification of their structure which facilitates higher yields but also decreases efficiency of their recycling processes and stimulates diffuse pollution in an entire region. Water and microclimate man

can influence the rates and direction of basic processes in the landscape, but can be accomplished only on a scale much larger than a single farm.

This analysis leads to an important conclusion. The farm is an autonymous or self-sufficient unit with respect to its economic processes. But a farmer's sovereignty over the natural processes is not great. Thus, there is a considerable functional discrepancy between a farm as an economic enterprise and its role in environmental protection programs. The activities within the farm may be on too small a scale to curb environmental deterioration. This does not mean, however, that environmentally friendly technologies are not important. On the contrary, they are very important, but their execution needs to be coordinated on a large spatial scale, including many farms, to achieve significant impact on a range of the natural processes in question. This situation calls for coordination of farmers' activities at the landscape level. In addition, the introduction of biogeochemical barriers may interfere with economic benefits obtained by individual farmers; therefore, the decision on their optimal localization should be undertaken from the perspective of protecting the interests of all of society (the societal pillar of sustainability) rather than those of the individual farmers. The problems and conflicts arising in such situations were analyzed well by Hardin (1968).

LANDSCAPE MANAGEMENT IN DEVELOPMENT OF SUSTAINABLE AGRICULTURE

The problems of spatial layout and design of various landscape elements can be addressed by spatial planning, which recently became one of the new dimensions of the EU policies of development. In order to reconcile the social and economic claims for spatial development with an area's ecological and cultural functions, the program for the European Spatial Development Perspective (ESDP 1999) was proposed by ministers responsible for spatial planning in the EU member states. With respect to agriculture, this political initiative aims not only at delimiting regions having different production intensity or specific problems (such as excessive specialization or concentration of production) but also indicates the need for recognizing the interactions between agriculture and other sectorial activities in order to find synergies that may result. The proposed ESDP policy based on those findings should lead to strategies for minimizing conflicts between different land users, for example between farmers and nature conservationists, by implementing the Natura 2000 program on a wide network of protected areas. According to the ESDP, "a broader land-use policy can provide the context within which protected areas can thrive without being isolated."

Recognition of the importance of spatial dimensions in the relationships between

One can expect that the development of the ESDP policy will also influence the CAP expanding arsenal of measures for controlling environmental deterioration from agricultural intensification. The discussion of the control measures of environment deterioration proposed by the CAP, presented at the beginning of this chapter, showed that regulations rely mainly on restricting the sources of environmental threats (e.g., Nitrate Directive). Methods for building resistance to threats at the landscape level are still not a broadly used operational option, although Recommendation Nr. R(94)6, on sustainable development and use of the countryside with a particular focus on safeguarding wildlife and landscape (issued by the Council of Europe in 1994) proposes structuring landscape with hedges, ponds, and buffer zones. The aim of those activities is the stimulation of natural processes of landscape self-purification and control of diffuse pollution. Nevertheless, to date, there is no EU directive for enhancing landscape resistance to environmental threats. Such regulations will rein-force the arsenal of current regulations with a new category of measures that allow some intensification of production with side effects that could be neutralized by natural processes enhanced through landscape management.

This new approach to control of environmental threats has a very important implication for the development of sustainable agriculture. More intensive agricul-tural production results in higher yields, which in turn provide opportunities for improved economic revenue. This aspect is of special interest to farmers for whom a positive link between income and environmental threat control measures would be more persuasive than would proposals restricting production.

This statement is supported by the results of a census carried out by the Research Center for Agricultural and Forest Environment in Poznań, Poland. Farmers were asked to identify what they considered the most important environmental threats. The majority indicated those threats that directly impact yields, while those threats with an indirect effect on production were either neglected or recognized only by those farmers with a higher education. For example, 71% of 168 respondents clearly recognized effects of drought, which decreased yields, while only 7% indicated existing water pollution problems as important. It seems, therefore, that control measures for environmental deterioration that do not restrict production and revenue constitute a better strategic approach than those that establish such restrictions.

The crucial point for applying a landscape approach to controlling environmental degradations is that side effects of agricultural production intensification should not exceed the buffering capacities of the landscape. Observation of that condition requires proper education of farmers and development of extension services qualified in landscape ecology.

Landscape ecologists showed some time ago that the mosaic structure of land-scape can stop disturbances (e.g., Forman and Gordon 1986). The review of recent ecological studies presented in this book clearly shows that many natural processes

Cultivated fields can be considered "landscape ovens," heating air, which could even influence mesoscale air circulation due to increased rates of heat convection. Shelterbelts in a simplified agricultural landscape are one of the best tools by which one can manage not only microclimatic conditions but also water cycling in the landscape. The tradeoff between those processes depends on the kind of biogeochemical barrier, its structure, and the pattern of the network.

A network of biogeochemical barriers can also efficiently control diffuse water pollution if combined with management practices aimed at the removal of accumulated plant debris, which rapid decomposition under favorable microclimatic conditions can turn a biogeochemical barrier from a sink to a source of contaminants. The biological processes of nutrient uptake, denitrification, assimilatory reduction of nitrates, and decomposition of organic matter play very important roles in the process of nitrogen cycling in the landscape. It seems that processes of biological recycling within the biogeochemical barrier play a substantial role in processes that determine the control of the barrier for modification of adverse matter cycles in landscapes.

A mosaic of refuge sites in the agricultural landscape houses much richer and more diverse plant and animal communities than does a uniform landscape that is composed of large cultivated fields and that largely lacks other nonagricultural landscape components. Analyses indicate that a more diversified landscape structure stimulates the appearance of a larger number of taxa as well as greater density and biomass of the animal communities. In nonarable habitats sheltered by the biogeochemical barriers, for example, mid-field ponds separated from fields by meadows or hedges, even protected or rare species can thrive in the agricultural landscape. The success of biodiversity protection by the biogeochemical barriers depends on farming operations. Thus, a tradeoff between farming intensity and pattern of refuge sites must be established in order to reconcile agricultural activities with biodiversity protection.

Elaboration of the methods of decision making in regard to proposed land uses constitutes the important step reported in this book to communicate landscape ecology achievements to policy makers. Development of computer tools to model the environmental consequences of proposed land-use changes can facilitate the incorporation of landscape ecology guidelines into agricultural policies. The same is true for elaboration of operational guidelines for managing biogeochemical barriers, e.g., instructions on management practices for shelterbelt maintenance. Such modeling can also help efficiently translate landscape ecology into common practice, such as attempts to introduce riparian forest buffers into the U.S. program for water quality protection.

CONCLUSION

348   LANDSCAPE ECOLOGY IN AGROECOSYSTEMS MANAGEMENT

It seems likely that agrolandscape ecology will be an important discipline that will influence agricultural practices in the 21st century. These tasks are not easy to accomplish. First, for example, restricting contamination of water reservoirs and building environmental resistance require considerable money. Hence, poor farmers will have difficulty implementing effective environment protection programs. Second, farmers are still not sufficiently educated to make profitable use of the achievements of contemporary ecology to restrict nonpoint pollution, save water, or control erosion processes through diversifying landscapes. Environmentally friendly cultivation technologies recently proposed by various proponents of alternative agriculture indicate a number of activities that are very sound from an ecological point of view. But the efficiency of environmental protection of rural areas could be enhanced even further if field practices could be supplemented by structuring landscapes with various biogeochemical barriers, which would not only augment control of nonpoint pollution, save water, and modify microclimatic conditions but also conserve biodiversity.

REFERENCES

Com 1999. Direction toward sustainable agriculture. Commission of the European Communities. Brussels. 30 pp.

ESDP. 1999. *European Spatial Development Perspective.* Office for Official Publications of the European Communities. Luxemburg. 82 pp.

European Environment Agency. 1998. *Europe's Environment: The Second Assessment.* Elsevier Science Ltd. Oxford. 293 pp.

Forman R. T. T., Godron M. 1986. *Landscape Ecology.* Wiley & Sons. New York. 619 pp.

Granstedt A. 2000. Increasing the efficiency of plant nutrients recycling within the agricultural system as a way of reducing the load to the environment — experience from Sweden and Finland. *Agric. Ecosys. Environ.* 80: 169–185.

Hardin G. 1968. The tragedy of the commons. *Science* 162: 1243–1248.

Ryszkowski L., Szajdak L., Karg J. 1998. Effects of continuous cropping of rye on soil biota and biochemistry. *Crit. Rev. Plant Sci.* 17: 225–244.

Stanners D., Bourdeau P. 1995. *Europe's Environment.* European Environment Agency. Copenhagen. 676 pp.

Index

A

Abandoning cultivars, 188
Aboveground insects, mosaic landscape, 203
Accipiter gentilis, in Turew shelterbelts, 208
Acidification, agriculture intensification, 16
Active surface
  defined, 35
  radiation balance, 32
Aerial photography, forest islets, 324
Afforestation, *See also* Forest islets, Forests
  aesthetic/recreational land value, 304
  ammonium cations, 123
  anti-erosion measure, 303
  biodiversity preservation, 308
  bird habitat, 207, 308
  clear cutting, 310
  crop protection, 310
  decreased, 306
  design and management guidelines, 307–311
  early recognition of benefits, 300
  economics, 310, 311
  evapotranspiration, 302
  extreme weather phenomena, 303
  factors affecting success, 305
  Hungary, 305
  insect habitat, 308
  landscape simplification, 258
  management, 299–312, 321
  mid-field, management of, 304–311
  networks, 19
  nitrogen pollution, 119–122
  past rationale for, 304–305
  persistence, 309–310
  plant diversity, 193
  plant litter and nitrogen leaching, 133–134

  refuge habitats, 300, 304
  set-aside programs, 300
  slope-width relationship, 308
  solar energy conversion, 23
  south Romanian plains, 305
  tree age and bird density, 208
  tree species, 300, 307
  underestimation of problems, 305
  water, 302
  wet conditions, 308
  wind-breaks, 308
Agricultural landscape
  analysis, functional approach, 1–7
  animal abundance and diversity, 199–209
  animal community impoverishment, 197
  as buffer, 147
  biological control of pests, 204
  diversification and plant diversity, 194
  Europe, 220
  increasing without ecological consideration, 67
  "leakiness," 147
  management, 10, 333–334
  protection in east and central Europe, 305
  rural settlement, 259
  sustaining, 333
Agricultural revolution, 10
Agricultural subsidies, European Union, 290
Agriculture, *See also* Agriculture intensification
  21st century challenges, 332–337
  beginnings, 10
  code of best practices, 343
  cultivation effect on biomass, 12–17, 194–199
  development of, 9–24, 112
  effect on Protozoa, Enchytraeidae, Collembola,
    198
  energy input, 11

B

H

INDEX   357

INDEX   361

INDEX   363